物联网与嵌入式技术及其在农业上的应用

◎ 马德新 著

中国农业科学技术出版社

图书在版编目（CIP）数据

物联网与嵌入式技术及其在农业上的应用／马德新著．—北京：中国农业科学技术出版社，2019.6
　　ISBN 978-7-5116-4189-2

　　Ⅰ.①物… Ⅱ.①马… Ⅲ.①互联网络-应用-农业②智能技术-应用-农业③微处理器-系统设计-应用-农业　Ⅳ.①S126

中国版本图书馆 CIP 数据核字（2019）第 089675 号

责任编辑　　白姗姗
责任校对　　贾海霞

出 版 者	中国农业科学技术出版社 北京市中关村南大街 12 号　邮编：100081
电　　话	（010）82106638（编辑室）　（010）82109702（发行部） （010）82109709（读者服务部）
传　　真	（010）82106650
网　　址	http://www.castp.cn
经 销 者	各地新华书店
印 刷 者	北京建宏印刷有限公司
开　　本	787mm×1 092mm　1/16
印　　张	29.25
字　　数	715 千字
版　　次	2019 年 6 月第 1 版　2019 年 6 月第 1 次印刷
定　　价	168.00 元

版权所有·翻印必究

作者简介

马德新，博士，青岛农业大学副教授，硕士研究生导师。近年来先后主持或参加国家自然基金、山东省重点研发计划、山东省高等学校科技计划、青岛市民生科技计划等各级课题 8 项，发表学术论文 40 余篇，其中 SCI/EI 论文 15 篇，申请/获得专利 20 余项，获得软件著作权 15 项，参编教材 1 部，获得 2016 年度山东省科技进步奖二等奖 1 项，获得 2018 年度山东省高等学校科学技术奖三等奖 1 项。

前　　言

随着网络技术和计算机技术的高速发展，嵌入式产业迅速崛起，嵌入式系统已经越来越多地应用在各个领域之中。嵌入式操作系统作为嵌入式系统的重要组成部分，发挥着越来越重要的作用。在工程实践中，嵌入式系统往往需要有较高的实时性，这就向嵌入式操作系统提出了更高的要求，本书第一篇主要讨论为满足嵌入式应用领域的需要，对μCOS-Ⅱ操作系统的实时性进行研究。首先概述了嵌入式系统、嵌入式实时操作系统和嵌入式μCOS-Ⅱ，随后对μCOS-Ⅱ的工作原理及进程管理等进行详细的介绍，并对μCOS-Ⅱ嵌入式操作系统的优势和在实时性方面的不足作了深入研究与分析。

物联网（Internet of Things，IoT）是新一代信息技术的重要组成部分，物联网可以看作是物物相连的互联网，通过感知、识别技术，与普适计算等技术融合应用，形成了继计算机、互联网之后又一次信息产业发展的浪潮。我们在分析物联网系统各部分功能与特点的基础上，从基于Web的物联网业务环境的基本原则出发，将物联网系统架构分为感知域和业务域。无线传感器网络作为感知域中的重要内容，具有特别重要的地位。本书第二篇对物联网关键技术进行研究，重点研究了基于Web的物联网体系结构和感知域中的拓扑管理与路由技术，取得了部分研究成果及创新。

随着各种嵌入式技术、物联网技术及装备的普及应用，面向感知过程的智能技术在农业生产中发挥越来越重要的作用。在以上研究的基础上，对嵌入式技术、物联网技术在农业上的研究与应用进行探索，具体包括嵌入式操作系统的实时性、物联网感知域关键技术、智能化设施园艺技术、水肥精准调控技术、信息化管理系统等。本书第三篇、第四篇对嵌入式和物联网技术在农业上的应用和项目分别进行研究和总结。

本书部分内容得到山东省重点研发计划项目（2019GNC106001）、青岛市民生科技计划项目（18-6-1-112-nsh）、山东省高等学校科技计划项目（J17KA154）、青岛农业大学高层次人才科研基金项目（663-1116017）的资助，在此一并表示感谢。

著　者

2019年4月

目　录

第一篇　嵌入式关键技术研究

第一章　绪　论 (3)
　第一节　嵌入式系统的概况 (3)
　第二节　嵌入式操作系统与实时操作系统 (5)
　第三节　常用嵌入式操作系统介绍 (7)

第二章　嵌入式实时操作系统 μC/OS-Ⅱ (10)
　第一节　μC/OS-Ⅱ的特点及结构 (10)
　第二节　选择 μC/OS-Ⅱ的原因 (12)
　第三节　μC/OS-Ⅱ工作原理及进程控制 (12)
　第四节　μC/OS-Ⅱ进程的创建和删除 (17)
　第五节　μC/OS-Ⅱ进程调度分析 (19)

第三章　μC/OS-Ⅱ在 PC 上的移植 (21)
　第一节　μC/OS-Ⅱ源码简介 (21)
　第二节　开发工具简介及安装 (21)
　第三节　移植及配置详细步骤 (22)

第四章　μC/OS-Ⅱ调度的研究与改进 (24)
　第一节　实时任务的调度 (24)
　第二节　μC/OS-Ⅱ内核调度算法的改进 (29)
　第三节　各种调度算法评估 (33)

第五章　优先级反转问题的研究与改进 (36)
　第一节　进程优先级反转现象的研究 (36)
　第二节　优先级反转的理论解决 (38)
　第三节　各种操作系统对优先级反转的问题的解决 (40)
　第四节　优先级反转的实验模型 (41)
　第五节　互斥量 Mutex 的研究和改进 (43)

参考文献 (46)

第二篇　物联网关键技术研究

第一章　绪　论 (51)
- 第一节　引　言 (51)
- 第二节　物联网 (51)
- 第三节　无线传感网络 (59)
- 第四节　主要贡献 (66)

第二章　基于 Web 的物联网体系结构 (69)
- 第一节　物联网系统体系架构 (69)
- 第二节　物联网发展存在的问题及需求 (73)
- 第三节　基于 Web 的物联网系统服务平台 (74)
- 第四节　基于 Web 的物联网系统的商业模式 (74)
- 第五节　应用服务中间件 (77)
- 第六节　应用与服务管理平台 (80)
- 第七节　物联网生态系统 (82)

第三章　一种基于虚拟坐标的动态自适应分簇协议 (83)
- 第一节　引　言 (83)
- 第二节　相关工作 (84)
- 第三节　系统模型 (84)
- 第四节　基于虚拟坐标的动态自适应分簇协议 (86)
- 第五节　算法性能评估 (90)

第四章　一种基于小生境粒子群优化的自适应分簇协议 (93)
- 第一节　引　言 (93)
- 第二节　相关工作 (94)
- 第三节　系统模型 (94)
- 第四节　无参数小生境粒子群优化算法的改进 (96)
- 第五节　一种基于小生境粒子群优化的自适应分簇协议 (101)
- 第六节　算法性能评估 (105)

第五章　一种基于粒子群蚁群优化的动态分簇路由协议 (109)
- 第一节　引　言 (109)
- 第二节　相关工作 (110)
- 第三节　系统模型 (110)
- 第四节　蚁群算法及性能分析 (111)
- 第五节　一种基于粒子群蚁群优化的动态分簇路由协议 (117)
- 第六节　算法性能评估 (118)

第六章　附　录 (121)

参考文献 (122)

第三篇 关键技术在农业上的应用

茶树种质及基因资源发掘与开发利用数据库系统构建 …………………………… (137)
茶园水肥一体化技术应用现状与发展前景 ………………………………………… (140)
滴灌在樱桃种植区的应用 …………………………………………………………… (143)
电子元器件在智能家居领域中的应用发展 ………………………………………… (146)
基于"互联网+"的农业信息传播应用现状与发展趋势 …………………………… (150)
基于彭曼-蒙特斯公式的温室茶树腾发量计算模型研究 ………………………… (154)
基于物联网的水肥精准配比调控技术 ……………………………………………… (162)
农村经济地理信息系统的构建 ……………………………………………………… (167)
青岛农业科技传播服务平台的构建 ………………………………………………… (172)
实时操作系统 μC/OS-Ⅱ 调度算法的研究 ………………………………………… (176)
实时操作系统 μC/OS-Ⅱ 的改进与应用研究 ……………………………………… (181)
数字茶园实验基地建设实践与探索 ………………………………………………… (185)
数字化农业平台科技传播服务系统的构建 ………………………………………… (189)
水肥一体化技术的发展现状与应用对策研究 ……………………………………… (194)
水肥一体化技术的应用发展建议 …………………………………………………… (198)
水肥一体化技术在设施温室中的应用分析 ………………………………………… (202)
水肥智能调控技术在设施温室中的应用 …………………………………………… (205)
网站创建与管理课程建设的探索与实践 …………………………………………… (208)
我国农业信息管理系统发展现状及趋势分析 ……………………………………… (212)
无线传感器网络在环境监测应用
领域的研究进展 …………………………………………………………………… (215)
智慧农业现状与发展建议 …………………………………………………………… (218)
作物腾发量计算方法比较与评价 …………………………………………………… (223)
A clustering protocol based on Virtual Area Partition using Double Cluster Heads
    scheme ……………………………………………………………………………… (227)
A clustering protocol with Adaptive Assistant-Aided Cluster Head using Particle
    Swarm Optimization ……………………………………………………………… (236)
An Adaptive Assistant-Aided Clustering Protocol using Niching Particle Swarm
    Optimization ……………………………………………………………………… (246)
An Adaptive Clustering Protocol using Niching Particle Swarm Optimization ………… (255)
An Adaptive Node Partition Clustering protocol using Particle Swarm
    Optimization ……………………………………………………………………… (269)
An Adaptive Virtual Area Partition Clustering Protocol using Particle Swarm
    Optimization ……………………………………………………………………… (278)

An Adaptive Virtual Area Partition Clustering Routing protocol using Ant Colony
　　Optimization ……………………………………………………………………（289）
An Efficient Node Partition Clustering protocol using Niching Particle Swarm
　　Optimization ……………………………………………………………………（297）
An energy distance aware clustering protocol with Dual Cluster Heads using
　　Niching Particle Swarm Optimization …………………………………………（305）
An Energy Efficient Clustering protocol based on Niching Particle Swarm
　　Optimization ……………………………………………………………………（315）
Energy-aware Clustering Protocol with Dual Cluster Heads using Niching Particle
　　Swarm Optimization for Wireless Sensor Networks ……………………………（321）
Solar-powered Wireless Sensor Network's Energy Gathering Technology ………（331）
Virtual Area Partition Clustering protocol with Assistant Cluster Head …………（336）
Wireless Sensor Networks' Application Research Progress ………………………（345）

第四篇　项目资料

茶园水肥一体自适应调控模式研究 …………………………………………………（353）
农村土地管理信息系统 …………………………………………………………………（372）
测土配方施肥专家系统 …………………………………………………………………（382）
农业物联网管理系统 ……………………………………………………………………（388）
农产品价格预警管理系统 ………………………………………………………………（410）
农产品电子商务服务平台 ………………………………………………………………（419）
农产品质量安全可追溯管理系统 ………………………………………………………（429）
智能农业机械物联网系统 ………………………………………………………………（442）
农民教育培训系统 ………………………………………………………………………（453）

第一篇

嵌入式关键技术研究

第一章 绪 论

第一节 嵌入式系统的概况

一、嵌入式系统的定义及组成

根据 IEEE（国际电器和电子工程师协会）的定义，嵌入式系统是"控制、监视或辅助设备、机器和车间运行的装置。"这个定义把嵌入式系统的概念定义得很宽泛。具体的我们也可以这样说，嵌入式系统是以应用为中心，以计算机技术为基础，并且软硬件是可裁剪的，适用于对功能、可靠性、成本、体积、功耗等有严格要求的专用计算机系统。嵌入式系统是现代多学科互相融合的产物，嵌入式系统无多余软件，并且以固化态出现，硬件亦无多余存储器，有可靠性高、成本低、体积小、功耗少等特点。嵌入式系统又是知识密集，投资规模大，产品更新换代快，且具有不断创新特征、不断发展的系统，系统中采用片上系统（SoC）将是其主要发展趋势。也有人认为，只要计算机（或处理器）是作为某个专用系统中的一个部件而存在的，那么这个计算机系统就可以称之为嵌入式系统。

所谓将计算机嵌入到系统中，一般并不是指直接把一台通用计算机原封不动地安装到目标系统中，而是指为目标系统构筑起合适的计算机系统，再把它有机地植入，甚至融入目标系统。常规的计算机系统是面向计算（包括数值和非数值）和处理的，而嵌入式计算机则一般是面向控制的。

整个嵌入式系统的体系结构可以分成四个部分：嵌入式处理器、嵌入式外围设备、嵌入式操作系统和嵌入式应用软件。如图 1-1 所示。

嵌入式系统的核心是各种类型的嵌入式处理器，嵌入式处理器与通用处理器最大的不同点在于，嵌入式处理器多工作在为特定用户群所专门设计的系统中，它将通用处理器中许多由板卡完成的任务集成到芯片内部，从而有利于嵌入式系统在设计时趋于小型化，同时还具有很高的效率和可靠性。嵌入式处理器的种类主要有：嵌入式微处理器（EMPU）、嵌入式微控制器（EMCU）、嵌入式 DSP 处理器（EDSP）、嵌入式片上系统（ESoC）。

在嵌入式系统的硬件系统中，除了中心控制部件以外，还用于完成存储、通信、调试、显示等辅助功能的其他部件，事实上都可以算作嵌入式外围设备。目前常用的嵌入式外围设备按功能可以分为存储设备、通信设备和显示设备三类。

为了使嵌入式系统的开发更加方便和快捷，需要有专门负责管理存储器分配、中断

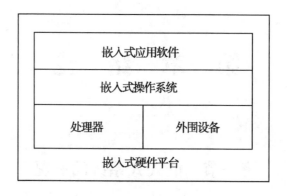

图1-1　嵌入式系统的组成

处理、进程调度等功能的软件模块，这就是嵌入式操作系统。嵌入式操作系统是用来支持嵌入式应用的系统软件，是嵌入式系统极为重要的组成部分，通常包括与硬件相关的底层驱动程序、系统内核、设备驱动接口、通信协议、图形用户界面等。嵌入式操作系统具有通用操作系统的基本特点，如能够有效管理复杂的系统资源，能够对硬件进行抽象，能够提供库函数、驱动程序、开发工具集等。但与通用操作系统相比较，嵌入式操作系统在实时性、硬件依赖性、软件固化性以及应用专用性等方面，具有更加鲜明的特点。

嵌入式应用软件是针对特定应用领域，基于某一固定的硬件平台，用来达到用户预期目标的计算机软件，由于用户任务可能有时间和精度上的要求，因此有些嵌入式应用软件需要特定嵌入式操作系统的支持。嵌入式应用软件和普通应用软件有一定的区别，它不仅要求其准确性、安全性和稳定性等方面能够满足实际应用的需要，而且还要尽可能地进行优化，以减少对系统资源的消耗，降低硬件成本。

二、嵌入式系统的现状和发展

嵌入式系统的出现至今已经有30多年的历史了，嵌入式技术也经历了几个发展阶段。进入20世纪90年代后，以计算机和软件为核心的数字化技术取得了迅猛发展，不仅广泛渗透社会经济、军事、交通、通信等相关行业，而且深入家电、娱乐、艺术、社会文化等各个领域，掀起了一场数字化技术革命。多媒体技术与Internet的应用迅速普及，电子、计算机、通信（3C）一体化趋势日趋明显，嵌入式技术再度成为一个研究热点。综观嵌入式技术的发展，大致经历了以下几个阶段。

第一阶段是以单片机为核心的可编程控制器形式的系统，同时具有与监测、伺服、指示设备相配合的功能。这种系统大部分应用于一些专业性极强的工业控制系统中，一般没有操作系统的支持，通过汇编语言编程对系统进行直接控制，运行结束后清除内存。这一阶段系统的主要特点是：系统结构和功能都相对单一，处理效率较低，存储容量较小，几乎没有用户接口。由于这种嵌入式系统使用简便、价格低廉，以前在国内外工业领域应用中较为普遍，但是已经远远不能适应高效的、需要大容量存储介质的现代化工业控制和新兴的信息家电等领域的需求。

第二阶段是以嵌入式 CPU 为基础、以简单操作系统为核心的嵌入式系统。这一阶段系统的主要特点是：CPU 种类繁多，通用性比较弱，系统开销小，效率高；操作系统具有一定的兼容性和扩展性；应用软件较专业，用户界面不够友好；系统要用来控制系统负载以及监控应用程序运行。

第三阶段是以嵌入式操作系统为标志的嵌入式系统。这一阶段系统的主要特点是：嵌入式操作系统能运行于各种不同类型的微处理器上，兼容性好；操作系统内核小、效率高，并且具有高度的模块化和扩展性；具备文件和目录管理、设备支持、多任务、网络支持、图形窗口以及用户界面等功能；具有大量的应用程序接口（API），开发应用程序简单；嵌入式应用软件丰富。

第四阶段是以基于 Internet 为标志的嵌入式系统，这是一个正在迅速发展的阶段。目前大多数嵌入式系统还孤立于 Internet 之外，但随着 Internet 的发展以及 Internet 技术与信息家电、工业控制技术等结合日益密切，嵌入式设备与 Internet 的结合将代表着嵌入式技术的真正未来。

当今嵌入式系统种类繁多、应用数量大、分布范围广，已经广泛应用于工业、农业、商业、金融、科研、国防、医疗、运输等一系列国民经济领域中。嵌入式系统目前已经成为通信和消费类产品的共同发展方向。在通信领域，数字技术正在全面取代模拟技术。在个人领域中，嵌入式产品将主要是作为个人的移动数据处理和通信工具，手写文字输入、语音拨号、收发电子邮件、以及彩色图形、图像都已在现实生活中得以实现。在自动控制领域，嵌入式系统技术不仅用于 ATM 机、自动售货机、工业控制等专用设备，还和移动通信设备、GPS（全球定位系统）相结合，发挥着巨大的作用。

21 世纪无疑将是一个网络的时代，随着网络技术的飞速发展，将嵌入式系统应用到各种网络环境中去的呼声自然也越来越高，嵌入式设备与 Internet 的结合已经成为嵌入式技术未来发展的趋势。

第二节　嵌入式操作系统与实时操作系统

一、嵌入式实时操作系统的特点

从嵌入式系统发展过程我们可以看到最初的嵌入式系统并没有操作系统。甚至直到现在，很多的嵌入式系统也不需要操作系统。首先是因为没有必要，功能简单的设备系统只要用监控程序就可以满足要求了；其次是因为当时的硬件条件不允许，在硬件条件很苛刻的情况下，根本没有操作系统生存的空间。但是随着硬件条件的改善以及嵌入式系统应用领域日益扩大，所需提供的功能越来越复杂，简单的监控程序已经不能很好地适应嵌入式系统的开发。这就促使人们在嵌入式系统中引入了操作系统。

用于嵌入式计算机的操作系统称为嵌入式操作系统。它已经成为嵌入式系统极为重要的组成部分，通常包括与硬件相关的底层驱动软件、系统内核、设备驱动接口、通信协议、图形用户界面等。

嵌入式操作系统具有通用操作系统的基本特点，如能够有效管理越来越复杂的系统资源；能够把硬件虚拟化，使得开发人员从繁忙的驱动程序移植和维护中解脱出来；能够提供库函数、驱动程序、工具集以及应用程序。

实时操作系统阶段嵌入式系统的主要特点是：操作系统的实时性得到了很大改善，已经能够运行在各种不同类型的微处理器上，兼容性好。操作系统内核精小、效率高，具有高度的模块化和扩展性。此时的嵌入式操作系统已经具备了文件和目录管理、设备管理、多任务、网络、图形用户界面等功能，并提供了大量的应用程序接口，从而使得应用软件的开发变得更加简单。

同时，嵌入式系统实时性要求高，所以嵌入式操作系统往往又是实时操作系统。许多嵌入式操作系统的内核是微内核结构，而不是宏内核。其他如在硬件相关性、软件固化性以及专用性等方面都具有较为突出的特点。

二、实时操作系统关键技术指标

由于实时操作系统在实时应用中的特殊地位，在实时操作系统的研究设计中对其性能指标的要求比通用操作系统严格。对于通用操作系统来说，其目的是方便用户管理计算机资源，追求系统资源的最大利用率。而实时操作系统追求的是实时性、可确定性和可靠性。

评价一个实时操作系统一般可以用以下几个技术指标来衡量。

1. 任务调度算法

实时操作系统的实时性在很大程度上取决于它的任务调度算法。

2. 上下文切换时间

在多任务系统中，上下文切换时间是当处理器的控制权由运行进程转移到另外一个就绪任务时所需要的时间。这是影响实时操作系统性能的一个重要指标。

3. 最大中断禁止时间

当实时操作系统运行在内核空间或执行某些系统调用时，是不会因为外部中断的到来而中断执行的。只有当重新回到用户态时才响应外部中断请求，这一过程所需的最大时间就是最大中断禁止时间。

三、实时操作系统的任务调度

在操作系统的多任务调度算法的设计上，要根据系统的具体需求来确定调度策略，实时调度策略可以按不同的方法分为：静态/动态；基于优先级/不基于优先级；抢占式/非抢占式；单处理器/多处理器。

各种实时操作系统的实时调度算法可以分为以下三种类别：基于优先级的调度算法（Priority-driven scheduling，PD）、基于CPU使用比例的共享式的调度算法（Share-driven scheduling，SD）以及基于时间的进程调度算法（Time-driven scheduling，TD）。本文主要介绍基于优先级的调度算法。

基于优先级的调度算法给每个进程分配一个优先级，在每次进程调度时，总是调度那个具有最高优先级的进程来执行。根据不同的优先级分配方法，基于优先级的调度算法可以分为静态优先级调度算法和动态优先级调度算法两种类型。

关于调度策略及其改进是本文研究的一个重点，在以后的章节中将会作更进一步的介绍。

四、可剥夺型内核及先级反转问题

一般认为只有可剥夺型内核的操作系统才可以称为实时操作系统。根据 IEEE 实时 UNIX 分委会对实时操作系统的定义，进程的可剥夺调度也是实时操作系统的基本特征之一。因此，绝大多数商业的实时内核都是可剥夺型内核。

可剥夺型内核是指最高优先级的进程一旦就绪，总能得到 CPU 的使用权。当一个比当前运行进程优先级高的进程进入了就绪态时，当前进程的 CPU 使用权就被剥夺了，或者说被挂起了，更高优先级的进程立刻得到了 CPU 的使用权。这样就可以保证就绪的最高优先级的进程，往往也是最重要的进程，能够及时得到运行。那么，采用了可剥夺型内核的操作系统是否一定就能使最高优先级的进程得到及时的运行呢？不是的。对于可剥夺型内核有个很常见的问题，那就是优先级反转。在这种特殊情况下，高优先级的进程可能会被低优先级的进程阻塞，结果会在比自己优先级低的进程之后运行，甚至有时还会得不到运行的机会，这在实时要求比较高的情况下是无法忍受的。

关于优先级反转问题及其解决也是本文研究的一个重点，在以后的章节中将会作更进一步的介绍。

第三节　常用嵌入式操作系统介绍

一、μC/OS 简介

μC/OS 可以说是最小的操作系统内核了。这里的 μ 表示 micro，所以 μC 就是指微控制器。其作者 Jean J. Labrosse 将第 1 版的源代码发表在 1992 年的"Embedded System Programming"杂志上，从而引起人们的注意和采用。在此基础上，后来又推出了 μC/OS 的第 2 版，即 μC/OS-Ⅱ。

现在 μC/OS-Ⅱ已经在世界范围内得到广泛使用，包括诸多领域，如手机、路由器、集线器、不间断电源、飞行器、医疗设备以及工业控制。这表明 μC/OS-Ⅱ具有足够的安全性与稳定性，能用于与人性命攸关的、安全性条件极为苛刻的系统。

二、Linux 简介

嵌入式 Linux 的开发和研究是操作系统领域中的一个热点。Linux 能够支持 x86、ARM、MIPS、ALPHA、Power PC 等多种体系结构，目前已经成功移植到数十种硬件平台，几乎能够运行在所有流行的 CPU 上。Linux 内核的高效和稳定性已经在各个领域内得到了大量事实的验证，Linux 的内核包括进程调度、内存管理、进程间通信、虚拟文件系统和网络接口五大部分，其独特的模块机制可以根据用户的需要，实时地将某些模块插入到内核或从内核中移走。这些特性使得 Linux 系统内核可以裁剪得非常小巧，很适合于嵌入式系统的需要。

　　Linux虽然已经被成功地应用到了PDA、移动电话、车载电视、机顶盒、网络微波炉等各种嵌入式设备上，但在医疗、航空、交通、工业控制等对实时性要求非常严格的场合中还无法直接应用，原因在于现有的Linux是一个通用的操作系统，虽然它也采用了许多技术来加快系统的运行和响应速度，但从本质上来说并不是一个嵌入式实时操作系统。Linux内核采用的是宏内核，整个内核是一个单独的、非常大的程序，这样虽然能够使系统的各个部分直接沟通，有效地缩短任务之间的切换时间，提高系统响应速度，但与嵌入式系统存储容量小、资源有限的特点不相符合。目前在基于图形界面的特定系统定制平台的研究上，与Windows CE等商业嵌入式操作系统相比还有很大的差距，整体集成开发环境有待提高和完善。

三、其他嵌入式操作系统简介

1. Windows CE

　　Windows CE是美国Microsoft公司专门为各种移动和便携电子设备、个人信息产品、消费类电子产品、嵌入式应用系统等设计的一种32位高性能操作系统。它具有一个简洁、高效的完全抢占式多任务操作系统内核，支持强大的通信和图形功能，能够适应广泛的系统需求。Windows CE是微内核结构的操作系统，它是从整体上为有限资源的平台设计的多线程、完整优先权、多任务的操作系统。操作系统由一些独立的模块组成，每个模块提供特定的功能，大的模块又可以分成几个组件。这种组件式结构能使系统变得非常紧凑，仅需使用很少的硬件资源就可运行。最主要的系统模块有四个，即内核、持久性存储模块、绘图窗口事件子系统和通信模块。内核是整个操作系统的核心部分，它负责基本的操作系统功能，包括内存管理、进程管理和必需的文件管理。它的模块化设计允许它对从掌上电脑到专用工业控制器的用户电子设备进行定制。操作系统的基本内核需要至少200K的ROM（只读内存镜像）。Windows CE不需要任何特定的硬件结构，实际的硬件系统完全由用户根据需要自由设计。

　　Windows CE的图形用户界面也是非常值得一提的。它兼容于Microsoft公司的Windows操作系统，支持超过1 000个常用的32位Windows应用程序结构函数，支持高分辨率真彩色。对于需要有较强人机界面的应用，特别是对于用惯了Windows界面的用户，这当然是个非常重要的因素。此外，其他Windows操作系统上的应用软件移植到Windows CE上也比较方便，因此，Windows CE所拥有的应用软件以及所支持的外设的数量是其他系统难以比拟的。

2. VxWorks

　　VxWorks是美国Wind River公司1983年设计开发的一种嵌入式实时操作系统的产品。因其良好的持续发展能力、高性能的内核、友好的开发环境、卓越的可靠性，被广泛应用于通信、航空、航天等实时性要求极高的领域中。VxWorks采用微内核的结构，支持多种硬件环境。另外，它还具有网络协议丰富、兼容性好和裁减性好等特点，具有程序动态链接和下载的功能。VxWorks是一种功能强大而且比较复杂的操作系统，包括了进程管理、存储管理、设备管理、文件系统管理、网络协议及系统应用等几个部分，为用户提供高效的实时多任务调度、中断管理、实时的系统资源以及实时的任务间通信，具有快速多任务切换、抢占式任务调度、任务间通信方式多样化等特点。VxWorks十分灵活，操作系统由400多个相对独立的、短小精炼的目标模块组成，用户可以根据需要选择适当的模块来裁减和配置系统，有多达1 800个功能强大的应用程序接口。此外，VxWorks支持广泛的工业标准，如POSIX1003.b实时扩展、ANSI C和TCP/IP网络协议等。VxWorks还提供了强大的网络支持，能与其他许多主机系统进行通信。

第二章 嵌入式实时操作系统 μC/OS-Ⅱ

第一节 μC/OS-Ⅱ的特点及结构

一、μC/OS-Ⅱ的特点

μC/OS-Ⅱ读做"micro COS2",意为"微控制器操作系统版本2",是一个实时内核,主要提供任务管理功能。它是由 μC/OS 升级而来的,并且做了很大的改进。μC/OS-Ⅱ的使用对象是嵌入式系统,并且很容易移植到不同构架的微处理器上。下面简单介绍一下它的特点。

1. 公开源代码

源代码清晰易读且结构协调,注解详尽,组织有序。

2. 可移植性

绝大部分 μC/OS-Ⅱ的源代码使用移植性很强的 ANSI C 写的,和微处理器硬件相关部分采用汇编写,并且压到了最低限度。只要该处理器有堆栈指针,有 CPU 内部寄存器入栈、出栈指令就可以移植 μC/OS-Ⅱ。目前 μC/OS-Ⅱ已经移植到绝大多数8位、16位、32位以及64位微处理器上。

3. 可固化

μC/OS-Ⅱ是为嵌入式应用而设计的,只要具备合适的系列软件工具(C编译、汇编、链接及下载/固化),可以将 μC/OS-Ⅱ嵌入到产品中作为产品的一部分。

4. 可裁剪

可以只使用 μC/OS-Ⅱ中应用程序需要的哪些系统服务。这种可裁剪性是靠条件编译实现的。

5. 占先式

μC/OS-Ⅱ完全是占先式实时内核,即总是运行就绪条件下优先级最高的任务。

6. 多任务

可以管理64个任务,但系统保留了8个任务,应用程序最多可以有56个任务。赋予每个任务的优先级必须是不相同的。

7. 可确定性

全部 μC/OS-Ⅱ的函数调用和服务的执行时间具有可确定性,即它们的执行时间是可知的,进而言之,μC/OS-Ⅱ系统服务的执行时间不依赖于应用程序任务的多少。

8. 任务栈

每个任务都有自己单独的栈，μC/OS-Ⅱ允许每个任务有不同的栈空间。

9. 系统服务

μC/OS-Ⅱ提供多种系统服务，如邮箱、消息队列、信号量、块大小固定的内存的申请与释放、时间相关函数等。

10. 中断管理

中断可以使正在执行的任务暂时挂起，中断嵌套层数可达255层。

11. 稳定性与可靠性

μC/OS-Ⅱ是基于μC/OS的，μC/OS自1992年以来已经有好几百个商业应用。μC/OS-Ⅱ与μC/OS的内核是一样的，只不过提供了更多的其他功能。

二、μCOS-Ⅱ的系统结构

图2-1说明了μC/OS-Ⅱ的软硬件体系结构。应用程序处于整个系统的顶层，每个任务都可以认为自己独占了CPU，因而可以设计成一个无限循环。μC/OS-Ⅱ处理器无关的代码，μC/OS-Ⅱ的系统服务，应用程序可以使用这些API函数进行内存管理、任务间通信以及创建、删除任务等。

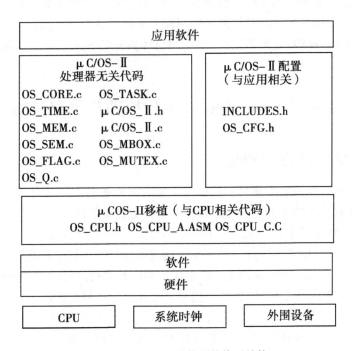

图2-1 μC/OS-Ⅱ软硬件体系结构

大部分的μC/OS-Ⅱ代码是使用ANSI C语言书写的，因此μC/OS-Ⅱ的可移植性较好。尽管如此，仍然需要使用C和汇编语言写一些处理器相关的代码。μC/OS-Ⅱ的移植需要满足以下要求：①处理器的C编译器可以产生可重入代码；②可以使用C调用进入和退出 Critical Code（临界区代码）；③处理器必须支持硬件中断，并且需要一个定时中断源；④处理器需要能够容纳一定数据的硬件堆栈；⑤处理器需要有能够在 CPU 寄存器与内存和堆栈交换数据的指令。移植 μC/OS-Ⅱ的主要工作就是处理器和编译器相关代码。

第二节 选择 μC/OS-Ⅱ的原因

本课题采用的是由 Jean J. Labrosse 编写的开放式实时操作系统 μC/OS-Ⅱ，主要是基于以下的考虑。

一是它的内核是完全免费的，用户不需支付任何费用，有利于降低系统开发成本。

二是它的源代码是公开的，并且仍在不断的升级，增加新功能。源代码的开放可以使得用户根据实际要求对源代码进行取舍，去掉不必要的变量和不使用的函数，提高系统性能。另外，由于对系统内核有源代码级的了解，用户可以添加自己的模块，与原有系统内核兼容，使得系统具有可扩展性。

三是系统内核实用性强、可靠性高。从最老版本的 μC/OS，以及后来的 μC/OS-Ⅰ，到最新版本的 μC/OS-Ⅱ，该实时内核已经走过了 10 年多的历程。10 多年来，世界上已有数千人在各个领域使用了该实时内核，如医疗器械、网络设备、自动提款机、工业机器人等等。这些应用的实践是该内核实用性、无误性的最好证据。

四是操作系统内核对处理器以及 ROM、RAM 资源的要求不高，有利于在各种处理器上的移植。

第三节 μC/OS-Ⅱ工作原理及进程控制

一、μC/OS-Ⅱ的工作原理

μC/OS-Ⅱ的工作核心原理是：近似地让最高优先级的就绪任务处于运行状态。首先，初始化 MCU，再进行操作系统初始化，主要完成任务控制块 TCB 初始化，TCB 优先级表初始化，TCB 链表初始化，事件控制块（ECB）链表初始化，空任务的创建等等；然后就可以开始创建新任务，并可在新创建的任务中再创建其他的新任务；最后调用 OSStart（）函数启动多任务调度。在多任务调度开始后，启动时钟节拍源开始计时，此节拍源给系统提供周期性的时钟中断信号，实现延时和超时确认。

操作系统在下面的情况下进行任务调度：中断［系统占用的时间片中断 OSTimeTick（）、用户使用的中断］和调用 API 函数（用户主动调用）。一种是当时钟中断来临时，系统把当前正在执行的任务挂起，保护现场，进行中断处理，判断有无任务延时到期，若有则使该任务进入就绪态，并把所有进入就绪态的任务的优先级进行比较，通过任务切换去执行最高优先级的就绪任务，若没有别的任务进入就绪态，则恢复现场继续执行原任务。另一种调度方式是任务级的调度即调用 API 函数（由用户主动调用），是通过发软中断命令或依靠处理器在任务执行中调度，如任务要等待信号量或一个正在执行的任务被挂起时，就需要在此任务中调度，找出目前处于就绪态的优先级最高的任务去执行。当没有任何任务进入就绪态时，就去执行空任务。

本课题采用的 μC/OS-Ⅱ，相对早期的 μC/OS，有了很大的改进，体现在以下方面。

1. μC/OS-Ⅱ 提供了对 64 个任务的管理

除了系统内核本身所保留了 8 个任务外，用户的应用程序最多可以有 56 个任务。由于 μC/OS-Ⅱ 是一个基于优先级的（不支持时间片轮转调度）实时操作系统，因此每个任务的优先级必须不相同。（μC/OS-Ⅱ 把任务的优先级当作任务的标识来使用，如果优先级相同，任务将无法区分；所以它只能说是多任务，不能说是多进程，至少不是我们所熟悉的那种多进程。因此在本文中对进程和任务未作严格区分，望读者注意。）系统中的每个任务都处于以下 5 种状态之一的状态中，这 5 种状态是休眠态、就绪态、运行态、等待态（等待某一事件发生）和被中断态。图 2-2 是 μC/OS-Ⅱ 控制下的任务状态转换图。

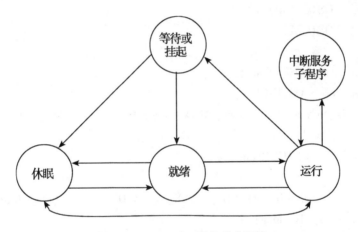

图 2-2　μC/OS-Ⅱ 任务状态切换

2. μC/OS-Ⅱ 是一个占先式（preemptive）的内核

即最高优先级的任务一旦就绪，总能得到处理器的控制权。当一个运行着的任务使一个比它优先级高的任务进入了就绪态，则当前任务的处理器控制权就被剥夺了，或者说被挂起，那个高优先级的任务立刻得到处理器的控制权。如果是中断服务（ISR）使一个高优先级的任务进入就绪态，中断处理完成后，被中断了的任务将被挂起，优先级高的那个任务则开始运行。

3. μC/OS-Ⅱ 提供了对信号量（Semaphore）、邮箱（MailBox）和消息队列（Message Queue）的支持

通过这 3 种方法可以完成任务与中断服务程序（ISR）之间的通信、任务与任务之间的通信以及多个任务对共享资源的互斥访问。

二、μC/OS-Ⅱ 进程控制块

我们知道，μC/OS-Ⅱ 是一个微内核结构的操作系统。但作为一个操作系统，进程

管理，特别是进程调度和进程状态的改变都是必不可少的。否则就不能称之为一个操作系统了。

　　进程控制块是进程管理中最重要的数据结构，是进程存在的必要条件。μC/OS-Ⅱ进程控制块是一个 OS_TCB 数据结构，它的字段包含了与一个进程相关的所有信息，是系统对进程进行控制的唯一手段。分析操作系统的进程控制块是分析进程管理的基础。同时通过对 μC/OS-Ⅱ进程控制块的分析，我们也可以从中了解一些 μC/OS-Ⅱ进程管理的特点。以下就是 μC/OS-Ⅱ进程控制块的程序源代码。

```
typedef struct os_ tcb {
    OS_ STK    * OSTCBStkPtr;
#if OS_ TASK_ CREATE_ EXT_ EN > 0
    void       * OSTCBExtPtr;
    OS_ STK    * OSTCBStkBottom;
    INT32U     OSTCBStkSize;
    TNT16U     OSTCBOpt;
    TNT16U     OSTCBId;
#endif
    struct os_ tcb * OSTCBNext;
    struct os_ tcb * OSTCBPrev;
#if ((OS_ Q_ EN > 0) && (OS_ MAX_ QS > 0)) || (OS_ MBOX_ EN >0) || (OS_ SEM_ EN > 0)
    OS_ EVENT * OSTCBEventPtr;
#endif
#if ((OS_ Q_ EN > 0) && (OS_ MAX_ QS > 0)) || (OS_ MBOX_ EN >0)
    void      * OS_ TCBMsg;
#endif
#if (OS_ VERSION >= 251) && (OS_ FLAG_ EN > 0)) && (OS_ MAX_ FLAGS >0)
# if OS_ TASK_ DEL_ EN > 0
    OS_ FLAG_ NODE    * OSTCBFlagNode;
# endif
    OS_ FLAGs    OSTCBFlagsRdy;
#endif
    INT16U    OSTCBDly;
    INT8U     OSTCBStat;
    INT8U     OSTCBPrio;
    INT8U     OSTCBX;
    INT8U     OSTCBY;
    INT8U     OSTCBBitX;
```

```
    INT8U      OSTCBBitY;
#if OS_ TASK_ DEL_ EN > 0
    BOOLEAN    OSTCBDelReq;
#endif
} OS_ TCB;
```

从进程控制块的组成可以大致上看出相应内核的许多性质和功能，我们对其中比较重要一些变量进行分析。

OSTCBStkPtr 是一个堆栈指针，指向当前进程堆栈栈顶的指针。这个堆栈指针记下了一个进程在运行被暂时剥夺时，即进程切换点上的堆栈的"顶"。μC/OS-Ⅱ允许每个进程有自己的堆栈，尤为重要的是，每个进程的堆栈的容量是任意的。

OSTCBEventPtr 是指向事件控制块的指针。它反应 μC/OS 对"事件"队列的支持。事实上，μC/OS-Ⅱ内核中所有的进程间通信机制都是通过这个队列实现的。在多进程的系统中，起码的进程间通信机制是必须要有的，否则便受到很大的局限，甚至失去实际的意义。

OSTCBStat 反映着进程的当前状态，如"就绪"或"睡眠"，或者正在"运行"。

OSTCBPrio 是进程的优先级，OSTCBPrio 的值越小，进程的优先级越高。在 μC/OS 中，所有进程的优先级都各不相同，每个优先级都只能有一个进程（或者没有），这对于 μC/OS 的进程调度算法是决定性的。另外，正是因为如此，就也可以用进程的优先级作为其唯一的标识。

OSTCBDly 用来记录需要延时的时间。当需要把进程延时若干时钟节拍时，或者需要把进程挂起一段时间以等待某事件的发生时，就要用到这个变量。它的存在表明内核支持 delay（time）、sleep（time）一类的功能。

OSTCBX、OSTCBY、OSTCBBitX 以及 OSTCBBitY，都是因为 μC/OS-Ⅱ特定的调度策略与算法而设的。这些值是在进程建立时计算好的，或者是在改变进程优先级时计算出的。它们的作用在分析其调度算法时将会看到。

OSTCBDelReq 是一个布尔量，用于表示进程是否要删除。

其余的成分，就都是因为 μC/OS 本身的实现所需要的了。

在操作系统初始化时，系统申请了一块 RAM 空间来存储空闲进程控制块。一旦进程建立，系统就会分配一个进程控制块给该进程。当进程的 CPU 使用权被剥夺时，μC/OS-Ⅱ用它来保存该进程的状态。当进程重新得到 CPU 使用权时，进程控制块能确保进程从被中断的那一点继续执行下去。OS_ TCB 全部驻留在 RAM 中，在进程建立的时候，OS_ TCB 被初始化。所有进程控制块都是放在进程控制列表数组 OSTCBTbl [ ] 中的。系统占用了两个任务，一个用于空闲任务，另一个用于任务统计。但是，μC/OS-Ⅱ作者建议用户不要使用优先级为 0，1，2，3 以及 OS_ LOWEST_ PRIO-1，OS_ LOWEST_ PRIO-2，OS_ LOWEST_ PRIO-3 及 OS_ LOWEST_ PRIO，因为未来的 uC/OS-Ⅱ版本中，可能会使用这些优先级。

在 μC/OS-Ⅱ初始化时，所有进程控制块被链接成空闲进程控制块的单向链表，如图 2-3 所示。当进程一旦建立，空闲进程控制块指针 OSTCBFeeList 指向的进程控制块便赋给了该进程，然 OSTCBFeeList 的值调整为指向链表中下一个空闲的进程控制块。一旦进程被删除，它的进程控制块就回到空闲进程控制块链表中。

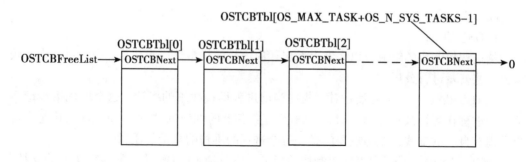

图 2-3 空闲进程表

三、μC/OS-Ⅱ进程的状态

在任一给定的时刻，任务的状态一定（只能）是以下 5 种状态之一。图 2-4 是 μC/OS-Ⅱ 控制下进程状态转换详细图。

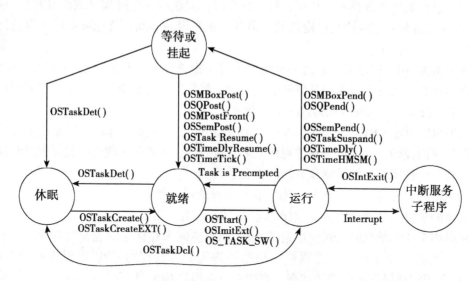

图 2-4 μC/OS-Ⅱ进程状态转换

1. 睡眠态

是指进程驻留在程序空间，还没有交给 μC/OS-Ⅱ来管理。把进程交给 μC/OS-Ⅱ，是通过调用 OSTaskCreate（）或 OSTaskCreateExt（）来实现的。这些调用只是用于告诉 μC/OS-Ⅱ，进程的起始地址在那里；进程建立时，用户给进程赋予的优先级是多少；进程要用多少栈空间等。

2. 就绪态

进程一旦建立，这个进程就进入了就绪态，准备运行。进程的建立可以在多任务运行开始之前，也可以动态地由一个运行着的进程建立。如果多任务已经启动，且一个进程是被另一个进程建立的，而新建立的进程的优先级高于建立它的任务的优先级，则这个刚刚建立的任务将立即得到 CPU 的使用权。一个进程可以通过 OSTaskDel（）返回睡眠态，或通过调用该函数让另一个进程进入睡眠态。

3. 运行态

调用 OSStart（）可以启动多进程。OSStart（）函数只能在启动时调用一次，该函数运行用户初始化代码中已经建立的、进入就绪态的优先级最高的进程。优先级最高的进程就这样进入了运行态。任何时刻只能有一个进程处于运行态。就绪的进程只有当所有优先级高于这个进程的进程都转为等待状态或者是被删除了，才能进入运行态。

4. 等待态

正在运行的进程可以通过调用 OSTimeDly（）或者 OSTimeDlyHMSM（）将自身延迟一段时间。这个进程于是进入等待状态，一直到函数中定义的延迟时间到时。这两个函数会立即强制执行进程切换，让下一个优先级最高的、并进入了就绪态的进程运行。等待的时间过去以后，系统服务函数 OSTimeTick（）使延迟的进程进入就绪态。

正在运行的进程可能需要等待某一事件的发生，可以通过调用以下函数之一实现：OSFlagPend（）、OSSemPend（）、OSMutexPend（）、OSMboxPend（）或 OSQPend（）。如果某事件并未发生，调用上述函数的任务就进入了等待状态，直到等待的事件发生了。当进程因等待事件被挂起时，下一个优先级最高的进程立即得到了 CPU 的使用权。当进程发生了或等待超时时，被挂起的进程就进入就绪态。事件发生的报告可能来自另一个进程，也可能来自中断服务子程序。

5. 中断服务态

正在运行的进程是可以被中断的，除非该进程将中断关闭，或者 μC/OS-Ⅱ 将中断关闭。被中断了的进程于是进入了中断服务态。响应中断时，正在执行的进程被挂起，中断服务子程序控制了 CPU 的使用权。中断子程序可能会报告一个或多个事件的发生，而使一个或多个进程进入就绪态。在这种情况下，从中断服务子程序返回之前，μC/OS-Ⅱ 要判定，被中断的进程是否还是就绪队列中优先级最高的。如果中断服务子程序使另一个优先级更高的进程进入了就绪态，则新进入就绪态的这个优先级更高的进程将得以运行；否则，原来被中断的进程将继续运行。

第四节　μC/OS-Ⅱ 进程的创建和删除

一、μC/OS-Ⅱ 进程的创建

如果要用 μC/OS-Ⅱ 管理用户的进程，就必须先创建进程。在 μC/OS-Ⅱ 中一个进程或者说一个任务就是一个无限的循环。μC/OS-Ⅱ 进程的创建是由 OSTaskCreate（）和 OSTaskCreateExt（）这两个函数完成的。

OSTaskCreate（）需要 4 个参数，task 是指向进程代码的指针，pdata 是进程开始执行时传递给进程的参数指针，ptos 是分配给进程的堆栈的栈顶指针，prio 是分配给进程的优先级。

使用OSTaskCreate()创建一个进程时,首先会检测分配给进程的优先级是否有效。因为在μC/OS-Ⅱ中有64个优先级,每个优先级最多只能对应一个进程。所以检测优先级是否有效就是检测所要创建的进程的优先级的值是否在0~OS_LOWEST_PRIO之间(常数OS_LOWEST_PRIO=63)以及该优先级是否已经被其他的进程占用。μC/OS-Ⅱ规定每个优先级最多只可以有一个进程存在。这样对于已经创建的进程,就可以把指向各个进程控制块的指针存放在一个以优先级为下标的数组OSTCBPrioTbl[ ]中,并通过这个数组实施进程管理。因此,给定一个优先级,如果该数组中的相应指针为0,就表示系统中尚无这个优先级的进程存在,因而可以创建;否则就表示已经有进程存在,因而不能再创建。在确认将要使用的优先级没有被其他进程占用以后,系统通过在数组OSTCBPrioTbl[ ]相对应的元素中放置一个非空指针,保留该优先级。然后调用OSTaskStkInit()建立进程的堆栈。内核中维持着一个空闲进程控制块的队列OSTCBFreeList,需要时就从队列的前端取下一个,并将其地址填入数组OSTCBPrioTbl[ ]中的相应位置上。同时,内核中也有一个已创建进程的控制块队列OSTCBList,顺着这个队列就可以找到系统中当前的所有进程,所以要把新创进程的控制块挂入这个队列。在完成堆栈的建立后,调用OS_TCBInit()从进程控制块链表中获取一个空闲的进程控制块。至此,一个进程的创建过程基本完毕。

OSTaskCreateExt()是OSTaskCreate()的扩展,使用OSTaskCreateExt()来创建进程会更加灵活,但是会增加一些额外的开销。

二、μC/OS-Ⅱ进程的删除

进程既然有创建就会有消亡。μC/OS-Ⅱ中所说的进程的删除是指进程将返回并处于休眠状态,进程的代码不再被μC/OS-Ⅱ调用。进程的删除是由函数OSTaskDel()实现的。在Linux中,一个进程是不能直接删除另一个进程的,它能做的只是向另一个进程发送一个软中断信号使其结束。所以,在直接意义上,一个进程的终结只能由其自己通过系统调用exit()完成。而在μC/OS-Ⅱ中,则允许一个进程直接删除一个进程,包括自身。由于在μC/OS-Ⅱ中进程的优先级同时也起着进程号的作用,所以OSTaskDel()的参数就是目标进程的优先级。如果要删除的进程是当前进程本身,则用一个特殊的优先级OS_PRIO_SELF,即常数0xFF。不过,系统中有一个进程是特殊的,那就是空闲进程。这个进程不允许删除,否则就有可能使进程调度无法进行。

OSTaskDel()首先检查参数的有效性,确保要删除的不能是空闲进程,参数中优先级的值也应该是有效范围内的,而且,占用该优先级的进程是应该存在的。另外,中断服务程序是不允许删除的,所以还要检查对函数的调用是否来自中断服务程序。如果是OS_PRIO_SELF就替换成当前进程的优先级。

根据给定的优先级找到目标进程以后,首先清除就绪进程组内代表着目标进程的标志位,如果这是本进程组中最后一个就绪进程,则还要进一步将就绪进程分组中的相应标志也清除。如果进程处于互斥型信号量、邮箱、消息队列或信号量以及事件标志的等待表中,还要把它从这些表中清除掉。然后将指向被删除任务的进程控制块的指针置为

NULL，就从优先级表中把 OS_TCB 清除了。这样，目标进程就不会再被调度了。接着把被删除进程的进程控制块从 OSTCBList 开头的进程控制块双向链表中去掉并加入空任务链表，以供建立其他任务时使用。

第五节 μC/OS-Ⅱ进程调度分析

对于严格按照优先级高低调度的系统而言，进程的调度只有在系统中的就绪队列进程集合发生变化的时候才有需要，而就绪进程集合的变动只可能发生在以下几种情况。

运行中的进程因操作受阻或其他原因而暂时放弃运行，这意味着一个就绪进程变成了非就绪状态。

创建了新的进程。

运行中的进程或者就绪进程被终止。

运行中的进程因某种原因唤醒了一个或几个进程。

某种外部事件的发生使中断服务程序唤醒了一个或几个进程。

进程的调度与切换是由 OSSched（）完成的，应用程序可以直接调用这个函数，所以也是一个系统调用。

进程的调度并不是任何时候都可以进行的，μC/OS-Ⅱ提供一对函数 OSSchedLock（）和 OSSchedUnlock（），让应用程序可以通过 OSSchedLock（）冻结调度，而调用这个函数的效果就递增一个计数器 OSLockNesting，使它变成非 0。另一个方面，在中断服务程序内部也不允许进程调度。发生中断时，中断相应程序先递增 OSLockNesting，然后在完成了中断服务程序以后，从中断返回之前再递减 OSLockNesting。这样，调度就不可能发生在中断服务程序内部，但是可以在中断返回前夕发生。

如果允许调度，那么第一件事当然是寻找系统中当前最应当得到运行机会的就绪进程，在 μC/OS-Ⅱ中就是优先级最高的就绪进程。每个就绪的进程都放在就绪表中（readylist）中，就绪表中有 2 个变量，OSRdyGrp 和 OSRdyTbl［ ］。

如前所述，数组 OSTCBPrioTbl［ ］是以优先级为下标的，其大小是 64，所以 μC/OS-Ⅱ内核允许 64 个优先级，从而可以有 64 个进程。设计者把这 64 个可能的进程按优先级大小划分成 8 组，每一组用一个 8 位的位图代表。OSRdyGrp 中的每一个位表示 8 组进程中每一组是否有进入就绪态的进程。根据具体进程的优先级就可以从数组 OSMapTbl［ ］中查找到它所属进程组的位图。进程优先级的低 3 位用于确定是在 OSRdyTbl［ ］中的所在位。接下去的 3 位用于确定是在 OSRdyTbl［ ］数组的第几个元素。进程进入就绪态时，就绪表 OSRdyTbl［ ］中的相应元素置 1，同时将在 OSRdyGrp 中将代表该组的位也置 1。这样，从这些位图就可以知道哪些进程已经就绪，而根据具体的编组标志位和组内标志位，就可以算出进程的优先级，从而找到目标进程的控制块。进程的优先级以及就绪表 OSRdyGrp 和 OSRdyTbl［ ］之间的关系如图 2-5 所示。

位图 OSRdyGrp 反映着哪些进程组中是就绪进程。另一方面，各个进程组的标志位在位图中的位置也是有规律的，位置靠右的标志位代表着优先级较高的进程（组）。这样，从原理上说，只要从右到左扫描位图 OSRdyGrp，碰到的第一个非 0 标志位就代表

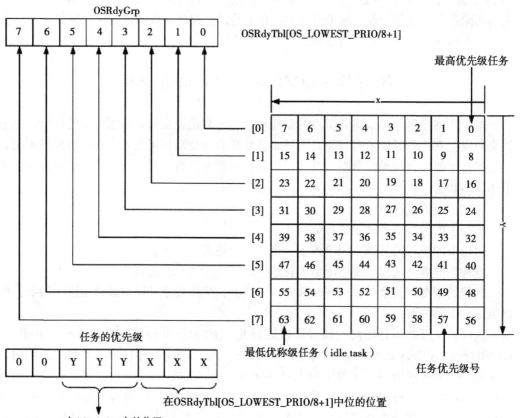

图 2-5 μC/OS-Ⅱ进程就绪表

着当前优先级最高的就绪进程所在的进程组。但是，预先编制好一个对照表，然后当需要时，以 OSRdyGrp 的数值下标访问该数组的方法显然速度更快些。这个数组就是 OSUnMapTbl[ ]。所以，以位图 OSRdyGrp 的数值为下标，直接就可以得到优先级最高者所属的组号。组号越小，组内进程的优先级就越高。

知道了组号 y 以后，就可以以此为下标在 OSRdyTbl[ ]中得到相应的组内位图。

同理，以这个位图的数值 OSRdyTbl[y]为下标，又可以在 OSUnMapTbl[ ]内查得该组内优先级最高的进程号（优先级的最低三位），即组内号。将组号与组内号编号拼合在一起，就得到目标进程完整的进程号，即优先级。再以此为下标，就可以从 OSTCBPrioTbl[ ]中得到指向目标进程控制块的指针 OSTCBHightRdy。

在找到就绪进程后，接下去的事情就是任务切换了。任务切换是由函数 OSSched( ) 完成的，应用程序可以直接调用。

第三章 µC/OS-Ⅱ在 PC 上的移植

µC/OS-Ⅱ是一个完整的，可移植、固化、裁减的占先式实时多任务内核，所有源代码都是开放的，而且没有系统空间和用户空间的区分，也就是说内核和应用程序之间是没有界线的。源代码可以在 DOS 环境下编译、链接和运行。这样，可以很方便地修改其操作系统的内核源代码。

第一节 µC/OS-Ⅱ源码简介

以 µC/OS2.52 为例，首先从 http://www.ucos-Ⅱ.com/获得源代码。以默认安装路径 C：\ SOFTWARE 为例，目录结构为：

C：\ SOFTWARE \ BLOCKS 子程序模块目录。可将用到的与 PC 相关的函数模块编译以后放在这个目录下。

C：\ SOFTWARE \ TO 这个目录中存放的是和范例 TO 相关的文件。

C：\ SOFTWARE \ uCOS-Ⅱ 与 µC/OS-Ⅱ相关的文件都放在这个目录下。

C：\ SOFTWARE \ uCOS-Ⅱ \ DOC 与 µC/OS-Ⅱ相关所有文档文件。

C：\ SOFTWARE \ uCOS-Ⅱ \ EX1_ x86L 这个目录里包括例 1 的源代码。

C：\ SOFTWARE \ uCOS-Ⅱ \ EX2_ x86L 这个目录里包括例 2 的源代码。

C：\ SOFTWARE \ uCOS-Ⅱ \ EX3_ x86L 这个目录里包括例 3 的源代码。

C：\ SOFTWARE \ uCOS-Ⅱ \ EX4_ x86L.FP 这个目录里包括例 4 的源代码。

C：\ SOFTWARE \ uCOS-Ⅱ \ Ix86L 这个目录下包括依赖于处理器类型的代码。此时是为在 80x86 处理器上运行 µC/OS-Ⅱ而必须的一些代码，实模式，在大模式下编译。

C：\ SOFTWARE \ uCOS-Ⅱ \ Ix86L-FP 这个目录不同于上一个目录是应用程序中的任务允许使用浮点处理单元（FPU），µC/OS-Ⅱ在做任务切换时保存 FPU 寄存器。

C：\ SOFTWARE \ uCOS-Ⅱ \ SOURCE 这个目录里包括与处理器类型无关的源代码。这些代码完全可移植到其他架构的处理器上。

第二节 开发工具简介及安装

以 Borland C/C++ V3.1 为例，默认安装路径 C：\ BC31，常用目录结构：

C：\ BC31 \ BIN 全部编译链接的开发工具都放在这个目录下；

C：\ BC31 \ LIB 全部链接库文件存放在这个目录下；

C：\ BC31 \ INCLUDE 全部头文件都存放在这个目录；

由于部分代码为汇编语言，还需要汇编器，采用 Borland Turbo Assember V3.0 的汇编器，默认安装路径 C：\ TASM。（若 BC31 中附带 TASM 则不需要）

第三节 移植及配置详细步骤

第一步 在 C：\ SOFTWARE \ uCOS-Ⅱ \ EX1_ x86L \ BC45 \ SOURCE \ 目录中，用写字板打开 TEST. lnk 文件 [注意 TEST. lnk 文件不是直接以文档文件形式存在的，而是一个名为 TEST 的一个快捷方式图标。因此不能被直接打开。只要 TEST 图标拖入文档编辑器（如记事本）中就可以看到下述代码]。

/v /s /c /P-   +
C：\ BC45 \ LIB \ C0L. OBJ   +
……
C：\ BC45 \ LIB \ EMU. LIB   +
C：\ BC45 \ LIB \ MATHL. LIB +
C：\ BC45 \ LIB \ CL. LIB

修改为：

/v /s /c /P-   +
C：\ BC31 \ LIB \ C0L. OBJ   +
……
C：\ BC31 \ LIB \ EMU. LIB   +
C：\ BC31 \ LIB \ MATHL. LIB +
C：\ BC31 \ LIB \ CL. LIB

第二步 进入 C：\ SOFTWARE \ uCOS-Ⅱ \ EX1_ x86L \ BC45 \ TEST \，编辑 test. mak 文件，在其中增加 TASM 的安装路径 BTASM = C：\ TASM 并将源文件中的 ASM = $（TASM）\ BIN \ TASM 修改为 TASM 的安装路径 ASM = $（BTASM）\ BIN \ TASM（若 BC31 中附带 TASM 则不需要）。

第三步 编辑同目录下的 maketest. bat 文件。将范例原文件的 C：\ BC45 \ BIN \ MAKE-f TEST. MAK 修改为 C：\ BC31 \ BIN \ MAKE-f TEST. MAK

第四步 运行 maketest. bat，在 C：\ SOFTWARE \ uCOS-Ⅱ \ EX1_ x86L \ BC45 \ SOURCE \ 及 C：\ SOFTWARE \ uCOS-Ⅱ \ EX1_ x86L \ BC45 \ WORK \ 下就能找到生成的 TEST. EXE 文件。

第五步 打开 C：\ SOFTWARE \ uCOS-Ⅱ \ EX1_ x86L \ BC45 \ SOURCE \ TEST. C 可以对其进行修改。修改后进入 DOS 环境。在 C：\ SOFTWARE \ uCOS-Ⅱ \ EX1_ x86L \

BC45 \ TEST \ 下输入命令 MAKETEST，就可以对程序进行编译和连接。然后输入 TEST 就可以运行 TEST. EXE。如图 3-1 所示。

第六步 运行 C：\ BC31 目录下 bin 文件夹中的 TD. EXE 文件，执行 File \ Open

命令，打开 C：\ SOFTWARE \ uCOS-Ⅱ \ EX1_ x86L \ BC45 \ WORK \ test.exe，就可以进行源码调试了。

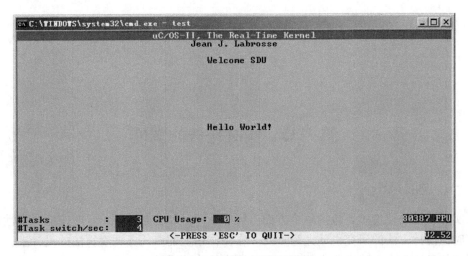

图 3-1　μC/OS-Ⅱ运行界面

第四章 μC/OS-Ⅱ调度的研究与改进

第一节 实时任务的调度

一、调度策略

实时软件的评价主要包括两个方面：可靠性评价和实时性评价，对应的可靠性设计和实时性设计方法的研究也就尤为重要。在操作系统的多任务调度算法的设计上，要根据系统的具体需求来确定调度策略，实时调度策略可以按不同的方法分为：静态/动态；基于优先级/不基于优先级；抢占式/非抢占式；单处理器/多处理器。其中，静态是指在任务的整个生命期内优先级保持不变，任务的优先级是当系统建立任务时确定的。动态是指在任务的生命期内，随时确定或改变它的优先级别，以适应系统工作环境和条件的变化。

非抢占式的调度算法要求每个任务自我放弃 CPU 的所有权，异步事件还是由中断服务来处理。非抢占式调度的优点是：响应中断快、几乎不需要使用信号量来保护共享数据，但是它存在一个非常大的缺陷是响应时间，高优先级的任务已经进入就绪态，但还不能运行，而且其等待时间是个不确定数，这对于实时系统的实时性要求来说是不容许的，所以在商业软件中几乎没有这类内核。

1. 静态优先级法则通常按照下述原则确定任务的优先级

（1）按任务的类型。例如，执行操作系统核心程序的任务，前台作业的任务可给予较高的优先级，这些任务或者运行频繁，或者要求快速的响应。

（2）按资源的使用要求。例如，要求 CPU 时间较短或要求内存量较少的进程可给予较高的优先级。这种算法称作短作业优先。

（3）按作业到达系统的时间次序，即先来先服务算法。

（4）按用户的要求指定优先级别。

静态优先级的特点是实现简单，但不灵活。

2. 动态优先级的改变也遵循一定的原则

例如，在 UNIX 系统中的进程调度采用的调度算法，就是动态优先级，进程的优先级根据一组参数周期性的重复计算，参数有：进程已用的 CPU 时间量、进程运行所需的内存总量等。该系统中优先级的原则如下。

（1）一个进程占用 CPU 的时间越长，优先级就越低；反之，一个进程离开 CPU 的时间越长，优先级就越高，调度的可能性就越大。

（2）执行紧急程序的进程优先。例如，执行操作系统核心程序进程的优先权要比执行用户程序的进程优先权高，因为操作系统核心程序是管理者程序，它常常涉及公共资源的使用，优先执行这些进程，可以减少其他进程为等待公共资源而排队的现象。

另外，我们需要说明一下时间片轮询调度法（Round-Robin Scheduling），该方式仅在同优先级调度时有效。时间片轮询调度通过限定任务时间片参数，以防止同优先级中某一任务独占 CPU。在一般情况下，实时系统应用中很少需要使用时间片调度方式（除了防止某个任务独占 CPU 之外），同时由于如下原因，这种调度方式很少使用。

一是任务运行时需要不停地减少时间片计数器的值，增加了系统开销。

二是一般情况下，将多个任务设计成共同优先级是因为它们之间共享某个不可重入的重要区域，防止这些任务间的相互抢占而设计的，若此时使用时间片的调度机制，会增加系统的不安全性。

本文重点讨论以下两种调度算法，速率单调调度算法 RMS（Rate-Monotonic Scheduling Algorithm）和截止期最早优先调度算法 EDF（Earliest Deadline First Scheduling Algorithm），其中 RMS 算法是当前静态优先级调度算法中的最优算法，而 EDF 是动态优先级调度算法中的最优算法。在实际的嵌入式实时操作系统设计中，主要就是利用这两种调度算法之一，而且通过研究这两种最优的经典算法，不仅可以学习任务调度算法中的基本知识，而且还能够可以让我们在这个基础上进一步完善和改进我们自己的算法设计。在此基础上我们还要进一步讨论另外一种现在较为流行和先进的调度算法：多级反馈队列调度算法，该算法是根据任务执行情况的反馈信息而对任务队列进行组织和调度的。

二、速率单调调度算法

在讨论 RMS 和 EDF 算法之前，首先需要申明以下三个假设。

假设 1：没有任务具有任何不可优先抢占段，而且优先抢占的耗费是可以忽略的。

假设 2：只有数据处理的需求是重要的；内存、I/O 和其他资源需求是可忽略的。

假设 3：所有任务都是独立的，同时不存在优先约束。

以上假设简化了我们将要讨论的调度算法。

假设 1 表明，我们可以在任何时候抢占任何任务并且在不付出任何代价的情况下恢复该任务，所以被抢占的任务次数并不会改变处理器总的工作量。

假设 2 可知，对于任务的可行性，我们只需要确保系统有足够的处理能力可以执行任务并且使任务满足其最终期限，而内存或者其他的拘束条件并不能使问题复杂化。

假设 3 中的不存在优先约束的意思是任意任务释放的次数并不依赖于其他任务的完成次数。

RMS 调度算法是一种在实际中被广泛研究和使用的算法之一。它是最理想的单处理器基于静态优先级的可抢占式调度算法，也就是说，当任意一种静态优先级算法可以产生可行的调度时，RMS 算法也可以使用。

在研究该算法之前，除了上述我们阐述的三个假设之外，还需另外提出两个假设。

假设4：任务集合中的所有任务都是周期性的。

假设5：一个任务的相对结束期限等于它的生命期。

假设4只是为了讨论方便所做出的假设。

假设5极大的简化了我们对RMS的分析，因为它确保任何时候，在多个生命期内的任务中至多有一个任务在运行。

在RMS算法中，一个任务的优先级和它的周期成反比，也就是说一个具有较小周期的任务可以分配到一个较高的优先级。同时具有较高优先级的任务可以抢占较低优先级的任务。

当存在一个任务集里的任意一个任务在任何时刻系统都可以满足它的时间要求，那么这个任务集可以被称为"可调度"的任务集。RMS调度算法就是这样一种算法，通过对它分配给每个任务一个固定的优先级来最大化任务集的"可调度性"。

静态优先级调度算法的一个最大的局限就是它不可能总是能够使得CPU完全被利用。虽然如此，RMS算法还是一种最优的固定优先级调度算法，它的处理器利用率公式如下：

$$W_n = n(2^{\frac{1}{n}} - 1) \tag{4.1}$$

式中，n是系统中任务的个数。从公式可以看出，当系统中只有一个任务时，最好情况处理器利用率是100%，但是随着任务数的增多，利用率随之降低，最终达到它的极限值69.3%（ln2）左右。所以可以得出：当系统中所有任务的总利用量不超过$n(2^{1/n} - 1)$时，RMS算法将可以调度所有的任务满足它们各自的时间要求。但这是充分不必要条件，在实际系统中，总有任务集具有较大并超过此限制的总利用量。

以下我们阐述对于RMS算法可调度性的充分必要条件：

$$W_i(t) = \sum_{i=1}^{t} e_i \left[\frac{t}{p_i} \right] \tag{4.2}$$

$$L_i(t) = \frac{W_i(t)}{t} \tag{4.3}$$

$$L_i = \min_{1 \leq i \leq t} L_i(t) \tag{4.4}$$

$$L = \max\{L_i\} \tag{4.5}$$

其中，$W_i(t)$是任务集$T_1, T_2, \cdots T_i$总的任务量，这些任务初始于时间段$[0, t]$。

RMS算法任务可调度性充要条件：对于给定的n个周期性任务（优先级分别为$P_1 \leq P_2 \leq \cdots \leq P_n$），当且仅当$L_i \leq 1$时，任务$T_i$可以被RMS算法合理调度。

三、截止期最早优先调度算法

EDF算法是一种动态优先级调度算法，任务优先级在其初始时并不固定，而根据它们的绝对最后期限来改变，所以遵照EDF算法调度的处理器总是执行所有任务中最早达到绝对最后期限的任务。

在讨论EDF算法中，我们将保留在讨论RMS算法时的所有假设，但除了所有任务都是周期性的这一假设。EDF 算法是一种最优的单处理器动态调度算法，也就是说，假如 EDF 算法在单处理器上对一个任务集不能合理调度，那么其他算法也不能够对该任务集合理调度。

EDF 调度算法的实现步骤如下：

检查任务队列中所有就绪任务；比较所有任务截止期；将具有最早截止期的任务给予最高优先级，即最先执行该任务。

EDF 算法的优点在于效率高、容易计算和推断，这是动态优先级调度研究的一个主要组成部分。EDF 的缺点在于，理论表明这种算法能对可调度负载进行优化，但是它不能解决过载问题。发生过载时，EDF 性能退化很快。

EDF 调度的实现相对容易。执行队列总是由下一个时限来分类。当一个任务处于激活状态时必须将该任务加入执行队列中，或者如果任务的时限在当前正在执行的任务的时限之前，那么该任务就会抢先占用当前的任务。

通常周期任务的最终期限是它下一个周期开始的时刻。Liu 和 Layland 证明，如果存在某一种调度能够满足所有的时限要求，那么每一次总是执行当前具有最早时限的未完成任务就可以满足所有时限的要求，并证明了对于一个 n 个周期任务的任务集 $T=\{t_1, t_2, t_3, \cdots, t_n\}$ 可以被 EDF 所调度，C_i 为任务执行的时间，当且仅当：

$$\sum_{i=1}^{n} \frac{C_i}{T_i} \leq 1 \tag{4.6}$$

这是一个 EDF 策略可调度的充分必要条件。这也意味着如果不满足此条件，则没有一个可行的调度算法来调度此任务集。

如果所的任务都是周期性的（这些任务的出现仅仅由于时间的流逝），那么抢先占用就没有必要。这样就简化了调度程序并且降低了任务切换的开销。如果任务可以在周期中的任何时刻执行，那么可实现性分析就简单可行。也就是说，早期的结果是可以接受的，并且任务准备好随时执行。从另一个角度来看，这样就允许任务的执行可以根据周期小幅变化。

EDF 算法在实际应用中也存在一些问题，如动态调度系统开销太大、在过载情况下会出现不稳定现象、优先级反转等。

四、多级反馈队列调度算法

Unix4.3BSD 中使用以多级反馈队列为基础的进程调度算法。所有可运行的任务都分配给一个调度优先级，这个优先级决定了它们将放入哪个运行队列中，在选择一个新的任务运行时，系统从最高优先级向最低优先级搜索运行队列，并选出第一个非空队列中的第一个任务，如果一个队列中有多个任务，则系统按先来先服务原则分配处理机来运行它们，即系统分配以相同的时间量，按照他们在队列中的顺序运行它们，如果一个任务被封锁，它不被放回任何运行队列中。如果一个进程用完分配给它的时间段，则把它放回它原来队列的末端，并选出该队列前端的进程投入运行。时间片越短，交互响应越好。然而，较长的时间片提供了较高的系统总体吞吐量，因为系统减少了用于环境切换的开销,并且处理器的快速缓存将较少刷新。

多级反馈队列调度算法过程如图 4-1 所示。

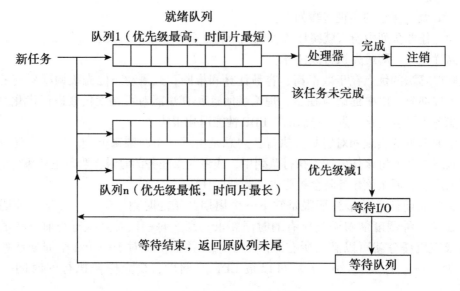

图 4-1 多级队列反馈调度算法

对图 4-1 有以下说明。

◆ 各级就绪队列中的任务具有不同的时间片,优先级最高的第一级队列中任务执行时间片最小,随着队列的级别增加,其任务的优先级降低了,但是时间片大小却增加了。队列提高一级,其时间片增加一倍。

◆ 各级队列均按先来先服务原则排序。

◆ 当第 n-1 级任务就绪队列为空后调度程序才去调度第 n 级就绪队列中的任务,调度方法相同。所有就绪队列中优先级最低级队列中的任务是采用时间片轮询调度。

◆ 当比运行任务更高级别的队列中到达一个新的任务时,它将抢占先运行任务的处理机,而被抢占的任务回到其原队列末尾。

Unix4.3BSD 使用的时间是 0.1 秒,这是一个在不影响编辑器之类交互作业所需响应时间的前提下最长时间片的经验值。为了反应对资源的要求和进程消耗的资源量,系统将动态调整进程的优先级,进程根据它们调度优先级的变化在各个运行队列间移动,也就是多级反馈队列反馈的意义。

UNIX 中优先级的计算公式:

$$进程优先级 = PUSER + \left[\frac{P\_CPU}{4}\right] + 2 \times CG \quad (4.7)$$

式中,PUSER = 50 是用户态下执行的基始优先级;

P_ CPU 是进程最近使用 CPU 的估计值;

CG 是一个用户可设置的常数因子,范围在 -20~20,正常值是 0。

计算所得大于 127 的值都定为 127。这种计算方法使得优先级随着最近 CPU 使用量线性降低。用户可控的常数因子起到了一个有限的加权因子作用。

对于参数 P_ CPU 可以根据下面的公式调整：

$$P\_CPU = \left[\frac{2 \times load}{2 \times load + 1}\right] \times P\_CPU + P\_nice \qquad (4.8)$$

式中，load 是前一次一分钟时间间隔的系统操作中运行队列长度的抽样平均值。

从上面的讨论来看，多级反馈队列调度算法在系统调度中根据任务每次完成情况的反馈来动态调整任务属性，从而达到最优的调度效果。该算法集合了各种调度算法的优点和反馈思想，是现代操作系统中较为先进的一种调度算法。

第二节　μC/OS-Ⅱ内核调度算法的改进

μC/OS-Ⅱ系统采用静态优先级分配策略，由用户来为每个任务指定优先级。虽然任务的优先级可通过 OSTaskChangePrio（）函数改变，但函数功能简单，仅以用户指定的新优先级来替换任务当前的优先级。随着嵌入式技术的发展，对嵌入式系统的要求越来越高，多样化的调度方法已成为一种趋势。在发展较成熟的实时系统中有多种任务调度方法可以借鉴，如截止期最早优先算法、速率单调算法以及可达截止期最早优先算法等。

一、速率单调算法改进任务调度

速率单调算法是一个经典的实时任务调度算法，主要用来处理周期性的任务。在此算法中，任务的优先级按任务的周期来分配，任务的周期越短优先级越高，反之亦然。

速率单调算法是一个静态调用的算法，需要事先由用户按任务的周期给每个任务分配固定的优先级。由于 μC/OS-Ⅱ本身采用的即是静态调度方法，因此只需由用户按照任务的周期确定其优先级并将系统中的 OSTaskChangePrio（）函数屏蔽即可，并不需要对内核代码做较大改动。

尽管这种改进对系统的改动最小并且 RMS 算法已被证明是静态最优的，但如前文所述这种算法本身存在局限性，因此在实际应用中常需要用动态的 RMS 改进。RMS 算法动态改进有很多种方法，考虑到 μC/OS-Ⅱ本身的特点，我们没有采用通用的做法，而是应用了存储系统中写回法的思想完成了改进。写回法是在存储系统的 Cache—主存层次中，为了保持 CACHE 与主存内容的一致，在 CPU 执行写操作时，信息只写入 Cache 中，当 Cache 内容需要替换时，才将改写过的 Cache 块先送回主存（写回），然后再调块。这种方法有利于省去许多将中间结果写入主存的无谓开销。在动态算法中，由于事先并不知道任务优先级间关系，建立或删除一个任务就有可能引起任务优先级的变化。这种优先级变化有可能很频繁，而且系统开销很大，而写回法的最大的优点就是节约系统开销，所以这里引入了写回法来实现动态的 RMS。这里引入了一个数组，在任务优先级改变的时候用来记录中间结果，当就绪任务排序结束的时候，根据数组中的记录更新就绪任务的优先级，从而节约了大量系统时间。为此，需为系统增加表 4-1 所示的项目。

第四章 μC/OS-Ⅱ调度的研究与改进

表 4-1 RMS 改进中增加的项目

项 目	功 能
INT8U cycal	在 OS_TCB 增加的数据项，用来记录任务的周期，cycal 的值在任务创建的时候由用户给出
INT8U csRec [64] [2]	公共数组，用来记录每个任务的周期并在就绪任务优先级排队的过程中记录任务优先级的变化

在 RMS 算法中，系统创建的时候需要由用户来提供任务的周期而不是优先级，因此将 OSTaskCreate () 中的参数 INT8U Prio 改为 INT8U cycal，并在函数内定义局部变量 INT8U Prio 以记录分配给任务的优先级。在本文所述 RMS 改进中，需要为μC/OS-Ⅱ系统编写 OSTaskPrioCreate () 函数，用来在创建或删除一个任务的时候处理任务优先级调整的工作。该模块流程如图 4-2 所示。在对就绪任务优先级进行调整的时候，该模块首先在数组中对任务的优先级完成调整并记录任务优先级的调整情况。在执行此函数后，就绪任务队列中的任务的优先级可能会有改变，因此还需要在 μC/OS-Ⅱ系统中添加 prio_adjust () 函数，该函数应用 μC/OS-Ⅱ系统原有的 OSTaskChangePrio () 函数来对就绪任务更新，代码如下所示：

```
vid prio_adjust () {
int i;
for (i=0; i<LOWEST_PRIO-8; i++) {
    if (csRec [i] [0] <0) {
OSTaskChangePrio (i, i+csRec [0] );
    }
  }
}
```

为防止多个任务同时调用 OSTaskPrioCreate () 函数造成混乱，这段代码应按临界资源来处里，需要在调用前关中断，调用后再开中断。

二、截止期最早优先算法改进任务调度

RMS 算法主要是针对周期性的任务，在实际应用中还有许多非周期性的任务，而且在未来的应用中这种需求会更普遍，这里引入已被证明是动态最优的截止期最早优先算法，以期对 μC/OS-Ⅱ系统进行改进。

表 4-2 截止期最早优先算法改进中添加的项目

项 目	功 能
INT8U deadline	在 OS_TCB 增加的数据项，用来记录任务的截止期。在任务调度时用当前系统时间 OSTime 和 cstime 的差更新 deadline
INT8U csRec [64] [3]	记录每个优先级任务的调用时间、截止时间、优先级附加值（记录任务优先级的变化）。任务创建或销毁时要根据其截止期分配合理的优先级并插入此数组，在任务发生调用时用数组来更新其相应项的值

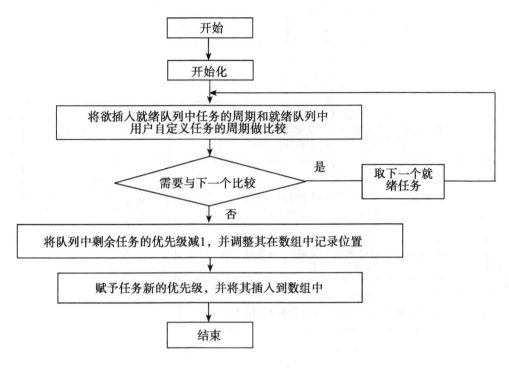

图 4-2 RMS 算法流程

在截止期最早优先算法中，系统按任务的截止期给每个任务分配优先级。任务的截止期越早其优先级越高，反之亦然。为此，本文采用写回法的思想，需要为系统增加表 4-2 所示的项目。

在截止期最早优先算法中，需要用户为任务指定其截止期。在本文所述改进中将 OSTaskCreate（）和 OSTaskCreateExt（）中的参数 INT8U Prio 改为 INT8U deadline，并在函数内定义局部变量 INT8U Prio 来记录分配给任务的优先级。与动态 RMS 改进相同，该算法改进也要在系统中增加 OSTaskPrioCreate（）函数，但函数的分配优先级的方法不同，不再按周期分配而是按任务的截止期分配，流程图如图 4-3 所示。

执行 OSTaskPrioCreate（）函数后，系统要同样调用 prio_ adjust（）函数更新就绪任务的优先级，其代码与 RMS 算法改进中相同，这些代码同样要用当作临界代码来处里。

三、可达截止期最早优先算法改进任务调度

截止期最早优先算法并不考虑任务的截止期是否已超时，在系统中调用超时的任务会造成系统时间的浪费，在时序要求严格的系统中还可能造成严重的后果。可达截止期最早优先算法是对截止期最早优先算法的一种改进，实质是使截止期最早的任务优先级最高，同时要保证在调度时超过截止期任务的不予调度。系统首先要根据系统当前的时

第四章 μC/OS-Ⅱ调度的研究与改进

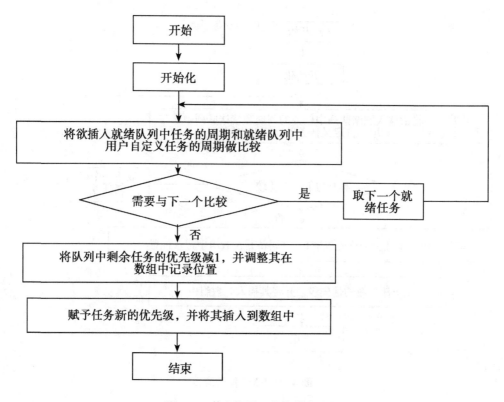

图 4-3 截止期最早优先算法流程

间、任务的调用时间来更新任务的截止时间。如果计算结果小于零，说明任务超时，就要删除这个任务。如果任务没有超时则与截止期优先算法中任务的处理方法相同。在本文所述可达截止期最早优先算法改进中，需在 μC/OS-Ⅱ系统中增加表 4-3 所示的项目。

表 4-3 可达截止期优先算法改进中添加的项目

项目	功能
INT8U deadline	在 OS_TCB 增加的数据项，用来记录任务的截止期。在任务调度时用当前系统时间 OSTime 和任务被调用的时间 cstime 的差更新 deadline
INT8U cstime	在 OS_TCB 增加的数据项，记录任务首次调用时的系统时间
INT8U csRec [64] [3]	记录每个优先级任务的调用时间、截止时间、优先级附加值（记录任务优先级的变化）。任务创建或销毁时要根据其截止期分配合理的优先级并插入此数组，同时用数组来更新其相应项的值

本文所述可达截止期最早优先算法的改进与截止期最早优先算法的改进多数工作相同，主要差异是可达截止期最早优先算法要判断任务的截止期是否超时，系统不会调度超时的任务。为此在截止期最早优先算法改进中所述 OSTaskPrioCreate（）中添加对任

务截止期是否超时的判断。流程如图 4-4 所示。

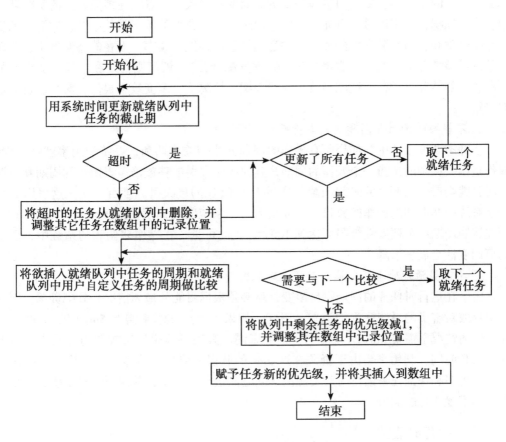

图 4-4　可达截止期最早优先算法流程

第三节　各种调度算法评估

一、静态算法评估

静态调度是在系统开始运行前进行调度，严格的静态调度在系统运行时无法对任务进行重新调度。静态调度算法的输入至少包含了所有任务的列表以及它们各自的运行时间。调度的目标是把任务分配到各个处理机，并对每一处理机给出所要运行任务的静态运行顺序。RMS（Rate Monotonic Scheduling），即速率单调调度，是许多实时系统所使用的一种静态调度方法。这种算法被广泛地应用于任务周期性执行并且各个任务之间不需要同步的场合。静态调度能够在较低的开销条件下保证任务运行时的优先级。然而，静态调度有以下缺点。

1. 对非周期任务的调度效率低

静态调度并不区分周期的和非周期的任务，而将它们等同视作周期性的任务，将最大可能发生的频率视作系统的 rate。为了降低调度延时，实时系统的资源被不定时（sporadic）的服务器直接或间接地分配给非周期性的操作。然而在多数典型的实时环境中，非周期操作并不以最大可能发生频率出现。一个典型的例子就是中断，它潜在的发生频率非常高，但实际发生的频率往往远远低于潜在发生频率。而在静态调度中，资源必须以最差情况来分配（根据潜在发生频率分配资源），调度则必须在中断可能出现的最高频率的情况下进行。这就导致了在非周期的情况下，调度效率很低，对系统资源浪费严重。

2. 不能有效的对非调和的周期性操作进行调度

在静态调度系统中，如果所有操作的周期并非相关可调和，系统利用率就会减少。操作是相关调和的是指一个操作的周期是其他操作周期的整数倍。当操作的周期互不调和时，就会产生未调度的时间间隙，这减少了 CPU 的最大可调度百分比。这里用 n 来表示系统中非调和的操作的数目，当 n 很大时，系统的可调度性的极限会比 69%稍大。而在调和周期中可调度性极限可达到 100%。由此可见，在操作非调和的系统中，系统的可调度性会显著下降。

3. 不能灵活地处理任务调用时资源需求的变化

由于在运行时任务的周期不易改变，就必须根据可能的最坏情况来分配资源，则有可能导致对宝贵的系统资源的浪费。因此，如果一个操作通常需要 5ms CPU 时间，而在特定的情况下需要 8ms CPU 时间，则静态调度策略必须假设每次调用此操作都需要 8ms CPU 时间，利用率会由于资源在这 3ms 的耽误而降低。总体而言，静态调度在资源的利用率和保证任务运行时对资源的访问之间进行了折中，限制了实时系统适应系统状态和配置变化的能力。

二、动态算法评估

1. 截止期最早优先算法（Earliest Deadline First）评估

EDF 是一种基于任务的截止期对任务的优先级进行排序的动态调度策略。该算法通过给距截止期较近的任务赋予较高的优先级来保证距截止期最近的任务在距截止期较远的任务之前被调用。只要有任务的调度请求，EDF 调度策略就被调用。新调度的任务可能会也可能不会优先占有当前执行的任务，这要看优先级部件是如何映射到线程优先级的。

EDF 最主要的局限是不管余留下的时间够不够在截止期之前完成此次执行，距离截止期最近的任务都会被调度。因此，若一个操作在截止期之前不能够完成，那么在截止期过去之前是检测不出的。若一个在截止期之前不能够完成的操作被调度了，本可以分配给其他任务的 CPU 时间就被浪费掉了。可达截止期最早优先算法是对截止期最早优先算法的改进，若一个任务在截止期之前不能够完成则系统不予调度。

2. 最小松弛度优先算法（Least Laxity First）评估

松弛度是用此刻到截止期的时间减去任务的剩余执行时间。LLF 通过综合考虑任务的剩余执行时间和任务的截止期来改善 EDF 算法。该算法先调度松弛度最小的任务。使用 LLF 可以检测出到最终期限之前不能完成的任务。若发生这种情况，在分配余下的 CPU 时间时，调度会管理这些任务。例如，一种策略就是简单的摘除那些松弛度不够的任务，这种策略会减小后面的任务错过截止期的概率，当系统暂时过载时效果更为明显。

3. EDF 算法和 LLF 算法比较

EDF 和 LLF 都是动态调度算法的典型代表。从调度的角度上而言，EDF 和 LLF 克服了 RMS 的利用率限制。EDF 和 LLF 也可以处理调和与非调和周期的情况，而且在资源需求中它们可以方便地反映调用的变化，允许将一个任务的 CPU 时间重新分配给其他任务，因此它们可以产生以 CPU 利用率为衡量标准的最佳调度。另外，EDF 和 LLF 都可以在单独静态优先级水平上调度任务而不需用频率来确定任务的优先级。但从执行的角度看，像 EDF 和 LLF 这样的动态调度策略需要较高的系统开销来在运行时对任务进行评估。另外，动态调度策略在调度量过大时对那些错过最终期限的任务不提供有效控制手段。当一个任务被加入调度队列以获得更高的利用率时，所有任务的实时安全性就降低了。因此对所有任务来说，当系统过载时，任务错过截止期的机率增加了。

4. 静态和动态调度算法的选择

选择用动态调度还是静态调度是很重要的，这会对系统产生深远的影响。静态调度对时间触发系统的设计很适合，而动态调度对事件触发系统的设计很适合。静态调度必需事先仔细设计，并要花很大的力气考虑选择各种各样的参数。动态调度不要求事先做多少工作，它是在执行期间动态地做出决定。

在资源利用方面动态调度比静态调度有更大的潜力。对于静态调度，系统通常需要过大范围的设计，以能够处理最不可能的事件。可是，在硬实时系统中，资源浪费通常是为保证满足系统所有的时限而付出的代价。

另一方面，若给定足够的处理能力，对静态系统一个最优或次优的调度策略可事先获得。例如，对一个反应堆控制的应用，花费数月的 CPU 时间来找出最好的调度策略是值得的。一个动态系统在运行期间无法承受复杂的调度计算花费。因此为了安全，就不得不在过大范围内设计，即使如此，也不能保证系统能够满足对它的指标要求，而是需要大量的测试。

第五章 优先级反转问题的研究与改进

第一节 进程优先级反转现象的研究

嵌入式实时操作系统一般都是可剥夺型内核,以保证最重要的进程(往往是优先级最高的进程)能够及时得到运行。但是如果用传统的信号量等机制对共享资源进行互斥操作,在某些不特定的时间里会出现高优先级的进程被低优先级的进程堵塞的现象。这种现象称为优先级反转。

Lampson 和 Redell 最早对优先级反转这一现象进行讨论。Lui Sha, John Lehoczky 和 Ragunathan Rajkumar 对这现象进行了研究和分析,并且提出了解决方法。

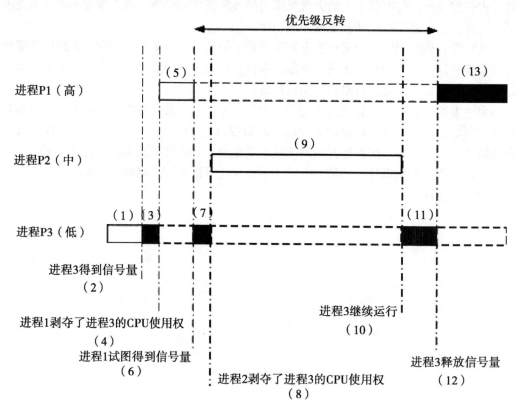

图 5-1 优先级反转问题

使用实时内核，优先级反转问题是实时系统中出现得最多的问题。图 5-1 解释优先级反转是如何出现的。图中进程 1 的优先级高于进程 2，进程 2 的优先级高于进程 3。

（1）进程 1 和进程 2 处于挂起状态，等待某一事件的发生，进程 3 正在运行。

（2）此时，进程 3 要使用共享资源。使用共享资源之前，先得到该共享资源的信号量。

（3）进程 3 得到了该信号量，开始使用该共享资源。

（4）由于进程 1 的优先级高，它等待的事件到来之后，剥夺了进程 3 的 CPU 使用权。

（5）进程 1 占用 CPU，开始运行。

（6）进程 1 需要使用进程 3 正在使用的资源。但由于该资源的信号量还被进程 3 占用，进程 1 只能进入挂起状态，等待进程 3 释放该信号量。

（7）进程 3 继续运行。

（8）由于进程 2 的优先级高于进程 3，当进程 2 等待的事件发生后，进程 2 剥夺了进程 3 的 CPU 使用权并开始运行。

（9）进程 2 占用 CPU，开始运行。

（10）进程 2 处理完毕，将 CPU 使用权还给进程 3。

（11）进程 3 继续运行，直到释放共享资源的信号量。

（12）高优先级的进程 1 得到该信号量，内核做任务切换，进程 1 得以运行。

（13）进程 1 得到信号量，继续运行。

在这种情况下，进程 1 的优先级实际上降到了进程 3 优先级的水平。因为进程 1 要等待，直到进程 3 释放占用的共享资源。由于进程 2 剥夺了进程 3 的 CPU 使用权，使进程 1 的状况更加恶化，并使进程 1 增加了额外的延迟时间。进程 1 和进程 2 的优先级发生了反转。

从软件的角度来说这个问题来自进程之间在运行上的依赖性，当一个优先级较高的进程的运行必须等待另一个优先级较低的进程完成其某一段的运行时，就可能会产生优先级反转的问题。而进程之间在运行上的这种依赖性则可以归纳为以下几类。

（1）因共享资源的互斥使用而造成的依赖。当优先级较低的进程在使用共享资源时，优先级高的进程不能将相应的共享资源剥夺过来，而只能等待。

（2）因数据的流通方式所造成的依赖。如果优先级较低的进程是某项数据的"生产者"（或提供服务者），而优先级较高的进程是此数据的"消费者"（接受服务者），则当然只能等待。

（3）因为在等待队列中的位置所造成的依赖。在一个先进先出的队列中，如果优先级低的进程排在前面，而优先级高的进程排在后面，就只好等待。

从前面的理论分析可以看出，一个没有解决优先级反转问题的可剥夺性内核在某些特殊场合下的实时性要差于不可剥夺内核，所以对于实时系统来说，内核的可剥夺性和优先级反转问题是两个密切联系的问题。因此解决优先级反转问题具有很高的理论意义和实践价值。

第二节 优先级反转的理论解决

Sha 等人提出了两种解决优先级反转的方案。即优先级继承协议（Priority Inheritance Protocol）、优先级上限协议（Priority Ceiling Protocol）。

一、优先级继承协议的原理

优先级继承协议的基本思想是：当一个进程阻塞一个或多个优先级更高的进程时，将该任务的优先级暂时提高到被它阻塞的所有进程中具有的最高优先级，从而使其能够抢占它所阻塞的所有进程而进入临界区，并且不影响与它所进入的临界区无关的其他高优先级的进程的执行，当它退出临界区时就恢复原来的优先级。

优先级继承协议的原理如图 5-2 所示，内核支持优先级继承。下面简要解释一下处理过程。

（1）进程 1 和进程 2 处于挂起状态，等待某一事件的发生，进程 3 正在运行。
（2）进程 3 正在运行，此时申请到信号量，获得共享资源的使用权。
（3）进程 3 使用共享资源，继续运行。
（4）进程 1 剥夺了进程 3 的 CPU 使用权。
（5）进程 1 开始运行。
（6）进程 1 申请共享资源信号量。内核知道该信号量被进程 3 占用，而进程 3 的优先级低，于是内核将进程 3 的优先级提升至与进程 1 相同。
（7）进程 1 等待共享资源的信号量，进程 3 使用该信号量，继续运行。
（8）进程 3 完成，释放共享资源信号量。同时内核恢复进程 3 本来的优先级，并把信号量交给进程 1。
（9）进程 1 得到继续执行。
（10）进程 1 完成，进行调度，进程 2 优先级高。
（11）进程 2 得到 CPU 使用权，开始运行。

这样就避免了高优先级的进程 P1 被比自己低优先级进程（如进程 P2）阻塞。

二、优先级上限协议的原理

在基本的优先级继承协议的基础上，Sha 等人提出了一个改进的方案，就时优先级上限协议。与优先级继承协议不同的是，优先级上限协议定义了信号量的优先级上限。一个信号量的优先级上限与可能锁定该信号量的所有进程中优先级最高的进程相等，调度的时候不仅比较进程之间的优先级，还要比较信号的优先级上限。

在上述例子中，如图 5-2，在时段（1）进程 P3 正在处在运行状态，而进程 P1 和进程 P2 处于挂起状态，等待某一事件的发生。在时刻（2）的时候进程 P3 要使用共享资源。在这种情况下进程 P3 只能独占共享资源，因此进程 P3 在使用共享资源前必须先得到该资源的信号量。在时刻（2）的时候进程 P3 得到了该信号量，并开始使用该共享资源。在时刻（3）的时候进程 P1 就绪了。由于进程 P1 的优先级比进程 P3 高，

第一篇　嵌入式关键技术研究

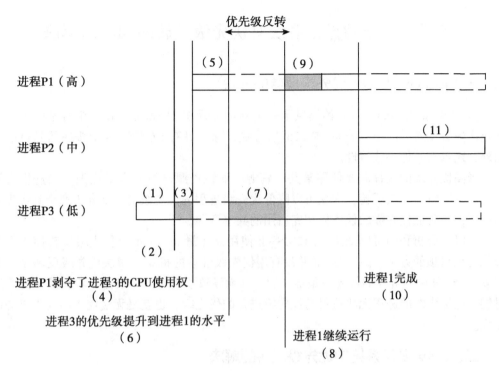

图 5-2　优先级继承协议的基本原理

在它就绪以后就剥夺了进程 P3 的 CPU 使用权开始运行。在进程 P1 运行了时段（5）后，也要使用进程 P3 正在占用着的共享资源，但由于进程 P3 还没释放该共享资源的信号量，所以进程 P1 无法获得信号量，在时刻（5）的时候只能进入挂起状态，等待进程 P3 释放该共享资源的信号量。这时候，由于占用信号量的进程 P3 的优先级低于被阻塞的进程 P1 的优先级，就说明可能会产生优先级反转。根据优先级上限协议的基本原理，将信号量的优先级提高到进程 P1 的水平并使进程 P3 继续运行。在时刻（8）的时候，进程 P3 使用完共享资源，释放占用的信号量，使进程 P1 得到信号量并剥夺进程 P3 的 CPU 使用权。时刻（10）的时候进程 P1 运行完毕，因为进程 P2 的优先级高于进程 P3，所以进程 P2 得到 CPU 使用权并开始运行。这样就避免了高优先级的进程 P1 被比自己低优先级进程阻塞。

三、非独占锁的优先级继承协议

前面讨论的都是独占锁的优先级继承协议，而未涉及非独占锁，如读锁的优先级继承协议。在 Linux 中提供了 read_lock（）和 write_lock（）两种锁机制，Write_lock（）为写锁，它需要保证写操作的一致性，一旦一个执行进程获得写锁，其他进程均被阻塞。read_lock（）为读锁，它允许多于一个的仅执行读操作的进程同时进入。Linux 中采用了一个 32 位的整型数作为锁状态标志，最高位是写锁，所以一个读锁可以同时容纳 $2^{31}-1$ 个不同进程进入。

第三节 各种操作系统对优先级反转的问题的解决

一、μC/OS-Ⅱ中对优先级反转的解决

μC/OS-Ⅱ从V2.04版开始提供了一个新的互斥机制mutex，在这个机制中采取了防止优先级反转的措施。mutex是二值信号量，除了具有μC/OS-Ⅱ中普通信号量的机制外，还具有其他一些特性。

当高优先级的进程需要使用某共享资源，而该资源已被一个低优先级的进程占用时，为了降解优先级反转，内核可以降低优先级进程的优先级提升到高于那个高优先级的进程，直到低优先级的任务使用完占用的共享资源。

从以上分析我们可以知道，μC/OS-Ⅱ使用互斥量mutex解决优先级反转使用了类似优先级封顶的方法。μC/OS-Ⅱ并没有用优先级继承的机制来解决优先级反转是因为在μC/OS-Ⅱ中，进程的优先级是唯一的。这就导致了当高优先级占用着它的优先级的时候，低优先级的进程无法获得高优先级进程的优先级，也就是无法继承高优先级的优先级。

二、Linux操作系统中优先级反转的解决

在Linux操作系统中也会存在优先级反转问题。在Linux内核中，进程的优先级虽然也是在动态地变化，但是这种变化只取决于进程对时间片的耗用，目的仅在于提高调度的公正性。没有"优先级继承"一类的措施，就不可能消除进程间因共享资源的互斥使用而造成的依赖以及因数据的流通方式所造成的依赖而产生的优先级反转。

另一方面，Linux操作系统进程间因为在等待队列中的位置所造成的依赖而产生优先级反转的情况似乎不是很明显。Linux内核的各种进程间通信机制都采用先进先出队列。这个问题的解决无需通过优先级继承一类的措施，只要把排队的规则从先进先出改成按级别排队即可。

随着Linux作为一种嵌入式操作系统，在嵌入式系统的领域中得到越来越广泛的应用，工程中对Linux内核的实时性的要求也越来越高，这就需要对内核在这些方面进行修改。而且Linux2.6版本中对进程的可剥夺性进行了修改，这也是为进一步解决Linux优先级反转问题提供了一个契机。

三、其他操作系统中优先级反转的解决

VxWorks采用优先级继承的算法来防止优先级反转。VxWorks与POSIX标准完全兼容，凡是在POSIX基础上作出了扩充改进的，就向用户分别提供两套函数，使用户在其他符合POSIX标准的系统上开发的软件用到VxWorks上，基本上只要重新编译、连接就可以运行。然后，如果有需要，如防止优先级反转，用户可以把已有的POSIX版本的函数调用换成经扩充修改后的VxWorks版本。

Windows CE 采用优先级继承的方法来解决优先级反转问题。当一个高优先级的进程试图获取一个由低优先级的进程占用的互斥量时，就提升低优先级进程的优先级。以此来解决优先级反转问题。从 Windows CE 3.0 开始，内核保证只处理优先级倒置到一个级别的深度。

第四节　优先级反转的实验模型

优先级反转是可剥夺型内核系统中一个常见的问题，由于操作系统的不同及解决方法不同，优先级反转现象的出现具有不确定性。为了更好的研究这一现象，本文根据优先级反转的基本理论模型在 μC/OS-Ⅱ 操作系统上建立了一个优先级反转的实验模型，模型程序的顺序图如图 5-3 所示。

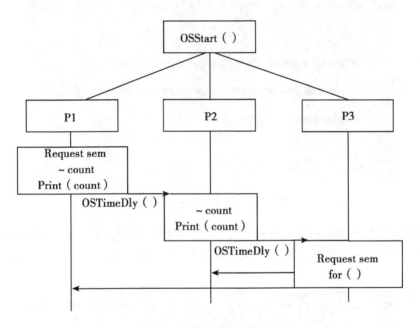

图 5-3　优先级反转模型顺序

在这个模型中，μC/OS-Ⅱ 操作系统创建了三个进程 P1、P2、P3，他们的优先级分别为 prio1、prio2、prio3，且 prio1 < prio2 < prio3，在 μC/OS-Ⅱ 中进程的优先级越高，其数值越小，故 P1 的优先级最高，P2 的次之，P3 的最小。由于 μC/OS-Ⅱ 是可剥夺型内核，当三个进程都就绪时，优先级最高的进程 P1 将获得运行，并且假设对进程状态不加以改变，进程 P1 将一直运行下去，优先级低的进程 P2、P3 将没有机会运行。模型中在 P1 运行之后调用 OSTimeDly () 函数，将自己延时等待，使优先级低的进程获得运行机会，同理，进程 P2 运行一次后也调用 OSTimeDly () 延时等待，使优先级最低的进程 P3 获得运行权。

假设进程 P1 和 P2 的延时时间相同，确保两个进程的运行次数相同，由于 P3 的优先级最低，在进程 P1 延时等待完后，自动被进程 P1 剥夺运行权，P1 和 P3 之间出现对共享资源进行竞争，而共享资源是由信号量 Sem 进行互斥操作的。在 μC/OS-Ⅱ 中信号量的创建是由函数 OSSemCreate（）实现的，对共享资源的互斥操作是由函数 OSSemPend（）和 OSSemPost（）实现的。在模型中，进程 P1 和 P3 对资源的竞争仅是在逻辑上存在的，是由函数 OSSemPend（）和 OSSemPost（）所保护的一段代码。在进程 P1 中，并没有对具体代码进行保护，仅仅是为了申请信号量。在进程 P3 中，我们在函数 OSSemPend（）和 OSSemPost（）所保护代码为一个 for 循环（延时），改变这个循环的次数可以控制进程 P3 对信号量的占用时间。

进程 P1 和 P2 对一个布尔数进行取反并打印，在正常情况下，因为进程 P1 的优先级高，所以进程 P1 打印的永远是 0，P2 打印的永远是 1，如图 5-4 所示。

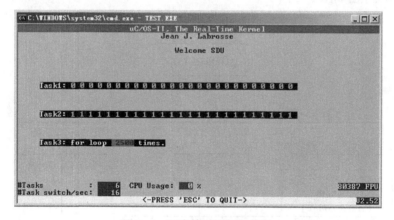

图 5-4　正常情况下运行结果

运行试验模型的结果看到，进程 P1、P2 发生了优先级反转，表现为 P1 打印 1，而 P2 打印 0。且经过一段时间之后，优先级反转现象还将出现，如图 5-5 所示。

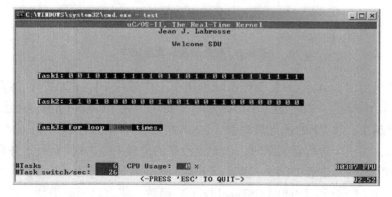

图 5-5　优先级反转时运行结果

第五节　互斥量 Mutex 的研究和改进

一、互斥量 mutex 的研究

μC/OS-Ⅱ从 V2.04 开始提供了一个新的互斥量 Mutex，在这个机制中采取了防止优先级反转的措施。互斥量 Mutex 中有个变量名为优先级继承优先级（Priority Inheritance Priority），从理论上讲，μC/OS-Ⅱ中的互斥量机制实现的是一个类似于优先级上限的算法而不是优先级继承。在 μC/OS-Ⅱ中实现优先级继承存在个特殊问题，由于 μC/OS-Ⅱ不允许两个进程有相同的优先级，所以高优先级进程的优先级就不能被低优先级进程继承。但是在系统中，那些进程要运行，以及这些进程的优先级是预知的。对每一个特定的互斥量，要使用这个互斥量的进程集合，以及这个集合中最高的优先级也是预知的，而这个互斥量的 PIP，应该比这个集合中最高的优先级 P 再高 1。当高优先级的进程申请共享资源的时候，就把占用共享资源进程的优先级提高到 PIP，这样在低优先级的进程释放共享资源以前就不会有其他优先级的进程剥夺它的运行。

互斥量的创建是由函数 OSMutexCreate（）完成的，参数 prio 就是互斥量的优先级继承优先级 PIP。函数 OSMutexCreate（）首先进行一些检查，其中最重要的是检查 PIP 的优先级数值是否已经被其他进程使用，如果没有就占用这个优先级，然后在空余事件控制列表中得到一个事件控制块 ECB，并将 ECB 设置成 OS_EVENT_TYPE_MUTEX。之后，OSMutexCreate（）置 mutex 的值为有效，同时 PIP 保存起来。最后调用 OSEventWaitListInit（），初始化等待任务列表。

创建了互斥量之后，需要访问共享资源的进程就可以通过函数 OSMutexPend（）进入受保护的临界区。函数 OSMutexPend（）的主要功能是通过对 OSEventCnt 低 8 位的判断，将进程分别处理。

当 OSEventCnt 的低 8 位等于 OS_MUTEX_AVAILABLE，即为 0xFF 时，这个 mutex 有效。OSMutexPend（）将 Mutex 赋予调用该函数的任务，且置 OSEventCnt 的低 8 位等于调用该函数的进程优先级。相当于使计数值变成 0，同时为这个进程以后可能的优先级变化做好了准备。

当 OSEventCnt 的低 8 位不等于 OS_MUTEX_AVAILABLE，则说明 mutex 被其他进程占用，相当于计数值变成了 0，所以当前进程不能进入临界区，那么调用进程应进入休眠状态，直到占用 mutex 的任务释放了 mutex。如果已在临界区中的进程优先级尚未提高到 PIP，并且又低于当前进程的优先级，就说明优先级将会反转，所以将其优先级提升至 PIP。当然，如果已在临界区中的进程处于就绪状态，就要将其从就绪队列中撤下来，在改变优先级后，再按新的优先级进入就绪队列。

如果当前进程不能进入临界区而要休眠等待，但是临界区中的进程已经提升了优先级，就是说它的优先级已经是 PIP 了，或者说本来就高于当前进程的优先级，就不需要改变其优先级了。

进程在退出临界区时需要通过调用函数 OSMutexPost（）释放互斥量。函数

OSMutexPost（）先查看占用 mutex 的任务的优先级是否已经升到了 PIP，因为有个高优先级的进程也需要这个 mutex。在这种情况下，占用 mutex 的进程优先级被降到原来的优先级，原来的优先级是从 OSEventCnt 得到。然后将调用函数的进程从就绪表中 PIP 位置上删去，放回到就绪表原来的优先级位置上，查看是否有进程正在等待 mutex。ECB 中的 OSEventGrp 不为 0，说明有进程在等待 mutex。调用函数 OSEventTaskRdy（）使等待列表中最高优先级的进程进入就绪状态，将新占用 mutex 进程的优先级保留在 ECB 中，之后调用 OSSched（）进行进程调度。如果没有等待 mutex 的进程，则 OSEventCnt 的低 8 位被置为 0xFF，表明 mutex 有效。

二、互斥量 mutex 的不足

从实验现象可以看出，互斥量机制起到了防止优先级反转的作用，但是这种机制存在不足之处。首先，使用互斥量机制需要传递一个优先级值的参数作为 PIP，这就要求在进行嵌入式系统设计时就安排好优先级，尽管这是系统设计的一部分，但是增加了系统设计的难度，也增加了系统设计错误的可能性。其次，μC/OS-II 支持 64 个优先级，而推荐使用的却只有 56 个，且每个优先级只能对应一个进程，按实现一个互斥量需要占用 3 个优先级来算，当系统中进程数很多，而都有可能会发生优先级反转时，这种机制显然是行不通的。

问题的根源来于在 μC/OS-II 中一个优先级只能对应一个进程。这是 μC/OS-II 进程管理的特色，也是进程管理算法实现的基础。μC/OS-II 的这种特点使得其在实现进程管理的时候相当简洁，但也造成了在解决优先级反转问题上的不足。因为要实现进程的优先级继承，关键在于进程的优先级可以动态变化，但是在 μC/OS-II 中很难实现这点。

三、互斥量 mutex 的改进

在 μC/OS-II 基础上对互斥量机制进行改进，具体思想是将 PIP 优先级作为一个跳板，让竞争共享资源的进程通过这个 PIP 优先级进行交换。如低优先级的进程 P3 阻塞了高优先级的进程 P1 时，把进程 P3 的优先级和 PIP 交换，把进程 P1 的优先级和 P3 的优先级交换，然后把 PIP 和进程 P1 交换，这样做的结果是进程 P1 和 P3 实现了优先级交换。当进程 P3 用完共享资源，释放信号量以后，用同样的方法恢复到原来的优先级。

在这个过程中，PIP 仅是被用来交换两个进程优先级的中间变量，在进程交换完毕之后，PIP 就不会被任何进程占用，通过这种方式可以减少使用 PIP 的数量，同时达到了优先级动态变化的目的。

实现这个机制中存在一个主要问题是在现有的 μC/OS-II 中没有现成的交换进程优先级的函数可用。我们可以对函数 OSTaskChangePrio（）进行修改，编写函数 OSTaskChangePrio_PIP（），在实验模型上实现这种机制，从实验结果来看（图 5-6），修改后的系统成功的避免了优先级反转问题。

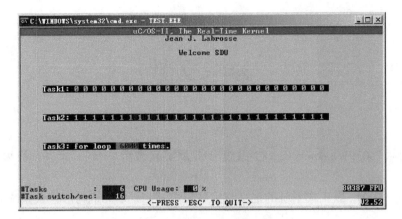

图 5-6 改进后的运行结果

参考文献

陈向群.2003.WindowsCE.NET 系统分析及实验教程［M］.北京：机械工业出版社.

何立民.2004.嵌入式系统的定义与发展历史［J］.单片机与嵌入式系统应用（1）：4-8.

蒋亚群，张春元.2002.ARM 微处理器体系结构及其嵌入式 SOC［J］.计算机工程（11）：4-6.

李小群，赵慧斌，叶以民，等.2003.一种基于时钟粒度细化的 Linux 实时化方案［J］.计算机研究与发展，40（5）：734-739.

陆松年.2000.操作系统教程［M］.北京：电子工业出版社.

唐寅.2002.实时操作系统应用开发指南［M］.北京：中国电力出版社.

王金刚.2004.基于 VxWorks 的嵌入式实时系统设计［M］.北京：清华大学出版社.

王学龙.2003.嵌入式 VxWorks 系统开发与应用［M］.北京：人民邮电出版社.

AndrewS Tanenbaum.1999.现代操作系统.陈向群等译［M］.北京：机械工业出版社.

Balaji Srinivasan.2003.A Firm Real-Time System Implementation using Commercial Off-The-Shelf Hardware and Free Software［D］.Department of Electrical Engineering and Computer Science：University of Kansas，Master Thesis.

Buttazzo G.1999.Optimal Deadline Assignment for Scheduling Soft Aperiodic Tasks in Hard Real-Time Environments［R］.IEEE Transactions on Computers.

Charles Crowley.2001.Operating Systems A Design-Oriented Approach［R］.McGraw-Hill Book Co.

Chenyang Lu，J A Stankovic.1999.Design and Evaluation of A Feedback Control EDF Scheduling Algorithm［J］.Journal of Assoc Compute Machinery（1）：55-69.

C.M.Krishna，Kang G.Shin.2001.Real-time Systems［M］.Beijing：Tsinghua Press.

Darren Cofer，Murali Rangarajan.2002.Formal Modeling and Analysis of Advanced Scheduling Features in an Avionics RTOS［J］.Lecture Notes in Computer Science（24）：138-152.

Dirk Desmet，Hugo De Man，D.Verkest.2003.Operating System based Software Generation for Systems on Chip［R］.NJ：Prentice Hall.

D.Verkest，J.da Silva.1999.A system-on-chip design methodology emphasizing dynamic memory management［J］. Journal of VLSI Signal Processing, 21（3）: 277-291.

Galli D. 2000. Distributed Operating Systems Concepts and Practice［R］. Upper Saddle River, NJ: Prentice Hall.

Goggin T. A. 2000. Windows CE 高级开发指南［M］. 北京: 电子工业出版社.

Jason Nieh, Chris Vaill. 2001. Virtual-Time Round-Robin: An O（1）Proportional Share Scheduler［C］. Proceedings of the 2001 USENIX Annual Technical Conference.

Jean J. Labrosse. 2002. MicroCOS-Ⅱ: The Real-Time Kernel［R］. USA: CMP Books.

Jean J. Labrosse. 2003. MicroC/OS-Ⅱ The Real-Time Kernel. Second Edition［M］. 北京: 北京航空航天大学出版社.

John A. Atankovic, Krithi Ramanuitham. 1991. The Spring Kernel: A New Paradigm For Real-Time Systems［R］. IEEE Trans.

J. B. Goodenough, Lui Sha. 1988. The priority ceiling protocol: A method for minimizing the blocking of high priority and a tasks［J］. ACM Ada Leters, 8（7）: 20-31.

Kartik Gopalan. 2001. Real-Time Support in General Purpose Operating Systems［R］. Tech Report.

Liu C, Layland J. 2002. Scheduling Algorithms for Multiprogramming in a Hard Real-time Environment［J］. Journal of the ACM（20）: 179-194.

Lui Sha, Ragunathan Rajkumar, John P. Lehoczky. 2001. Priority Inheritance Protocols: An Approach to Real-Time Synchronization［R］. IEEE Transactions on Computers.

Michael Barabanov. 2004. A Linux-based Real-Time Operating System［R］. New Mexico Institute of Mining and Technology, Master Thesis.

Naghizadeh M, Kim KH. 2002. A Modified Version of Rate-monotonic Scheduling Algorithm and its efficiency Assessment［J］. IEEE Trans. Software Engineering（1）: 285-299.

Nima Homayoun, Parameswaran Ramanathan. 2004. Dynamic priority scheduling of periodic and aperiodic tasks in hard real-time systems［J］. Real-Time Systems（6）: 207-232.

Ramamritham K, Stankovic J A. 1994. Scheduling Algrithms and Operating Systems Support for Real-time Systems. Software Engineering Journal, 83（1）: 55-67.

Ramamritham K, Stankovic J. 1994. Scheduling Algorithms and Operating Systems Support For Real-Tim Systems［R］. Proceedings of the IEEE.

R. K. Abbot, H. Garcia-Molina. 2002. Scheduling Real-time Transactions: a Performance Evaluation［J］. ACM Trans on Database Systems, 17（3）: 513-560.

Shui Oikawa, Raj Rajkumar. 2001. Linux/RK: A Portable Resource Kernel in Linux［R］. In IEEE Real-Time Systems Symposium Work-In-Progress, Madrid.

参考文献

Tilborg A, Koob G. 1991. Foundations of Real-Time Computing: Scheduling and Resource Management [R]. Boston: Kluwer Academic Pubishers.

WayneWolf. 2002. 嵌入式计算系统设计原理. 孙玉芳等译 [M]. 北京: 机械工业出版社.

William Stallings. 2001. Operating Systems Internal and Design Principles [M]. Fourth Edition, Prentice Hall.

Yu Chung Wang, Kwei Jay Lin. 1999. Implementing a General Real-time Scheduling Framework in the RED-Linux Real-time Kernel [J]. Journal of Software (1): 23-49.

Yu-ChungWang, Kwei-Jay Lin. 1999. Implementing a General Real-Time Scheduling Framework In the RED-Linux Real-Time Kernel [R]. IEEE Real-Time Systems Symposium.

Y. Mizuhashi, M. Teramoto. 1999. Real-time operating system: RX-UX 832 [J]. Microprocessing and Microprogramming (27): 533-538.

第二篇

物联网关键技术研究

第一章 绪 论

本章首先概述了物联网的定义及组成、物联网的应用现状及存在的问题,随后对物联网感知域中无线传感器网络相关技术的研究现状进行归纳和总结。

第一节 引 言

物联网被看作是继计算机、互联网和现代通信之后的一次信息产业革命,必将成为带动中国经济发展的一支重要力量。20世纪末美国等已经开始进行无线传感器网络的研究,最近几年来,欧盟、美国、日本等发达国家都针对物联网的发展和推进制定了本国的战略发展规划。2009年9月,欧洲物联网研究项目工作组(CERP-IoT)在欧盟委员会资助下制定了《物联网战略研究路线图》等报告。特别是奥巴马就任美国总统后,对IBM首席执行官彭明盛提出的"智慧地球"概念给予了积极的回应,并上升至美国的国家战略,推动了物联网全球范围的发展。在2010年全国"两会"期间物联网已经被写入政府工作报告,确立为国家五大战略性新兴产业之一。2013年2月,国务院正式出台了《关于推进物联网有序健康发展的指导意见》,该指导意见的出台,将极大推动物联网的有序健康发展。

物联网运营商通过与政府合作,整合产业内的设备提供商、运营商、系统集成商、软件开发商等,在各方互惠互利的前提下为各行各业搭建公共平台,并以平台为基础设施,以信息服务为产品,面向社会提供服务。因此构建物联网运营平台及组建物联网运营商,为各行各业提供基础设施建设,形成物联网现代化服务业与物联网信息终端制造业,促进大量人员就业,推动我国成为创新型国家,为国民经济发展创造新优势。

第二节 物联网

物联网,英文名称"The Internet of Things,IoT"是新一代信息技术的重要组成部分。物联网字面层次上的意思是物物相连的互联网,有两层含义,第一,物联网的基础和核心是Internet互联网,是在Internet基础上的扩展和延伸的网络;第二,用户端扩展和延伸到了任何物品之间进行信息交换与互联。物联网通过感知、识别等技术以及普适计算、云计算等技术融合应用,被称为继计算机、互联网之后信息产业发展的又一次浪潮。

The Web of Things(WoT)是受IoT在日常生活中设备与物体的启发,例如物体中包含一嵌入式设备或者计算机,与Web紧密联系在一起。智能设备和对象如无线传感器网络、环境设备、家用电器、RFID 标记对象等。与 IoT 中的很多系统不同的是，WoT 强调重用 Web 标准来快速互联生态系统中的嵌入式设备，例如 URL、HTTP、REST、ATOM 等技术和标准用于访问智能设备。

嵌入式设备世界经历了快速的变化，家电、工业机械、汽车以及日常生活中的物品越来越多的与微型计算机、传感器、网络接口等互联。各种计算设备和物体（如无线传感器网络、手机、嵌入式计算机、RFID 标签等），形成一个在线业务系统，在网上进行检索与数据交互。这种融合了传感技术、计算机、互联网的应用带来了新的机遇与挑战，数字通信网络越来越多的包含真实世界设备以及直接读写交互设备。物联网已经成为普适计算和泛在计算的重要应用领域，存在各种挑战与应用环境的复杂性。WoT 的下一个逻辑目标是朝着传感器、控制器、新设备的全球性网络发展，提供新的机会，探索事务的连通性并致力于快速原型、数据集成和对象的交互性。

由于网络无处不在，灵活多变，已经有一些协议和嵌入式设备在应用。WoT 的一个愿景是所有的设备无缝接入 Web 中，不仅仅是通过自定义的基于 Web 的用户界面，而且通过可重用的架构来与设备进行交互，我们希望应用设计者在传感器网络和 Web 应用之上进行考虑，去想象、设计、构建、评估与分享他们的想法，到底未来的网络会如何，应该如何。

一、物联网概述

1995 年，微软创始人比尔·盖茨在《未来之路》中首先提到物联网（Internet of Things，IoT）的理念，书中描述了很多未来的智能生活，例如"用户丢失的相机可自动发回信息，告诉用户目前它所处的位置"等。进入 21 世纪之后，芯片技术、传感器、网格计算、分布式计算、移动计算、云计算等技术得到了飞速发展，为物联网的数据处理提供了理论与技术支撑。2005 年以后，物联网发展浪潮席卷全球，很多国家从长远的角度提出了物联网发展的国家战略，并在一些方面已经从构思阶段步入了实施阶段。北京邮电大学张平教授给出了 MUSE（Mobile Ubiquitous Service Environment）的发展趋势，如图 1-1 所示。2012 年，《物联网"十二五"发展规划》由工信部发布，指出我国到 2015 年，要在关键标准研究与定制、核心技术研发与产业化、重大应用示范与推广、产业链条建立与完善等方面取得明显成效，在我国初步形成"应用牵引、创新驱动、安全可控、协调发展"的物联网发展格局。

物联网这个词语已出现多年，但是一直没有权威的定义，不同的国家和研究组织对物联网的理解不一样。如在美国，学术界提出的普适计算（Pervasive Computing）和企业界提出的智慧地球（Smart Planet），在欧洲，学术界提出的泛在计算（Ubiquitous Computing）和企业界提出的传感网（Sensor Networks）等都属于物联网的范畴。在英文维基百科（Wikipedia）里，物联网定义为"Internet of Things refers to uniquely identifiable objects (things) and their virtual representations in an Internet-like structure"，这个定义不是很全面，重点强调将物体连接起来的网络。在工业和信息化部报送中央人民政府的《关于支持无锡建设国家传感网信息中心情况的报告》中，将物联网定义为：物联网实现物与物、人与人、人与物全方面互联的网络，该网络以感知为目的，以运用

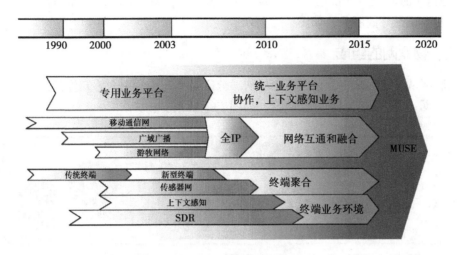

图 1-1 MUSE 发展趋势

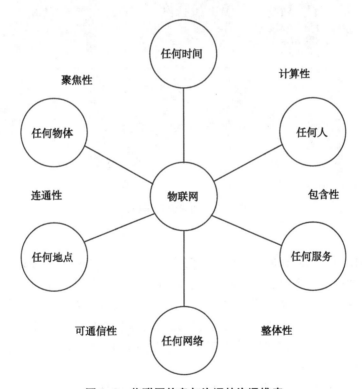

图 1-2 物联网信息与沟通的沟通维度

各种感知方式来获取各种物质世界的信息为突出特征，集成移动通信、互联网等进行信息的交互与传递，采用各种智能技术对感知信息进行处理分析，从而使人们对物质世界的感知能力得到提升，决策与控制实现智能化。我们可对物联网从信息与通信沟通的维度来理解：任何时间、任何地点、任何物体、任何人、任何服务、任何网络（Any

Time，Any Place，Any Things，Any One，Any Service，Any Network）都可以直接保持最佳联系和提供最好的服务，如图 1-2 所示。

二、物联网的组成

物联网从宏观上来看，包含三个层次，分别是感知层，用来感知世界；网络层，用来传输数据；应用层，用来处理数据。如图 1-3 所示。

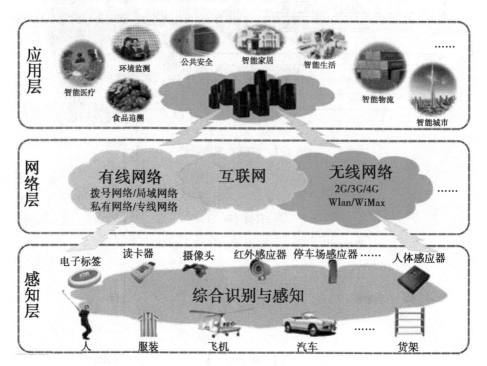

图 1-3 物联网三层结构

1. 感知层

感知层的作用是感知和采集信息，从仿生学来看，感知层为"感觉器官"，可以感知自然界的各种信息。感知层包含传感器、RFID 标签与读写器、激光扫描器、摄像头、M2M 终端、红外感应器等各种设备和技术。传感器及相关设备装置位于物联网的底层，是整个产业链中最基础的环节，解决人类世界与物理世界数据获取问题，首先通过传感器、RFID 等设备采集外部物理世界的数据，然后通过蓝牙、红外、工业总线、条码等短距离传输技术进行传输。

在 2009 年温家宝总理提出"感知中国"后，国家对传感器的研发投入加大，江苏省无锡市建成了我国首个传感中心，通过国家高层次海外人才引进，纳米传感器在医学上已经应用到临床。传感器是一门多学科交叉的工程技术，涉及信息处理、开发、制造、评价等许多方面，制造微型、低价、高精度、稳定可靠的传感器是科研人员与生产单位的目标。RFID 应用广泛，如身份证、电子收费、物流管理、公交卡、高校一卡通等，且 RFID 标签可以印刷，成本低廉，得到广泛的应用与普及。

2. 网络层

网络层的任务是将感知层的数据进行传输，将感知层获取的数据通过移动通信网、卫星通信网、各类专网、企业内部网、小型局域网、各种无线网络进行传输，尤其是互联网、有线电视网、电信网进行三网融合后，有线电视网也能提供低价的宽带数据传输服务，促进了物联网的发展。

网络层的研究开发主要有高校和大企业主导，在学术界，随着 IPv6 的诞生与实际应用，我国的科技工作者做出了很多贡献，如为了解决多网融合问题，北京邮电大学教授张平所在的泛在网络研究中心已经对网络的异构问题进行了大量的研究，完成了相关的"973"和"863"计划项目。在企业界，华为、中国移动、大唐电信等已经加入 3GPP 长期演进（Long Term Evolution，LTE）项目，LTE 是以频分多路复用/频分多址联接（FDM/FDMA）及多输入多输出（Multiple-Input Multiple-Output，MIMO）为核心的准 4G（Forth Generation of Mobile Phone Mobile Communications Standards）技术，在速率、功耗、延迟、高速移动性等性能指标方面取得了突破性进展。

3. 应用层

应用层的任务是对网络层传输来的数据进行处理，并与人通过终端设备进行交互，包括数据存储、挖掘、处理、计算，以及信息的显示。物联网的应用层涵盖医疗、环保、物流、银行、交通、农业、工业等领域，物联网虽然是物物相连的网络，但最终需要以人为本，需要人的操作与控制。应用层的实现涉及软件的各种处理技术、智能控制技术、云计算技术等。

物联网的应用层属于软件范畴，目前已经实现的有智能操控、安防、电力抄表、远程医疗、智能农业等，另如清华大学提出的物联网中间件，通过基于负反馈和指数平滑预测的负载均衡技术，能有效的提高吞吐量；清华同方公司已开发完成了"M2M 物联网业务基础平台"；以及南京理工大学提出的物联网高层架构。

三、物联网的应用现状

物联网应用涉及的领域很广，从简单的个人生活应用到工业现代化，到城市建设、军事、金融等领域。物联网的应用设计由传感器技术推动的各个产业领域，包括智能家居、智能农业、智能环保、智能医疗、智能物流、智能安防、智能旅游、智能交通等等。物联网的发展最终将现有各种产业应用聚集成为一个新型的跨领域的应用领域，下面简单介绍几种物联网应用（图 1-4）。

1. 智能家居

智能家居（Smart Home）利用物联网平台，以家居生活环境为场景，将网络家电、安全防卫、照明节能等子系统融合在一起，为人们提供智能、宜居、安全、舒适的家居环境。与传统家居相比较，智能家居为人们提供宜居舒适的生活场景，安全高效利用能源，生活工作方式得到优化，家居环境变成智慧、能动的生活工作工具，从而达到环保、低碳、节能。

我国的智能家居经过市场发展培养，智能家居发展迅速，从 2012 年开始，随着 4G 技术、云计算技术的应用推广，手机、平板等智能终端设备的普及，价格下降迅速，以

图 1-4 物联网的应用领域

及各物联网相关技术的发展,智能家居进入快速发展通道。

青岛海尔推出的"U-Home"系列家居智能产品,通过无线网络,以智能家电为载体,实现各种数字媒体、家电产品以及智能家居各子系统的管理与互联。从功能上可分为 U-Service、U-Play、U-Shopping、U-Safe 等覆盖服务、娱乐、安防等功能的数字化家庭解决方案。清华同方"e-Home"数字家园计划,核心是对家居环境进行管理,重点实现服务信息化和控制智能化,推动了智能家居的发展。上海世博会展示的"让城市生活更美好"更加多样化,有各种智能家居的应用场景。

2. 智能农业

传统农业主要依靠自然资源和劳动力,成本低廉、效率低下、劳动强度大、难度高,已不能满足现代农业的高产、高效、优质、安全的需要。随着物联网技术被引入农业中,农业信息化程度得到明显的提高。智能农业通过实时采集温湿度、二氧化碳浓度、光照强度、土壤温湿度、pH 值等参数,自动开启或者关闭控制设备,使农作物处于最优生长环境中。同时通过追踪农产品的生长监控信息,探索最适宜农产品生长的环境,为农业的自动控制与智能管理提供科学依据。传统农业中的灌溉、打药、施肥等农民靠感觉,凭经验,在智能农业中,这些通过相关设备自动控制,实行精确管理。

中国农业大学在山东省寿光市进行农业物联网应用实验，试点"农业管家"等农业信息化应用，将物联网技术融入农业生产中，使农作物处于适宜的生产条件下，帮助农户对农业种植环境进行科学管理，降低人工成本。"十二五"国家 863 计划项目"农业物联网与食品质量安全控制体系研究"，旨在建立开发共享的农业物联网应用支撑服务平台，构建基于农业物联网的食品安全质量控制标准体系和规范，提高农产品品质，保障国家食品安全。

3. 智能环保

随着社会的进步发展，环境污染变得更加严重，且出现了一些新情况，同时伴随着人们生活水平的提高，环保意识在不断增强。我国的环境保护方面的信息化程度较低，实现环保工作的自动化、智能化是未来工作的重点。

美国部署的"CitySense"系统，用于实时监测城市环境污染数据的，以及在大鸭岛部署的监测系统，用于监测海鸟栖息地信息；澳大利亚部署的生态监测系统用来监测蟾蜍的分布情况；我国在无锡部署的太湖水环境监测示范系统也初步完善。这些系统的部署与实现，弥补了传统监测手段的不足，提高了环保的效率，节省了开支，实现了环保的智能化与自动化。

4. 智能医疗

我们可利用物联网技术实时的感知的各种医疗信息，实现全面互联互通的智能化医疗。通过智能医疗系统，对病人和药品进行智能化管理，例如病人佩戴 RFID 设备，实时跟踪病人的活动范围；病人佩戴各种传感器，对重症病人进行全方位实时监控，特殊情况及时报警，节省了人力开支，提高了信息的准确性和及时性。智能医疗还能通过家庭医疗传感设备，实时监控家中老人或者病人的各项健康指标，并将各项指标数据传输到健康专家，给出保健或护理建议。但是也存在标准不统一、成本高、隐私保护难度大以及国内医疗相关企业竞争力弱等问题。

5. 智能物流

物联网在物流行业已得到广泛应用，智能物流运用传感器技术、RFID、GPS 等技术，对物品的运输、配送、仓储等环境进行跟踪管理，达到配送物品的高效、智能，减少了人力资源的浪费。智能物流实现了物流配载、电子商务、运输调度等多种功能的一体化，成为运用物联网技术较成熟的行业。

6. 智能安防

我国的安防体系存在安防设备智能化不足、功能单一、可靠性差以及服务范围窄等问题。物联网技术的快速发展，给安防行业带来了技术创新，通过把物联网的快速感应、高效传输等特点应用到安防领域，实现安防系统的智能化，提高安防系统的自适应能力、自学习能力，最终实现能针对不同的应急情况自动采取各种针对性的措施来保证安全。如上海世博会的各种安防系统，车辆安全监控系统实现对世博会园区 600 余万辆汽车的安检；智能火灾监控系统，在发现烟雾时能及时采取有效措施并报警。

四、物联网产业存在的问题

当前物联网产业发展存在的问题主要如下。

一是存在大量互不兼容的应用解决方案和产品，缺乏统一概念与实施办法，导致应用和架构碎片化。

二是各种示范工程不能实现互联互通，建设重复，地区与行业壁垒严重，同时缺乏可复制性、开放性，只是内部封闭的尝试，应用开发商和服务提供商介入困难。

三是应用开发商和服务提供商对设备制造商依赖，源于对硬件进行系统集成与开发，造成研发和实验成本过高，导致较高的产业化门槛，无法为大量中小企业创造机会，缺少创新性的应用，难以满足用户应用物联网的需求。

四是众多应用开发商、广大科研机构及高校，缺乏一个合适的实验、测试及验证平台来验证最新研究成果的有效及可行性。

五是产业链多样化且比较长，导致复杂的利益矛盾，需要一个系统来保证并连接产业链上各方的利益。

六是缺乏可持续发展的商业模式与应用开发团队，物联网规模化、产业化发展推动困难。

因此，当前物联网发展中，各类机构角色都希望形成自己的一套规范，导致开发的应用大部分为封闭的孤岛应用，缺少和谐融合的物联网服务平台，从而难以促进物联网产业的跨越式发展。以如此模式推进物联网产业，必将会导致基础资源重复和浪费、用户需求无法得到满足、物联网应用支撑平台建设困难、终端成本无法下降、信息安全难以保障、技术与规范无法统一、物联网服务及应用商用困难等问题。我们针对这些问题，提出了基于 Web 的物联网体系结构，并在此基础上构建物联网生态系统，可有效地解决这些问题。

五、物联网应用中的关键技术

物联网是以应用为核心的网络，应用创新是物联网发展的核心，强调用户体验为核心的创新是物联网发展的灵魂。下面介绍物联网应用中的几种重要技术。

1. 传感器技术

物联网能做到物物相连，进行感知识别离不开传感器技术。目前通常采用无线传感器技术，大量传感器节点部署在感知区域内，构成无线传感器网络。无线传感器网络作为感知域中的重要组成部分，有很多关键技术需要研究，如路由技术、拓扑管理技术等。

2. RFID 标签

RFID 本质上来说也是一种传感器技术，融合了无线射频技术和嵌入式技术，在物流管理、自动识别、电子车票等领域有广阔的应用前景。

3. 嵌入式技术

综合了集成电路技术、电子应用技术、传感器技术以及计算机软硬件技术，经过多年的发展，基于嵌入式技术的智能终端产品随处可见，从普通遥控器到航天卫星，从电子手表到飞机上的各种控制系统。嵌入式系统已经完全融入人们的生活中，也改变着人们的生活，推动工业生产以及国防技术的发展。

4. 应用软件技术

通过各种各样的应用软件技术提供不同的服务，满足我们的需求。我们应充分利用丰富的应用软件提供的各种功能，将物联网 Web 化，物联网应用融合到 Web 中，借助 Internet 物联网，为用户提供各式各样的服务。

虽然当前国内外在物联网领域已经取得了大量理论研究成果和部分应用示范，但问题仍较为突出，如封闭的内部尝试，缺乏开放性、示范性与可复制性；不能互联互通，存在严重的地区和行业壁垒，大量示范工程重复建设；产品、解决方案互不兼容，缺乏统一的概念，导致大量碎片化的框架和应用等。针对这些问题，在分析物联网系统各部分功能与特点的基础上，从基于 Web 的物联网业务环境的基本原则出发，将物联网系统架构分为感知域和业务域，提出了基于 Web 的物联网体系结构，将物联网 Web 化。构建基于 Web 的物联网系统服务平台，汇聚产业链上的设备和平台，引进国内外先进的技术和理念，形成物联网应用设备商店，为用户提供全方位的体验与服务，最终形成物联网应用服务云，构建物联网生态系统。

无线传感器网络作为感知域中的重要内容，具有特别重要的地位，是物联网感知获取信息的前提和基础，具有举足轻重的地位和作用。本文针对无线传感器网络中的拓扑、路由等关键技术进行研究，下面先简单介绍一下无线传感器网络相关情况。

第三节 无线传感网络

物联网系统感知域中的无线传感器是物联网中信息获取的重要元器件，无线传感器网络通过部署大量廉价的微型传感节点来完成以往昂贵的高精度仪器才能完成的监测任务。无线传感器网络（Wireless Sensor Networks，WSNs）结构如图 1-5 所示，包括管理节点、传感器节点（Sensor node）和汇聚节点（Sink node）。在监测区域附近或内部，部署大量的传感器节点，节点监测感知数据，以自组织的方式构成网络，通过多跳或单跳进行数据传输，到达汇聚节点，然后通过有线网络或者卫星等方式接入互联网，方便用户进行管理。

无线传感器网络在军事目标定位与跟踪、自然灾害救援、生物健康监测、危险环境探险与地震监测等场景中广泛应用。作为物联网应用的重要基础，物联网各应用系统中，无线传感器网络广泛存在。

一、无线传感器节点

无线传感器节点具备一定存储、通信与处理能力，通常是一个微型的嵌入式系统，一般通过电池供电，携带能量有限。每个传感器节点需具有本地信息的收集处理、路由数据的融合与存储，以及协作其他传感器节点来完成特定任务等功能。而汇聚节点的处理、通信与存储能力通常较强，将无线传感器网络连接至外部 Internet，把传感器网络感知汇聚的数据发送至管理节点，同时管理节点可发布监测任务等。

无线自组网（Mobile ad-hoc network）是由成十上百个节点组成无线、动态、多跳、移动、对等网络，通过移动管理和动态路由技术传输一定服务质量要求的多媒体信息，

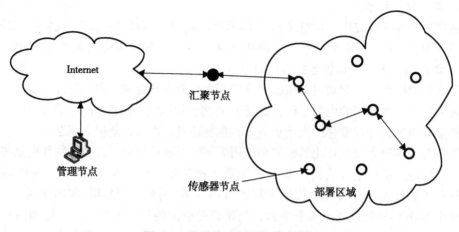

图 1-5 无线传感器网络体系结构

这种网络节点通常有持续能量供给。而无线自组网与无线传感器网络相似之处较多,但是差异也较大,网络中节点的数量更多,节点集成了控制、监测以及无线通信等系统,节点的能力有限,且能量供给受限,存在下面约束限制。

1. 电池能量有限

传感器节点微小,采用锂电或干电池供电,所携带的能量比较有限。潜在能源如太阳能、振动、声音、移动充电设备等可从周边环境来换取能源,但是并不通用,由于受成本、体积等方面的限制,通常还是采用电池供电。传感技术节点数量多、成本低,通常部署在环境恶劣复杂,以及人员不宜到达的区域,传感器节点进行能量补充或更换电池不能实现,因此必须通过高效利用能量来使网络寿命最大化。

无线传感器节点能量消耗主要在无线通信模块、传感模块、处理器模块,随着集成电路技术的迅速发展,传感模块和处理模块的能量消耗已经较低,图 1-6 为无线传感器节点能量消耗情况,从图中我们可看出能耗主要集中在无线通信模块上。

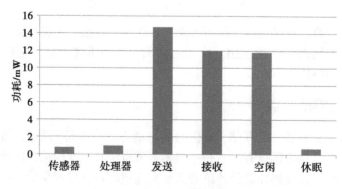

图 1-6 节点能量消耗情况

2. 通信能力受限

传感器节点的无线通信能耗与通信距离的关系为:$E = kd^n$,$2 \leq n \leq 4$,其中参数 n

的取值受多种因素的影响。从公式可知，传感器节点的能耗随通信距离的增大而迅速增大，因此需要在保证通信连通的前提下，采用多跳通信的方法，减少单跳通信距离。

3. 存储处理能力受限

传感器节点体积小、价格低、功耗小，这些必然限制其处理能力以及存储能力。为了完成各种监测任务，传感器节点需要进行数据的采集、处理、转发路由以及融合等工作，因此需充分利用传感器节点有限的存储处理能力协同完成。目前随着低功耗电路技术的发展进步，超低功耗的微型传感器已开发出相关产品，推动传感器的应用普及。

二、无线传感器网络特点

无线传感器网络具有如下特点。

1. 自组织网络

在大量无线传感器网络的应用中，传感器节点被随机放置在监测区域中，位置不能事先确定，通常在节点周围也没有基础设施，节点间的邻接关系也不能预先获知，还有随着有些节点的能量耗尽以及新节点的补充，这就要求传感器节点具有自组织能力，进行自动的管理，通过拓扑管理与网络协议自组织成多跳无线网络。

2. 网络动态性

网络中的节点受环境的影响，可能造成无线通信链路不畅通；有些传感器节点具有移动性；另传感器节点能量耗尽或者故障，新传感器节点的加入等，这使传感器网络具有网络动态性。

3. 大规模网络

通常在监测区域部署大量的传感器节点，规模成百上千甚至更多，大量传感器节点感知采集数据能提高精度，提高网络的容错性，对单个传感器节点精度要求降低，减少盲区。

4. 数据为中心

由于传感器节点通常随机部署，节点间的关系是动态的，用户查询事件时，直接将查询事件告知无线传感器网络而不是某个确定的传感器节点，无线传感器网络在获取指定的事件信息后报告给用户，因此是任务型网络，是一个以数据为中心的网络。

5. 可靠性网络

传感器节点通常部署在环境恶劣的地方，有些地方需通过飞机炮弹等散播发射过去，这要求传感器网络具有可靠性，节点不容易损坏，另外大量传感器节点导致维护困难，需要节点通信安全可靠，具有容错性与保密性。

三、无线传感器网络的关键技术

无线传感器网络涉及多学科交叉的研究领域，关键技术包含部署、定位、拓扑、路由、安全、同步、数据管理、数据融合等技术，我们选择网络拓扑控制技术和路由协议进行相关研究，无线传感器网络拓扑控制和路由技术及现状如下所述。

（一）网络拓扑控制技术

网络拓扑控制技术具有非常重要的意义，尤其是对自组织特性的无线传感器网络来说意义巨大，这是因为拓扑控制产生良好的拓扑结构，能够提高路由协议和 MAC 协议的效率，为目标定位、时间同步和数据融合等奠定良好的基础，提高能量利用率，减少优化网络的能量消耗，延长网络的寿命。

1. 无线传感器网络拓扑结构的优化意义

（1）减少节点间通信干扰，提高网络通信效率。网络中传感器节点通常在监测区域中大量密集部署，节点间通信必然会造成相互干扰，降低通信效率，从而造成能量耗费，若采用优良的拓扑结构，节点按照调度开启或者关闭，能减少干扰，节点能量得以高效的利用，通信效率提高。

（2）影响网络的生存周期。无线传感器节点通常采用电池供电的方式，故携带有限的能量，拓扑控制在保证网络连通性和覆盖度的前提下，使无线传感器节点的能量得到合理高效利用，从而延长生存周期。

（3）帮助路由协议。在无线传感器网络中，有些传感器节点可能随时失效，因此路由必须选择活动节点，拓扑控制能提供节点间的邻接关系，并确定那些节点处于活动状态。

（4）影响数据融合。数据融合指把采集的感知信息发送给传感器网络中的骨干节点，骨干节点进行数据融合，然后把融合后的数据发送至汇聚节点。其中骨干节点的选择是拓扑管理研究的一项重要内容。

2. 几种拓扑控制算法

无线传感器网络通过功率控制和骨干节点的选择，在满足网络连通度和覆盖率的前提下，剔除节点间非必要的通信链路，形成一能高效转发的拓扑结构，可分为节点功率控制和层次型拓扑结构两类。层次型拓扑控制选取一些节点作为骨干节点（簇头），利用分簇机制，由骨干节点（簇头）来组织产生一处理转发的主干子网，其他节点暂时关闭通信模块，进入休眠状态，从而节能能量消耗。节点功率控制在满足网络连通度的前提要求下，调节每个传感器节点的发射功率，从而使得均衡单跳可达邻居节点的数量。下面我们简单介绍几种拓扑控制算法。

（1）基于节点度的算法。节点度定义为距离该节点一跳的所有邻居节点的总数量，算法给出节点度的下上限要求，每个传感器节点通过动态调整发射功率，使得每个节点度在上下限范围之内。该算法节点间的链路具有一定的冗余性，利用局部信息达到整个网络邻居节点的连通性。基于节点度的算法可分为本地相邻平均算法 LMN（Local Mean of Neighbors Algorithm）和本地平均算法 LMA（Local Mean Algorithm），主要区别在于计算节点度的策略不同，但都是依据传感器节点度，通过动态调整传感器节点的发射功率来实现。但是该算法未考虑邻居节点信号的强弱，同时缺少严格的推导。

（2）基于临近图的算法。基于临近图算法中，每个传感器节点均使用最大功率发射形成拓扑图 G，然后按照规则 q 计算出邻接图 G'，然后 G' 中每个传感器节点以距离自己最远的通信距离决定发射功率。图 $G=(V, E)$，其中 V 是顶点的集合，E 是边的集合，集合 E 中的每条边 $l=(u, v)$，$u, v \in V$，由图 G 导出的邻接图 $G'=(V, E')$，对于一个结点 $v \in V$，给出其判定条件 q，E 中满足 q 的边 $(u, v) \in E'$。邻接图模型有

GG（Gabriel Graph）、MST（Minimum Spanning Tree）、RNG（Relative Neighborhood Graph）、YG（Yao Graph）等。

该算法的确定节点的邻居节点集合，通过调整节点的发射功率来达到能量节省，建立一个连通网络，有 DLSS（Directed Local Spanning Subgraph）算法和 DRNG（Directed Relative Neighborhood Graph）算法等。在 DRNG 算法中，如图 1-7 所示，若节点 u、v 满足 $d(u, v) \leq r_u$，且不在节点 p 满足 $w(u, p) < w(u, v)$，$w(p, v) < w(u, v)$ 和 $d(p, v) \leq r_p$，节点 v 被选为 u 的邻居节点，为节点 u 确定了邻居集合。

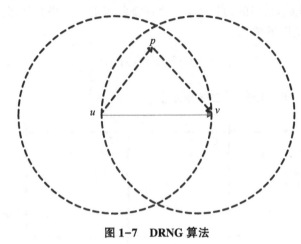

图 1-7　DRNG 算法

（3）LEACH 算法。LEACH（Low Energy Adaptive Clustering Hierarchy）属于层次型拓扑控制算法，将网络中的节点分成两类，成员节点和簇头节点，簇头节点协调簇内成员节点信息的感知采集，并对数据进行汇聚融合与转发，所以消耗的能量较多，而普通节点仅需在簇头节点的协调安排下进行数据采集，其他时间进入休眠，节省了能量，同时簇头按照一定的规则进行轮换，来达到网络中节点能量消耗的均衡。

LEACH 算法是周期性执行的，每个周期分为簇的形成阶段与数据通信阶段。在簇的形成阶段，传感器节点按照一定的规则产生簇头，形成分簇，在数据通信阶段，簇头节点汇聚簇内成员节点感知采集的信息，进行信息融合并把结果发送出去。由于簇头节点负责数据的汇集、融合、转发等工作，需要耗费更多的能量，因此簇头节点需轮换，以免过早耗尽能量而形成网络感知"盲区"。

LEACH 算法产生簇头过程为：每个节点随机产生一个在 0 到 1 范围内的数，如果这个数不大于阈值 $T(n)$，则该节点被选中为簇头节点。在每轮运行过程中，若该节点已当选过簇头，就置 $T(n)$ 为 0，该节点在本轮中不能再次当选簇头。随着网络的运行推移，节点当选簇头的数量增加，剩余的其他节点当选簇头的阈值 $T(n)$ 将增大，随机数在该范围的概率也增大，更有可能当选为簇头。

$$T(n) = \begin{cases} \dfrac{P}{1 - P \times [r \bmod (1/P)]}, & n \in G \\ 0, & 其他 \end{cases} \quad (1-1)$$

其中 P 是簇头数占总节点数的百分比，r 是选举轮数，$r \bmod (1/P)$ 代表本轮中当选过簇头节点的个数，G 为本轮中未当选过簇头的节点集合。

（4）GAF 算法。GAF（Geographical Adaptively Fidelity）算法及改进以节点所处的位置作为分簇依据，将监控区域分成若干单元格，每个单元格选出一个簇头，节点按照自己所处的位置加入相应的单元格，簇头负责簇内管理，其他节点根据安排大部分时间处于休眠状态，从而节省能量。GAF 算法的执行分为两步，第一步虚拟单元格的划分，根据节点所处位置和通信半径，将网络划分成多个虚拟单元格，需保证每个虚拟单元格中的节点间均能进行通信，每个节点根据位置信息加入相应的单元格。假设所有节点的通信半径为 R，虚拟单元格的边长为 r，为保证临近的单元格内任意两个传感器节点能进行通信，需要满足以下关系：

$$r^2 + (2r)^2 \leqslant R^2 \Rightarrow r \leqslant \frac{R}{\sqrt{5}} \tag{1-2}$$

初步划分的虚拟单元格如图 1-8 所示。

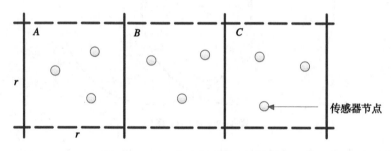

图 1-8　GAF 算法划分的虚拟单元格

GAF 算法第二步是各虚拟单元格内簇头的选择，每个传感器节点周期性进入休眠和工作状态，节点在进入工作状态时感知采集信息，并与其他节点交换信息，确认是否当选簇头。每个节点处于睡眠（Sleep）、活动（Active）与发现（Discovery）三种状态，如图 1-9 所示。在网络运行初始阶段，传感器节点均处于发现状态，节点发送信息通告自己的位置、标识等，这样各虚拟单元格内的节点均相互知道节点所处的状态；节点将自己的定时器设置为某区间内的随机值 Td，若超时，则发表声明进入活动状态，当选为簇头节点；若在超时之前收到其他节点的声明，竞选失败，则进入睡眠状态；节点在睡眠状态下设置定时器 Ts，在超时后重新进入发现状态；在活动期内，簇头节点发送广播包抑制其他节点成为簇头的请求，并设置定时器 Ta，在定时器超时后进入发现状态。

（二）网络路由技术

在网络中，路由负责把数据包从源点转发到目的点，首先需要寻找源点到目的点的路径，然后沿着路径对数据进行转发。无线传感器网络作为一种应用相关型的网络，路由协议在不同应用场景中差别较大，另无线传感器节点能量有限，有时需要和数据融合、拓扑管理等技术结合运用，来减少网络通信量，从而达到能耗减少的目的，这些因素对路由技术提出了一些要求。

（1）快速收敛。网络拓扑动态变化，节点能量和带宽有限，这要求路由机制适应

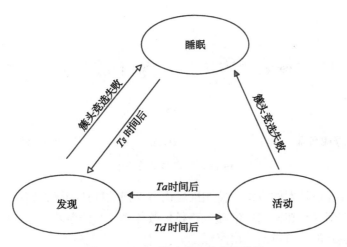

图1-9 GAF算法节点状态转换图

网络的动态变化，能快速收敛，来提高系统的效率。

（2）能量高效。路由协议不仅考虑单条路径的能耗，还需考虑整个网络的均衡与减少，简单高效的路由是能量和资源受限的传感网的首选。

（3）鲁棒性。部分传感器节点能量耗尽或者失效后，以及无线通信链路质量等因素要求具有一定的容错能量。

（4）扩展性。网络拓扑发生动态变化时，如新节点的加入、节点失效以及节能能量耗尽时，要求路由机制适应网络的变化，具有扩展性。

无线传感器网络是应用相关性网络，与应用密切相关，针对不同的应用场景，设计不同的路由协议，主要有以下几种。

（1）能量感知路由。无线传感器网络能量有限，因此节点可用能量（即节点剩余能量）作为选择路由路径的参数最容易被考虑到。能量多路径路由在目的点和源点之间建立多跳路由路径，根据每条路径上的通信量以及节点的剩余能量，赋予一选择概率，均衡数据传输过程中的能量消耗，达到延长网络生存周期的目的。该协议包含三个阶段，路由建立阶段、数据传输阶段与路由维护阶段，在路由建立阶段，传感器节点需要知道到达目的的下一跳节点，并计算选择下一跳节点的概率；在数据传输阶段，节点根据选择概率从多个下一跳节点中选择一个，并进行转发；路由维护阶段可周期性的从目的点到源点发送一查询消息来维持路径的活动性。

（2）基于查询的路由。汇聚节点（基站）发布查询消息，然后采用洪泛方式在网络中传播该消息，在传播过程中，在每个传感器节点上逐跳建立从数据源到汇聚节点的反向路由路径。当节点感知采集到数据需要发送时，沿着已建立的路由路径发送至汇聚节点，该过程分为三个阶段，消息扩散、路径建立以及路径加强，如图1-10所示。其中在消息扩散阶段，汇聚节点向网络中周期性发送广播消息；在路径建立阶段，节点从保存的下一跳信息中选择节点并转发；选择的路径加强了源点和汇聚节点间的路由路径。

谣传路由是另一种查询路由，该机制并不是采用洪泛的方式来建立转发路径，引入

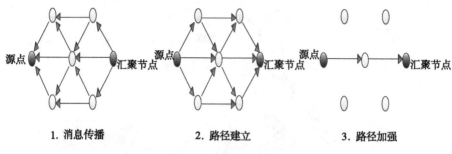

图 1-10 定向扩散路由

了查询消息单播随机转发的方法。监测区域内节点把数据选择随机路径扩散，同时汇聚节点发送的查询消息也在网络中沿随机路径扩散传播，当两条路径交叉时，就得到了一条从传感器节点到汇聚节点的路由路径。

（3）地理位置路由。在一些应用场景中，传感器节点需知道感知数据的位置信息，这样数据才具有意义，我们利用已知的地理信息，按照一定的路由方法策略将数据发送至目的节点。GEAR（Geographical and Energy Aware Routing）根据节点位置信息，建立汇聚节点到传感节点的路由路径，避免了洪泛广播消息，减少了系统开销，提高了路由效率。GEAR 路由分两步，第一步汇聚节点发布查询命令，根据事件所处区域将命令发送至在事件区域内，且离汇聚节点最近的传感节点，然后该传感节点将查询消息传播发送至区域内其他节点；第二步传感节点的感知监测数据沿着查询消息的反向路径向汇聚节点进行数据转发路由。

第四节　主要贡献

一、基于 Web 的物联网系统结构

当前国内外在物联网领域已经取得了大量理论研究成果和部分应用示范，但问题仍较为突出，如封闭的内部尝试，缺乏开放性、示范性与可复制性；不能互联互通，存在严重的地区和行业壁垒，大量示范工程重复建设；产品、解决方案互不兼容，缺乏统一的概念，导致大量碎片化应用和框架。各类角色都在或希望建立自己的一套规范，产生的应用多为封闭孤岛式应用，缺少和谐的物联网系统服务平台，这样很难推动物联网产业的发展。以这种模式推动物联网产业，必将会导致基础资源重复和浪费、用户需求无法得到满足、无法统一规范和技术、应用和服务商用困难、难以保障信息安全、各物联网终端成本无法下降，以及各物联网应用服务支撑平台建设困难等问题。我们针对这些问题，在分析物联网系统各部分功能与特点的基础上，从基于 Web 的物联网业务环境的基本原则出发，可以将物联网系统架构分为感知域和业务域，提出了基于 Web 的物联网体系结构，将物联网 Web 化。构建基于 Web 的物联网系统服务平台，汇聚产业链上的设备和平台，引进国内外先进的技术和理念，形成物联网应用设备商店，为用户提供全方位的体验与服务。物联网系统应用平台和服务管理平台构建在互联互通、标准化

接口框架之上，将具有多种异构平台的能力，如 M2M 平台、传感网平台、智能手机平台、WoT 平台等，通过云计算、物联网接口为应用开发者提供全方位服务，最终形成物联网应用服务云，构建物联网生态系统。

二、一种基于虚拟坐标的动态自适应分簇协议

在无线传感器网络中，分簇是一种流行有效的拓扑控制方法，层次结构清晰，兼顾负载均衡，能增强网络可靠性，具有较强的扩展性与鲁棒性。目前很多分簇协议存在簇头分布不均匀，簇头节点的选举未充分考虑剩余能量与通信距离的影响，以及分簇的数量不随网络情况的变化而变化等问题。针对这些问题，我们设计了一种基于虚拟坐标的动态自适应分簇（a Virtual Coordinates Dynamic Adaptive Clustering，VCDAC）协议，该协议按照虚拟分区的思想将网络分簇，并根据网络的状态动态决定分簇的数量，从而降低了能耗；在每个簇中，根据节点的剩余能量与结点间的距离，选择最合适的节点当选簇头，负责数据的收集汇聚与转发等任务，簇头分布均匀，选择合理，从而均衡了整个网络的能量消耗。仿真结果显示该协议能明显延长网络的生存时间，均衡网络负载，发送更多的数据至汇聚节点，具有优良的性能。

三、一种基于小生境粒子群优化的自适应分簇协议

无线传感器网络的拓扑结构对网络性能有较大的影响，良好的拓扑结构能提高 MAC 协议以及路由协议的效率。无线传感器网络是一种应用相关性网络，针对应用场景的不同，须采拓扑控制协议也不同。分簇方法作为拓扑管理中一种层次清晰的方法，能延长网络的生存时间，经常被转化为优化问题求来最优解。传统分析优化方法需要大量的计算，计算量随问题规模成指数增长，对于微型传感器来说，少的计算量与资源需求，并能产生可接受结果的优化方法是适宜的，粒子群算法就是一种计算效率高的仿生智能优化方法。本文首先提出了一种无参数小生境粒子群优化算法，并进行验证，然后利用该算法，设计了一种基于小生境粒子群优化的自适应分簇协议（an Adaptive Clustering Protocol using Niching Particle Swarm Optimization，ACP-NPSO）协议，该协议形成的分簇分布均匀，均衡了系统的负载，提高了系统的吞吐量。我们给出了 ACP-NPSO 的详细设计与仿真结果，实验结果表明，该协议具有良好的效率，能明显延长网络的生存周期，提高系统的吞吐量。

四、一种基于粒子群蚁群优化的动态分簇路由协议

无线传感器网络具有很强的应用相关性，是一种与应用密切相关的网络，不同应用场景中路由协议存在很大差别，没有通用的路由协议。根据逻辑结构的不同，路由协议可分为分簇型路由协议和平面型路由协议，同平面型路由协议相比较，分簇型路由协议结构清晰，能高效利用节点资源。本书设计了一种基于粒子群蚁群优化的动态分簇路由协议（a Dynamic Clustering Routing Protocol using Particle Swarm Optimization and Ant Colony Algorithm，DCRP-PSOACA），减少在数据传输过程中节点的能量消耗，达到节点能量高效利用的目的，从而实现网络的高效通信，延长了网络的生命周期。我们充分利用

粒子群优化算法优点，首先将网络分簇，然后利用蚁群算法的优点，在簇头节点与汇聚节点间形成多跳最优路径的选择。我们给出了 DCRP-PSOACA 的详细设计与仿真实验，结果表明该协议能提供稳健的路由路径，减少节点能量消耗，延长网络生存时间，实现高效通信。

第二章　基于 Web 的物联网体系结构

当前国内外在物联网领域已经取得了大量理论研究成果和部分应用示范工程，但问题仍较为突出，如封闭的内部尝试，缺乏开放性、示范性与可复制性；不能互联互通，存在严重的地区和行业壁垒，大量示范工程重复建设；产品、解决方案互不兼容，缺乏统一的概念，导致大量碎片化应用和框架。各类角色都在或希望建立一套自己的规范，形成封闭的孤岛式应用，缺少融合和谐的物联网服务平台，导致难以推动物联网产业跨越式发展。按照这种模式推动物联网产业，必将会导致基础资源重复和浪费、用户需求无法得到满足、物联网应用支撑平台建设困难、终端成本高企、难以保障信息安全、无法统一规范和技术、物联网应用和服务商用困难等问题。针对这些问题，我们提出了基于 Web 的物联网体系结构，将物联网 Web 化，针对上述问题进行解决，并在此基础上构建物联网生态系统。

第一节　物联网系统体系架构

在分析物联网系统各部分功能与特点的基础上，从基于 Web 的物联网业务环境的基本原则出发，可以将物联网系统架构分为感知域和业务域，如图 2-1 所示。

感知域：由无线传感器网络及包含的传感器、控制器等组成，用于读取感知信息（如传感器）和执行控制命令（如控制器）。

业务域：由网络、网关、平台和应用等组成，用于 Web 化传感器等设备的访问方式，通过 Web 业务网关或业务平台提供针对传感器/控制器的物联网应用。

一、感知域

感知域主要由无线传感器网络及包含的传感器、控制器等组成，用于读取感知信息（如传感器）和执行控制命令（如控制器）。

1. 传感器节点

传感器（transducer/sensor）能感知（感受）被测量的信息，并按照一定的规律将被感受到的信息转换成电信号等形式输出的一种检测装置，信息的输出满足传输、处理、记录、存储、控制、显示等方面的要求。传感器是实现自动检测和自动控制的首要环节，是物联网的最基础的感知单元，传感器节点由传感器/控制器、变送器和通信模块组成。

传感器具有微型化、智能化、数字化、系统化、网络化、多功能化等特点，在微电子机械系统技术的基础上获得了突飞猛进的发展。

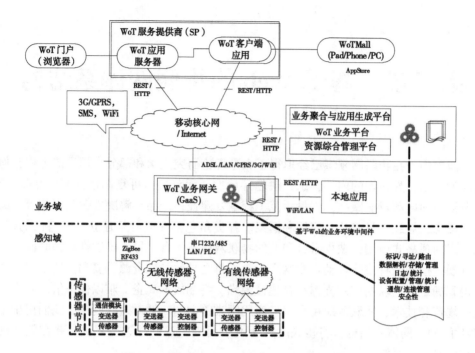

图 2-1　基于 Web 的物联网业务环境体系架构

随着社会步入信息时代，我们在想利用信息，首先必须准确的获取信息，传感器就是我们获取自然与工农业生产领域中信息的主要工具与手段。在现代工农业生产中，需要采用各种传感器来保证设备处于正常工作状态，监测控制生产过程中的各参数，从而保证产品质量。所以我们可以认为，没有各种各样的传感器做保障，现代化生产与生活失去了基础保证。

传感器已广泛应用于环境保护、生物工程、工农业生产、海洋探测、医疗诊断等领域，可以说生活生产中的各个环境，都离不开各式各样的传感器。因此传感器技术在推动经济社会发展作用也非常明显，各国都积极发展，促进了传感技术的进步。

传感器的功能可与人类的五大感觉器官类别，如光敏传感器——视觉；气敏传感器——嗅觉；声敏传感器——听觉；化学传感器——味觉；压敏、温敏传感器——触觉，可见传感器延伸了人的感觉感知功能，使人们更加准确地感知感觉被测物。

传感器/控制器，实现感知、控制功能，涵盖了智能农业、视频监控、位置服务、绿色社区、智能校园中的各式各样的电子产品、设备。

变送器主要用于将传感器和控制器提供的原始模拟信号或者特定接口数字信号转换为常用的数字接口，使之符合常用的网络传输需求。传感器的输入通常是非电量信号，通过变送器转换成电信号并同时放大，通常和传感器一块构成自动控制的监测源。变送器的种类很多，不同的物理量需要不同的传感器及相应的变送器，在工业控制仪器仪表方面，比较常见的变送器有电流变送器、流量变送器、温度变送器、电压变送器、压力变送器等。

通信模块提供远程/本地的有线和无线网络通信能力。通信模块根据通信能力可分为：
  远程无线能力：3G/GPRS、WiFi、SMS 等。
  本地无线能力：WiFi、ZigBee、RF433、蓝牙等。
  本地有线能力：串口 232/485、以太网 LAN、电力线通信 PLC 等。
  传感器中的通信模块一般来说功能要求不是太高，满足基本要求即可，当然针对一些特殊应用场合，通信能力要求不同。

2. 传感器网络

传感器网络通过传感器节点的有线/无线网络通信能力，使得多个传感器节点可以组成传感器网络。传感器网络提供的能力是：将同一逻辑或物理范围内的传感器节点连接在一起，使之能够通过相应的网络协议（例如，ZigBee，Ad hoc）进行相互通信。

传感器网络中的特定（或任意）节点能实现与网关的有线/无线通信，使得网关能通过特定的网络协议与一个或多个传感器节点进行通信，由网关汇聚传感器节点的网络通信需求。

在实际应用中，由于无线传感器网络的便捷性、部署的方便性等优点，我们通常采用无线的形式，当然在有些场合下，有线传感器网络更具有优势。

二、业务域

1. 物联网业务网关

物联网业务网关（注：能够提供物联网业务能力的物联网终端也可以被称为物联网业务网关）是实现 GaaS（Gateway as a Service）的核心，物联网业务网关能提供 Web 服务，使物联网应用直接以 Web 的方式访问网关，与传感器等进行通信，物联网业务网关提供南向和北向接口。

物联网业务网关的南向接口：通过有线/无线传感器网络与传感器/变送器通信，实现数据和控制信息的格式转换和交互流程。

物联网业务网关的北向接口：通过移动核心网络（Internet）实现与 Web 业务平台和/或物联网应用的交互。

物联网业务网关通过在业务网关上部署和运行业务网关中间件来提供业务能力，并实现业务网关的南向和北向接口。

物联网业务网关中间件提供的业务能力集主要包括以下方面。

（1）标识与寻址。包括标识、寻址和路由，实现传感器/控制器的可识别性和可达性。

（2）资源适配/资源抽象/存储与同步/缓存。包括数据解析/存储/管理，实现业务数据在网关上的存储。

（3）设备配置/管理/统计。实现对传感器/控制器的网管能力（不包括对传感器/控制器业务数据的管理）。

（4）协同与调度。包括通信/连接管理，实现网关对多个可用通信链路的选择，使得网关可以按照特定场景和策略有效地与传感器/控制器进行通信。

（5）发现与搜索。实现传感器/控制器数据和信息的开放与搜索发现。

(6) 使用统计。包括日志和统计。

(7) 安全性。包括接入控制和安全，实现网关和传感器/控制器数据和控制信息的安全访问。

2. 物联网业务平台

物联网业务平台处于移动核心网上层，包括业务聚合与应用生成平台、资源综合管理平台、业务环境中间件，如图 2-2 所示。

服务（业务）聚合与应用生成平台要能提供业务生成能力（如使用 BPEL 来实现通用的物联网业务流程），包括应用生成、应用登记、应用发布、应用签约计费等。通过平台提供的服务、应用生成能力来实现。

资源综合管理平台的功能包括如下。

(1) 资源标识分发与管理。

(2) 抽象出的海量资源提供基于 Web URI 的统一资源标识和管理，支持多级资源的映射，保证资源标识的唯一性、结构化和层次化。

(3) 支持资源被系统、应用、用户可发现，为海量异构资源提供高效的搜索。

(4) 资源接入管理，并能充分考虑泛在网众多能力受限设备和实时性业务的特性。

(5) 根据业务需要，对资源进行统一协同与调度。并提供基于上下文感知的异构资源动态切换与调度机制。

(6) 根据运营需要，分用户角色、应用类型、资源粒度的签约计费以及资源使用共享机制。

(7) 实现分布式物理平台间，平台与网关间的资源同步和分布式存储。

(8) 对资源实现使用统计。

(9) 引导和路由。对资源进行引导注册，并完成跨平台、平台与网关间的资源存储查找的路由功能。

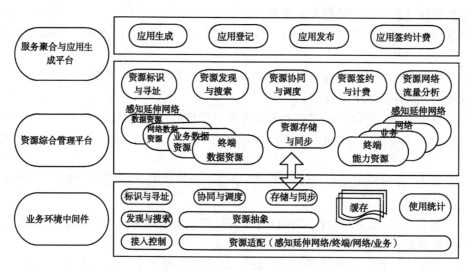

图 2-2　物联网业务平台功能模块

3. 物联网应用服务提供商

物联网应用服务提供商是物联网应用的提供者，包括公司或个人提供者，可以以应用服务器或客户端应用的方式提供。

物联网应用服务器：物联网应用服务器提供基于 Web 的服务能力，用户可以通过浏览器的访问 Web 应用，应用的生成需要通过物联网业务平台调用各类业务、资源。

物联网客户端应用：物联网客户端应用通过安装各类应用客户端的方式向用户提供各类应用，物联网客户端应用可以由物联网应用服务提供商（包括公司和个人）放到 IoTMall 的应用商店中，以供用户下载使用。

由此可见，基于 Web 的物联网体系结构强调将物联网 Web 化，采用 Web 的技术标准将物联网中的各种设备资源无缝接入到互联网中，如通过 HTTP 获取物联网应用中的各种感知元器件数据，以及进行控制指令的下达，而传统物联网体系未采用 Web 的方式。

第二节 物联网发展存在的问题及需求

经过多年发展，中国物联网在技术研发、标准研制、产业培育和行业应用等方面已初步具备一定基础，但也存在核心关键技术有待突破、产业基础薄弱、物联网产业链松散、存在网络信息安全隐患和产业聚合能力较低的现状，导致产业链上各类角色孤立化和技术低的局面。当前我国物联网产业发展存在的问题主要如下。

存在大量互不兼容的应用解决方案和产品，缺乏统一概念与实施办法，导致应用和架构碎片化。

各种示范工程不能互联互通，重复建设，存在行业和地区壁垒，缺乏可复制性、示范性与开放性，工程只是封闭的内部尝试，应用和服务提供商介入困难。

应用开发商和服务提供商依赖设备制造商，通常基于硬件进行系统集成和开发，导致产业化门槛高，研发和实验成本高，无法为大量中小企业创造机会，缺少创新性的应用，难以满足用户应用物联网的需求。

众多应用开发商、广大科研机构及高校，缺乏一个合适的实验、测试及验证平台来验证最新研究成果的有效及可行性。

产业链多样化且比较长，导致复杂的利益矛盾，需要一个系统来保证并连接产业链上各方的利益。

缺乏可持续发展的商业模式与应用开发团队，物联网规模化、产业化发展推动困难。

因此，当前物联网发展中，各类机构角色都希望形成自己的一套规范，导致开发的应用大部分为封闭的孤岛应用，缺少和谐融合的物联网服务平台，从而难以促进物联网产业的跨越式发展。以如此模式推进物联网产业，必将会导致基础资源重复和浪费、用户需求无法得到满足、物联网应用支撑平台建设困难、终端成本下降困难、难以保障信息安全、无法统一规范和技术、物联网应用与服务商用困难等问题。

从移动互联网产业的爆炸性发展中可以看出，苹果公司和 Google 公司构造的 iOS

和 Android 系统服务平台是其成功的重要原因。苹果以及 Google 通过其生态平台吸引了数以万计的第三方开发者，这些开发者也是苹果和 Google 的支持者。大量移动平台上的创新应用不但满足用户的需求，也为大量的中小企业带来了机会，为产业链的各方都创造了巨大的利益，推动了整个产业的良态发展。

因此，当前阻碍物联网产业链进一步发展的重要原因之一就是缺少融合统一的物联网系统服务平台。没有物联网系统服务平台，导致我们物联网产业链间无法进行高效的融合，所产生的孤岛应用普遍，一盘散沙，既无法满足用户需求，也无法满足产业链的各方的利益。对于物联网的发展来说，我们应当借鉴移动产业的成功经验。用户的需求才是物联网持续发展的动力，我们应该不断开发与发展创新型应用来满足用户需求和更好的体验。而满足产业链各方的利益是物联网产业化发展的重要因素，为大量中小企业以及少数大企业创造机会的能力对物联网系统平台来说是必要的，通过大力发展推动物联网系统服务平台，应用开发者开发创造更多更好的满足用户需求的创新应用，从而增加就业、促进经济迅速发展、推动物联网产业发展。只有建立无缝连接的统一的物联网系统平台，才能满足上述需求，整个物联网产业链才能真正迎来爆炸式的发展。

第三节　基于 Web 的物联网系统服务平台

基于 Web 的物联网系统平台为应用开发者、应用实验者、参观者和实际使用者提供最好的实际工作、生活和开发体验的环境。该系统汇聚各种开发平台，为应用开发者打造多元化的应用开发平台；构建产业园规模的示范平台和实验平台，为应用示范上线试运行创造环境；提供测试和评测设备，对平台上运行的应用和设备进行兼容性和互联互通性测试，使之符合国际标准。基于 Web 的物联网系统汇聚产业链上主要平台与设备，引进国内外先进的理念与技术，构建集上述三位一体的物联网生态系统服务平台，形成物联网应用设备商店，为用户提供全方位的体验与服务。基于 Web 的物联网系统服务平台如图 2-3 所示。

第四节　基于 Web 的物联网系统的商业模式

基于 Web 的物联网系统把政府、平台提供商、应用开发者、设备提供商、评测机构、最终用户的利益紧密结合在一起，如图 2-4 所示。通过各种标准接口与应用服务中间件，最终构建应用设备商店，兼顾各方利益，形成完善、各方共赢、多赢的商业模式。

1. 政府

支持产业化模式创新，引导并推动物联网发展应用服务创新；示范工程、产业园建设或平台建设模式创新；平台建设和初期运营阶段，作为产业孵化器；避免产业链上的利益矛盾，促进产业良性循环发展。

2. 平台提供商

集成多元设备组建服务平台；连接产业链各方利益，特别是连接设备制造商和应用

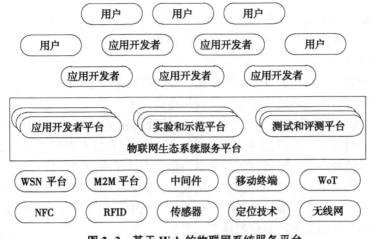

图 2-3 基于 Web 的物联网系统服务平台

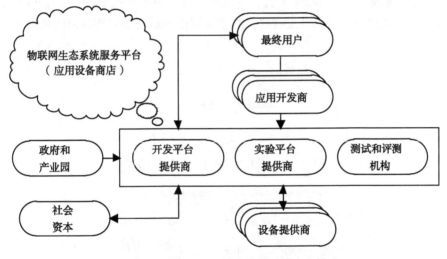

图 2-4 基于 Web 的物联网系统商业模式

开发商的利益，解决当前两者之间的矛盾和依赖关系；构建物联网的应用和设备商店，从应用开发商处收取合理服务费用。通过平台服务，积累各类创新应用和解决方案，为应用开发商、设备提供商和最终用户提供解决方案和技术咨询服务。

3. 应用开发者

降低开发者门槛，可专注应用开发，可用很低的成本创造出更多的新应用；吸引大批开发者，开发者投入低，形成聚集效应；当把开发成功的应用交付给最终用户时，可以得到最合适的和得以验证的设备平台；应用也得到了认证和验证。

4. 设备提供商

为平台提供实验设备，宣传设备商设备；可让应用开发商发掘出新的应用；通过实验成功应用的商业部署为最终客户提供设备，创新的设备推广和销售渠道；设备提供商得到评测机构的验证达到市场准入。

5. 评测机构

为应用及设备商提供评测和认证，同时保证设备和应用间的兼容性、互通性、安全性和可靠性，既对应用进行评测又对设备进行评测。

6. 最终用户

获得经实验和评测机构验证的可靠产品；在物联网系统平台中，有更多的设备和平台选择，根据验证后的产品进行采购，降低最终的成本。

基于 Web 的物联网系统整合产业链上各主要利益相关者的利益，减少政府的重复投资，提高投资回报率，注重形成规模效应；减少应用开发者对硬件和实验部署的投入，减少开发风险和周期，降低应用开发者门槛；专注于应用创新的开发，加速商业模式创新的推出，吸引更多的应用开发者，促成良性循环；推广被广泛采纳的平台以及标准化技术，打造有序竞争环境；连结中小企业、研究中心和大学等主要利益，孵化中小企业发展；加强对应用服务产业化的推进和引导。

物联网系统应用平台和服务管理平台构建在互联互通、标准化接口框架之上，将具有多种异构平台的能力，如 M2M 平台、传感网平台、智能手机平台、WoT 平台等，通过云计算、物联网接口为应用开发者提供全方位服务，最终形成物联网应用服务云，如图 2-5 所示。

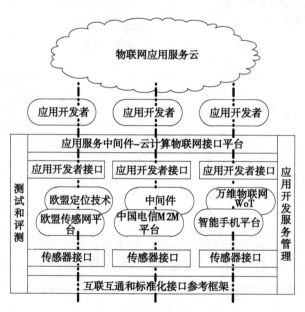

图 2-5　物联网应用和服务管理平台架构

物联网生态系统的管理实施，包括商务和技术方面，其中商务管理包括宣传推广、产业联盟、产业招商、政府公关、资金引进等；技术管理包括整体框架、国际合作、技术引进、整合和标准化等。通过示范应用开发、示范应用实验、示范应用展示、示范应用体验连接各参与方；通过应用开发服务管理，连接标准化的平台与接口，在完善的通信和云计算基础设施上为用户提供完美的体验与服务，如图 2-6 所示。

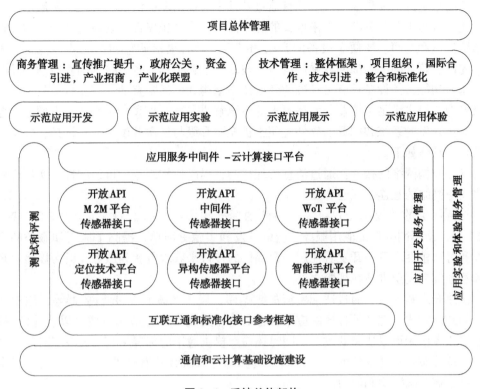

图 2-6 系统总体架构

第五节 应用服务中间件

在物联网应用环境中，存在各种各样的软、硬件，各种传感器、RFID 以及其他感知设备，多样的网络环境以及数据库系统，复杂的异构环境导致开发部署成本增加，应用软件的重用存在重重障碍。为了解决系统的异构性问题，我们引入应用服务中间件技术，来屏蔽各种软硬件以及应用服务环境的差异。

一、应用服务中间件功能

在物联网环境中，大量的应用和服务需要在异构的网络环境、异构的数据库环境、异构的资源平台、异构的操作系统上运行。复杂的环境加大了开发部署的成本，为应用的可重用带来了巨大障碍。为了解决异构问题，亟须一个统一的应用服务中间件来屏蔽各种硬件及服务环境的差异，使应用和服务能共享跨网功能，不同分布式数据的软件，能够帮助用户灵活并且高效的开发、部署、集成、管理大型分布式的 Web 应用、网络应用、移动应用和数据库应用，这是物联网生态链的一个有机组成部分。在应用服务中间件的帮助下，应用服务开发者可以从复杂烦琐的分布式计算资源管理问题中解脱出来，通过利用统一的管理引擎和开发平台，访问中间件提供的各种资源和服务。

通过应用服务中间件将使得物联网的研发、生产单位能够紧密结合在一起，提高物联网应用系统的开发集成速度，并且通过共性平台提高便捷的二次开发、扩展应用。平台在建设过程中要充分考虑未来增长的业务需求，设计扩展性强的系统结构，为后期建设预留接口和空间。要便于设备的升级和扩容，具有灵活调整和扩充的特性。采用成熟、先进、符合国际标准的软硬件设备与技术，能够根据业务、技术的发展动态随时增加、调整功能，使设备具有可扩展的开放体系结构和灵活配置的工作流程。具有扩展和升级能力，提出满足系统业务发展要求的系统解决方案，提高应用水平，方便应用范围的扩大；开发的系统和配置的软件应便于扩充、升级与二次开发，易于维护，并支持多种接口。具体如下。

支持异构网络的接入：通过消息代理机制，提高异构服务的可访问性，实现应用资源之间跨网络的互连、互通和互操作。

支持标准的协议和接口：支持主流的标准和协议，如 J2EE、EJB、XML、WML、REST 等，使不同类型的应用服务系统的开发和实施更简单，提高了应用和服务的可重用性和可移植性。只要开发者使用平台提供的接口来获取资源和服务，就可以将物理实现和管理细节完全屏蔽在视野之外。

快速开发和部署：通过适配主流的数据库、操作系统和 Web 服务器等，用中间件提供的开发方式，把各种各样的应用整合起来，真正使这些应用变成可以快速的拼插，使系统或者应用可以快速的改变。同时还需要提供通用的部署工具、开发工具等。

支持分布式和并发访问：支持海量资源的分布式计算和透明访问，为用户提供计算、存储等基础资源和能力。

中间件可以进一步细分为数据通信中间件和服务中间件，其目的是向下屏蔽底层设备之间的差异，向上以统一的规范提供对外服务访问接口。一方面，对于上层应用而言，底层传感器节点都是逻辑上无差异的设备资源，通过传感器节点适配器将来自上层应用的指令解释成与设备类型相关的命令，同时以通用的数据传输格式、数据存储格式将来自不同类型传感器节点的数据进行标准化处理，从而使得开发者不用过多关注底层的实现细节而更专注于业务逻辑。

为了实现对异构物联网络中异构节点资源以及服务资源的统一管理，我们定义了如图 2-7 所示的结构。从系统部署的角度来看，设备分为应用服务器、节点适配器以及传感器节点。节点适配器是一种能够适配不同类型传感器节点的通用数据处理模块，该模块可以支持设备的即插即用、支持数据的标准化和加解密操作。从软件层次的角度来看，中间件自上而下可以分为表示层、业务逻辑层、持久化层、通信层、设备驱动层。其中表示层、业务逻辑层、持久化层以及通信层属于服务中间件的逻辑单元，而设备驱动层属于传感器节点适配器的逻辑单元。

二、各层主要功能

表示层包含两部分的功能：一是接收客户端用户的请求，并提交给业务逻辑层进行相应的处理；二是接收业务逻辑层处理后的结果，进行相应整理布局后展现给用户。用户无须了解异构物联网络的技术细节，只需要使用简单的用户可理解的脚本命令或者图形界面就可完成对现实中的传感器进行管理和控制。

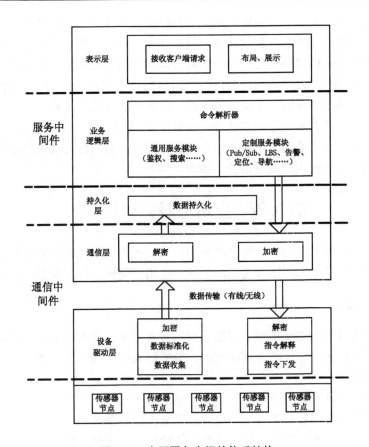

图 2-7　应用服务中间件体系结构

业务逻辑层的功能概括起来包括：调用相应的业务逻辑对表示层所提交的用户请求进行相关处理；对经过持久层获取的原始数据进行加工处理，并将结果发送到持久层进行加工处理；处理的最终结果反馈至表示层。本平台提供脚本解析器，对用户脚本进行解释并调用相应的服务模块来执行用户的命令。开发者只需要修改应用相关的参数就可以使用服务模块提供的功能，既不需要修改服务的细节，也不需要从底向上开发服务。出于安全性的考虑，对传感器节点与服务器之间的通信进行鉴权尤其重要。因此，需要保证只有认证的用户才可以读取传感器的数据，其命令才会被传感器接收和执行。

持久化层封装了所有与数据库有关的操作和细节，具有新增数据、删除数据、修改数据、查询数据等基本功能以及连接管理、性能管理、事务管理、缓存管理等功能。

通信层负责服务器与传感器节点适配器之间的数据交互。并根据节点对应的标识符选取对应的密钥对数据进行解密，将解密后的数据提交到数据库中，同时接收上层的控制命令，并对控制命令进行加密、传输。

设备驱动层运行于传感器节点适配器，通过传感器节点的对外访问接口进行数据采集以及指令下发。为了访问不同的传感器节点，本层需要封装适配不同传感器节点的驱动模块，通过调用对应的驱动模块可以对传感器节点进行数据采集或者配置管理。对于采集到的原始传感器数据，按照统一的标准进行数据格式化处理，使得上层服务看到的数据是设备无关的。而对于上层服务单元下发的配置管理指令，驱动模块需要根据设备类型将设备无关的指令信息解释为设备相关的指令信息下发到对应的设备。当将一个异构节点接入现有的传感网络中时，只需要将节点和中间件提供的标准适配器连接，平台就可以自动地识别新加入的传感器节点，并给传感器节点一个全网唯一的标识。此后，新节点就能像网络中已有的传感器节点一样，向服务器发送标准格式的数据，也能够通过标准的传输协议接受来自服务器的统一控制命令，同时在工作异常时也可以自动向服务器报警。为了保证通信的安全性，每个传感器节点所对应的处理进程都有唯一对应的标识符和密钥，由该密钥完成通信的加解密。

第六节 应用与服务管理平台

随着各种不同技术架构的应用和服务的引入，各种技术之间的异构性和不兼容问题逐渐显现。那么对这些采用不同技术的服务和应用的配置、使用等都需要进行管理。

一、基本功能

应用与服务管理的主要功能是根据用户所提的要求，依据各种已有的功能子系统，对物联网系统平台进行相应的管理和配置，最终形成物联网应用服务云。具体的功能主要包括以下方面。

对应用服务的开发进行管理：管理与开发相关的各种接口、访问控制策略的制定等，通过云计算物联网接口为应用开发者提供全方位服务。

应用实验和体验服务管理：管理实验和体验的相关流程，管理实验平台上运行的应用或服务，使用户的实验或者体验能够顺利的进行，达到预期的目的。

管理控制工具：支持图形化的管理控制工具，用于远程监视和更新应用服务状态，从单一的远程控制台可以安全方便的管理应用服务。同时，需要提供信息采集、内容管理、报表分析、事务服务、日志服务、安全管理、权限管理、质量服务等基础服务。

支持标准安全性原则：将标准安全性原则引入应用开发、集成、部署和管理当中，提供一个可靠安全的软硬件的平台。

依据管理对象的不同，服务管理主要包括开发服务管理和应用实验和体验服务管理。开发服务管理的主要功能是管理与开发相关的各种接口、访问控制策略的制定等，尽可能使试验床生态系统中各个部分保持良好运行状态，为用户提供更有效的服务，主要包括各种系统参数的配置、各个功能模块运行的监督、故障管理以及安全管理等。

二、开发服务管理

开发服务管理子系统的各类管理功能支持开发者的各种功能的实现，满足开发者和管理人员对于物联网生态系统操作、维护和管理的需要。管理人员可以通过相关的操作接口与服务管理子系统进行交互。常见的管理功能可以划分为如下的几个功能域。

性能管理（Performance Management）：主要是对物联网生态系统中的各个功能子模块的性能和网络单元的有效性进行评估，并给出相应的评价报告，包括各个模块的性能分析及性能控制。

配置管理（Configuration Management）：主要提供传感器网络、M2M、WoT 等功能模块的状态以及安装功能。依据用户应用程序的不同，给其提供各种不同的配置方案和组合，以期达到使用尽可能少的成本满足用户的各种功能需求。主要包括各个业务模块的开通和关闭、对网络的状态进行管理等。

故障管理（Fault Management）：主要是对传感器网络、M2M、WoT 等功能模块的运行异常情况和相关的各种设备安装环境异常进行管理。同时也包括对各个功能模块之间的网络连接以及与外界的接口的状态进行管理。

计费管理（Accounting Management）：主要功能是计算用户对整个物联网生态系统中的各个功能模块以及与之相关的硬件的使用情况，依据不同的使用情况计算提供服务的应收费用。同时提供对用户使用各种功能或者业务的收费过程的支持。

三、应用实验和体验服务管理

应用实验和体验服务管理的主要功能是对实验和体验各种不同的应用进行管理和控制，并管理实验和体验的相关流程，使用户的实验或者体验能够顺利的进行，以确保能够达到预期的目的。主要包括以下部分：实验和体验环境的设置管理、服务状态与部署管理、服务过程安全管理与故障管理等。

实验和体验环境的设置管理：该部分功能和开发服务管理中的配置管理类似，主要是根据实验或者体验者的需要，对传感器网络、M2M、WoT 等功能模块进行相关的配置和组合，使之能够达到实验者或者体验者的需求。该部分也提供各个不同的功能模块的开通和关闭的功能。

服务状态管理：设计针对异构网络的通用服务管理信息库，记录当前运行的服务状态，以及不同服务所占用资源等信息。

服务部署管理：负责服务的下发、部署及终止。根据网络中的情况，将新加入的服务根据网络资源分配部署到网络中，并记录相应的状态。同时负责服务的更新、下发、终止等工作。

实验过程的安全管理：在实验的过程中，实时监控各个功能模块以及与之相关的硬件设备的工作状态，确保实验的过程中，各个功能模块和相关的硬件设备在正常的工作状态下运行。同时，该部分还提供实验者的权限管理，对不同级别的用户授予不同的操作权限，剔除不合格的实验者，以保证整个试验床的安全、有效地运行。

故障管理：同开发服务管理子系统中的故障管理子系统的功能类似，对运行中的各个功能模块以及相关的硬件设备进行状态监控。同时也包括对各个功能模块之间的网络连接以及与外界的接口的状态进行管理。

四、安全管理

安全管理提供对各个功能模块以及与之相关的硬件产品进行安全保护的能力，对各种设备的工作状态进行安全检测，同时也提供用户权限的管理，安全审查及安全告警等功能。

物联网业务的发展，安全权限尤为重要。与传统的互联网不一样的是，物联网的终端大多数是一些无人控制的终端设备，这些设备在发生通信行为时，不能人为地进行鉴权权限管理。而物联网的应用业务的安全权限要求很高，故需要一套完整的从"设备-服务-用户"的三维权限矩阵。

生态系统平台内部，将推出面向用户、设备的统一身份认证。对于部分设备接入，将使用 RADIUS（Remote Authentication Dial In User Service）方式进行认证，在接入服务器一侧布置 NAS 等具有代理认证功能的设备，RADIUS 服务器和 RADIUS 客户端通过共享密钥进行交互信息的相互认证，用户密码等信息在网络上传输采用密文的方式，增加了安全性。RADIUS 在响应报文中携带授权信息，合并了授权和认证过程。

第七节　物联网生态系统

目前在国内，有少量以政府和运营商为主导的物联网公共平台进入试运行阶段。中国电信、中国移动、中国联通都已经开始构建 M2M 综合平台，支持应用、开放接口、运营支撑、承载和感知层技术，以行业专家系统、物联网信息管理中心、云计算平台为核心，提供对物联网终端监测和控制。上述平台均以运营商为主导建立，而不是从用户的真正需求角度出发，是否能够真正被市场接受还有待检验。

我们可以借鉴苹果公司、Google 公司、微软公司构造的 iOS、Android、Windows 系统服务平台，通过吸引数以万计的第三方开发者，为中小企业带来机会，为产业链各方创造巨大的利益，创造出更多的创新应用，满足用户的需求，推动产业的良性、健康发展。

物联网产业的良性发展除了产业需求、科技发展、政府推动外，还需要务实的物联网生态系统与商业模式、运营体系，并突破所需关键技术。当前物联网生态系统平台的设计与开发仍面临着许多挑战，目前很多研究仍处于起步阶段，主要面向科研与教学，许多问题仍需进一步研究，离实用还有很大的距离，更无法适应现代物联网产业应用多样性、产业化和多网互联与融合的战略需求。国内相关平台研发起步慢，投入少，研发水平相对滞后。为解决当前现有系统的不足之处，本项目旨在建立统一融合的物联网生态系统平台，攻克异构统一接入、服务应用管理、异构平台安全、不同行业应用等技术难题，突破新型物联网生态运行模式，通过系列创新技术突破，最终形成实际物联网生态系统平台，有效促进物联网技术的可持续性发展，为构建物联网产业化奠定基础。

第三章　一种基于虚拟坐标的动态自适应分簇协议

无线传感器网络应用广泛，如灾难救援、环境监测、军事跟踪等，在设计无线传感器网络时，我们需考虑应用场景、目标、代价、硬件等因素。由于无线传感器通常采用电池供电，所携带的能量受限，因此必须均衡、减少网络中传感器节点的能耗，达到延长网络的生命周期的目的。分簇协议是一种高效利用节点能量，层次结构清晰，兼顾负载均衡，增强网络可靠性的拓扑控制方法。本文设计了一种基于虚拟坐标的动态自适应分簇（Virtual Coordinates Dynamic Adaptive Clustering，VCDAC）协议，该协议能延长网络的生存周期，提高系统的吞吐量，均衡网络负载。针对 LEACH（Low-Energy Adaptive Clustering Hierarchy）、SEP（Stable Election Protocol）协议中簇头选择的随机性与分布不均匀性，借鉴基于地理位置分簇协议 GAF（Geographical Adaptive Fidelity）的虚拟单元格划分的思想，对监测区域进行动态划分，簇头的选择考虑簇内节点数量、节点间的平均距离以及节点与基站的距离等因素，选择最优节点担当簇头节点，负责簇内的收集、汇聚、转发任务，簇头的选择更加合理，分簇分布更均匀，进一步均衡了能量消耗。本章给出了 VCDAC 的详细细节与仿真实验，仿真结果表明，该协议能明显延长网络的生存时间，延缓节点的死亡速度，发送更多的数据至基站。

第一节　引　言

在无线传感器网络中，分簇拓扑控制用来将无线传感器网络监测区域划分成一个个簇，每个簇中包含若干成员节点和一个簇头节点，簇内成员节点直接与簇头节点进行交互，簇头负责簇内管理，收集节点采集的数据，进行数据融合后把结果发送至基站。在分簇拓扑结构中，所有簇内成员节点与簇头通信，避免了直接与基站通信，使得通信距离大大减小从而节省能量，簇内成员节点在其余时间进入睡眠状态，进一步节省能量；由于簇头负责收集簇内成员传感器节点的感知数据，对数据进行融合并路由转发，故承担较多责任，消耗能量相对较多，比成员节点耗费能量多。因此必须按照一定的规则进行簇头的轮换，将一些符合条件的节点调整为簇头节点，使网络内所有节点剩余能量保持均衡，延长网络生命周期，避免过早出现能量"空洞"。无线传感器网络中基于地理位置信息的分簇算法也被研究人员所关注，节点的地理位置可以通过 GPS 等低成本、体积小的设备来获取，也可以根据网络中少数已知节点位置，按照某种定制机制来得到，比较常见的定位方法有三角测量法、三边测量法、极大似然估计法、基于接收信息强度指示 RSSI 的定位、基于到达角 AOA 的定位、基于到达时间差 TDOA 的定位、基于到达时间 TOA 定位等。

目前很多分簇协议存在簇头分布不均匀，簇头节点的选举未充分考虑剩余能量与通信距离的影响，以及分簇的数量不随网络情况的变化而变化等问题。为此我们设计了一种基于虚拟坐标的动态自适应分簇（a Virtual Coordinates Dynamic Adaptive Clustering，VCDAC）协议，该协议按照虚拟分区的思想将网络分簇，并根据网络的状态动态决定分簇的数量，从而降低了能耗；在每个簇中，根据节点的剩余能量与结点间的距离，选择最合适的节点当选簇头，负责数据的收集汇聚与转发等任务，簇头选择合理，分布均匀，从而使整个网络的能量消耗较均衡。仿真实验结果验证，该协议能明显延长网络的生存时间，均衡网络负载，发送更多的数据至汇聚节点，具有优良的性能。

第二节　相关工作

在层次拓扑控制方面，TopDisc 算法仅考虑在保证网络连通度的前提下，形成尽可能少的分簇数量，而未考虑节点的剩余能量以及如何提高网络的健壮性问题；基于地理网格分簇 GAF 算法及其改进算法需要知道网络内节点的精确位置，未考虑簇内节点间距离对数据汇聚的影响；LEACH 协议中簇头的选择具有一定的随机性，且未充分考虑影响系统性能的其他因素，比如传输距离，网络的动态性等。HEED 协议是一种固定簇半径的分簇协议，HEED 协议中簇头的选择主要依据主次连个参数，通过这两个参数来衡量簇内通信代价，主参数依赖于剩余能量，衡量簇内通信成本标准是簇内平均可达能量（Average Minimum Reachability Power，AMRP），各节点以不同的初始概率发送竞选消息，初始概率为 CH_p 可根据公式（3-1）来确定。

$$CH_p = \max(C_p + E_{re}/E_{\max}, p_{\min}) \tag{3-1}$$

其中，C_p 和 $p_{\min}$ 为统一网络参数，对收敛速度有影响；$E_{re}/E_{\max}$ 为节点初始能量和剩余能量的比例。该协议较 LEACH 在成簇速度上有改进，同时考虑成簇后簇内通信开销，但是簇头的选择仍然是随机的。

PEGASIS 协议通过构造接近最优的链来减少能量消耗，减少簇重构的开销，节点仅需要和距离自己最近的邻居节点通信，但需要动态调整网络的拓扑结构，仍然需要知道各自邻居节点的能量状态。VAP-E 算法将网络监测区域进行虚拟分区，根据节点剩余能量多少来决定成簇的概率，未考虑节点间的距离等因素，且簇头的选择是随机的。

第三节　系统模型

一、网络模型

假设 N 个无线传感器节点，任意随机部署在 $M \times M$ 的感知监测区域内，传感器节点集合 $S = \{s_1, s_2, s_3, \cdots, s_n\}$，其中 s_i 代表第 i 个传感器节点，且 $|S| = N$，传感器网络部分结构如图 3-1 所示。我们对无线传感器网络做如下设定。

汇聚节点在监测区域的中央。

节点电池供电，能量受限，具备一定的初始能量。
节点的发射功率根据通信距离的远近可进行调整，双向通信且链路可靠。
所有传感器节点没有移动性。
每个节点知道自己的位置信息。
节点周期性的采集数据。
任意两个节点间的距离小于门限距离，即节点在自由信道模型下工作。

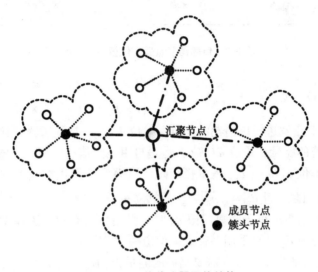

图 3-1　无线传感器网络结构

该网络模型运行 VCDAC 算法，簇头节点负责簇内管理协调，成员节点发送感知数据至簇头，这种结构设计具有如下优点。

每个成员节点传输感知数据至最近的簇头，节省了能量消耗。

仅簇头节点进行数据融合工作，进一步减少了能耗。

采用虚拟坐标的动态分簇，分簇情况由网络状态动态决定。

选择最合适的节点担任簇头，避免能量即将耗尽的节点成为簇头。

我们把每个周期定义为一轮，在每一轮中，首先形成分簇，簇头公布簇内每个成员节点的传输时隙；成员节点采集感知信息后，在各自的时隙中发送数据至簇头，其他时间进入休眠状态；簇头节点汇聚成员节点感知采集的数据，并将数据直接发送至基站（汇聚节点），基站将接收到的数据接至 Internet，为用户所利用。

二、能耗模型

我们采用文献中相同的能量消耗模型，接收器消耗能量运行接收电路，发射器消耗能量运行发送放大器和发送电路。根据接收器和发射器间不同距离，能量控制模块工作在多径衰减信道模型或自由空间信道模型。传感器节点通信能量消耗模型如图 3-2 所示。

发送器发送 l 比特数据经过距离 d，节点需要消耗的能量为：

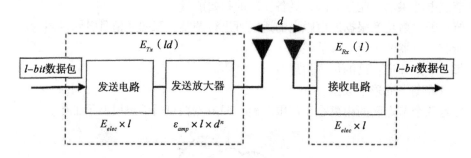

图 3-2 传感器节点通信能量消耗模型

$$E_{Tx}(l, d) = E_{elec}(l) + E_{amp}(l, d) == \begin{cases} l \times E_{elec} + l \times \varepsilon_{fs} \times d^2, & d \leq d_0 \\ l \times E_{elec} + l \times \varepsilon_{mp} \times d^4, & d > d_0 \end{cases} \quad (3-2)$$

其中，E_{Tx} 是节点发送模块消耗的能量，单位 J；E_{elec} 是发送器（接收器）电路收发每比特数据所耗费的能量，单位 nJ/bit；ε_{fs} 是自由空间信道模型下传感器节点发送放大器的功耗参数，单位 pJ/bit；ε_{mp} 是多径衰减信道模型下发送放大器的功耗参数，单位 pJ/bit；d_0 是距离门限，且 $d_0 = \sqrt{\varepsilon_{fs}/\varepsilon_{mp}}$，单位 m。

当节点间的通信距离小于距离门限 d_0 时，采用自由空间信道模型；当节点间的通信距离大于距离门限 d_0 时，采用多径衰减信道模型。

节点接收 l 比特的数据，节点需要消耗的能量为：$E_{Rx}(l) = l \times E_{elec}$，其中 E_{Rx} 是节点接收模块消耗的能量，单位 J。节点汇聚数据所需能量为 E_{DA}。

第四节 基于虚拟坐标的动态自适应分簇协议

本节详细设计了基于虚拟坐标的动态自适应分簇协议（a Virtual Coordinates Dynamic Adaptive Clustering，VCDAC），VCDAC 在 LEACH、SEP、VAP-E 等协议的基础上，借鉴了 GAF 算法虚拟单元格的思路，根据网络的状态，动态决定分簇的数量，并根据节点的剩余能量、节点间的距离以及与基站的距离情况等选择合适的节点担任簇头，进一步减少、均衡能量消耗，平衡网络负载，提高系统的性能。

一、算法的基础

VCDAC 在分簇和簇头选择时，充分考虑网络的状态与节点的剩余能量以及距离关系，针对汇聚节点（基站）位于感知区域内部，如上节所定义的网络模型，假设在第 r 轮选举中，选出 k 个簇头，则每个簇内包含 $(N/k) - 1$ 个成员节点和一个簇头节点，根据能量消耗公式，我们可得到在一轮中簇头节点消耗的能量为：

$$E_{CH} = (\frac{n}{k} - 1) l \cdot E_{elec} + \frac{n}{k} l \cdot E_{DA} + l \cdot E_{elec} + l \cdot \varepsilon_{fs} d_{toBS}^2 \quad (3-3)$$

其中 k 是簇头节点的数量，E_{DA} 是簇头节点进行数据融合时所耗费的能量，d_{toBS} 是簇头节点与基站间的平均距离，且：

$$d_{toBS} = \int_A \sqrt{x^2 + y^2} \frac{1}{A} dA = 0.765 \frac{M}{2} \tag{3-4}$$

非簇头节点在一轮中消耗的能量为：

$$E_{nonCH} = l \cdot E_{elec} + l \cdot \varepsilon_{fs} \cdot d_{toCH}^2 \tag{3-5}$$

其中 d_{toCH} 是簇内节点到簇头的平均距离，假设簇内节点均匀分布，且分布密度为 $\rho(x, y)$，则有：

$$d_{toCH}^2 = \int_{x=0}^{x=x_{\max}} \int_{y=0}^{y=y_{\max}} (x^2 + y^2)\rho(x, y) dx dy = \frac{M^2}{2\pi k} \tag{3-6}$$

所以单个簇内能量消耗大约为：

$$E_{cluster} = E_{CH} + \frac{n}{k} E_{nonCH} \tag{3-7}$$

在一轮运行中，整个网络的能量消耗为：

$$E_r = l(2nE_{elec} + nE_{DA} + \varepsilon_{fs}(kd_{toBS}^2 + nd_{toCH}^2)) \tag{3-8}$$

我们将 E_r 对 k 求导数并令结果为 0，即：

$$E_r' = l\varepsilon_{fs}\left(d_{toBS}^2 - \frac{N}{2\pi}\frac{M^2}{k^2}\right) = 0，得 k_o = \sqrt{\frac{n}{2\pi}}\frac{M}{d_{toBS}}，将 d_{toBS} 带入得：$$

$$k_o = \sqrt{\frac{n}{2\pi}} \frac{2}{0.765} \tag{3-9}$$

可见在我们设定的场景下，最优簇头数量仅与簇内节点数量有关，随着网络的运行，需动态调整监测区域内的分簇数量，延长网络的生存时间。另簇内节点信息比如剩余能量、位置等信息可以通过捎带的方式发送至汇聚点，以便获取准确信息，且不会增加系统的开销，在其他算法中，通常采取估算的方法，极不准确。

二、最优簇头数量的论证

在基于分簇的拓扑控制方法中，簇的大小与数量对系统性能影响较大，无论簇的数量过多或者过少，都会引起网络中能耗的不均衡和系统能耗增加。针对无线传感器能量受限考虑，运行中每轮最优簇头数量应使网络在该轮中总能耗最小，在我们设定的场景下，$k_o = \sqrt{\frac{n}{2\pi}}\frac{M}{d_{toBS}}$，下面我们验证 k_o 即为最优簇头节点数。

我们定义 $f(k) = (k - k_o) \times E_r'$，我们知道 $(k - k_o)^2 \times (k + k_o) \geq 0$，证明过程如下：

$(k - k_o)^2 \times (k + k_o) \geq 0$

$\Rightarrow (k - k_o) \times (k^2 - k_o^2)^2 \geq 0$

$\Rightarrow \left(k - \sqrt{\frac{N}{2\pi}}\frac{M}{d_{toBS}}\right) \times \left(k^2 - \frac{N}{2\pi}\frac{M^2}{d_{toBS}^2}\right) \geq 0$

$\Rightarrow \left(k - \sqrt{\frac{N}{2\pi}}\frac{M}{d_{toBS}^2}\right) \times \left(d_{toBS}^2 - \frac{N}{2\pi}\frac{M^2}{k^2}\right) \geq 0$

$$\Rightarrow \left(k - \sqrt{\frac{N}{2\pi}}\frac{M}{d_{toBS}^2}\right) \times l\varepsilon_{fs}\left(d_{toBS}^2 - \frac{N}{2\pi}\frac{M^2}{k^2}\right) \geq 0$$

$$\Rightarrow (k - k_o) \times E'_r \geq 0$$

$$\Rightarrow f(k) \geq 0$$

我们假设 $k \neq k_o$，所以 $f(k) > 0$，满足连续函数极小值存在的充分条件，k_o 即为所求的最优簇头数。

三、网络区域虚拟坐标动态自适应划分

我们说的网络区域虚拟坐标，指的是采用虚拟极坐标的形式，将网络区域划分成若干小的分区，每个分区为一簇。由于传感器节点在监测区域内均匀分布，所以每个分簇内节点数量大体相当，在每个簇内选择最合适的节点成为簇头节点，从而有效避免分簇过大或者过小情况的出现。我们采取的划分方法是以监测区域的中点作为虚拟极坐标的原点，以 $2\pi/k_o$ 为步长，将监测区域划分为 k_o 个分簇。每个节点相对于区域中心点所对应的虚拟极坐标系计算自己所处的分簇，并加入相应的分簇。随着网络的运行，网络中有些节点能量耗尽，此时网络中存活节点数量减少，最优簇头数量也会发生相应的变化。图3-3是一个在 100×100 区域内随机部署 100 个传感器节点，基于虚拟坐标分成 5个区域、4个区域的情况。

在监测区域中，汇聚节点（基站）位于监测区域的中心，在运行过程中，节点的剩余能量、位置等信息通过"捎带"的方式汇总至汇聚点；汇聚节点根据当前的网络状态，动态决定最优簇头数量，并选择合适的簇头节点，将结果发布至监测区域中；簇头节点与汇聚节点进行握手，表示确认成功，进行新一轮的监测任务。

四、虚拟坐标动态自适应分簇协议

无线传感器网络被分成若干个簇之后，需要选出合适的节点担任簇头节点，我们考虑簇头主要负责收集簇内数据，并与汇聚节点进行通信等任务，要求簇头具有充足的能量，在分析文献的基础上，经过认真分析和多次实验，我们构造簇头概率函数为：

$$f(i) = \alpha f_1(i) + \beta f_2(i) + \gamma f_3(i) \tag{3-10}$$

$$f_1(i) = E(i) / \frac{1}{m}\sum_{i=1}^{m} E(i) \tag{3-11}$$

$$f_2(i) = \frac{1}{N}\sum_{k=1}^{N} d(i, k) / \frac{1}{m}\sum_{k=1}^{m} d(i, k) \tag{3-12}$$

$$f_3(i) = \frac{1}{N}\sum_{k=1}^{N} d_{toBS}(i) / d_{toBS}(i) \tag{3-13}$$

$$\alpha + \beta + \gamma = 1, \quad 0 \leq \alpha, \beta, \gamma \leq 1$$

其中，$f_1(i)$ 是节点剩余能量与簇内节点平均剩余能量之比，该值越大表明该节点能量充足，更胜任担当簇头节点；$f_2(i)$ 是感知区域内节点与当前节点的平均欧氏距离与簇内节点与当前节点的平均欧氏距离之比，该值越大，表明簇内节点间欧氏距离相对较

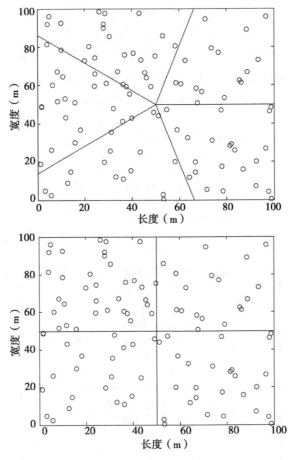

图 3-3 VCDAC 将监测区域基于虚拟坐标分区情况
（上图分成 5 个区域，下图分成 4 个区域）

小，簇内通信时消耗的能量越小，该阶段更适合充当簇头节点；$f_3(i)$ 是感知区域内节点到汇聚节点的平均距离与该节点到汇聚节点的距离之比，该值越大，表明该节点离汇聚节点距离越近，越适合充当簇头节点。

在每轮中，我们在虚拟分区内选择簇头概率函数值最大的节点充当簇头节点。我们可以看出，$f_1(i)$ 是出于节点剩余能量方面的考虑，让剩余节点更多的节点充当簇头；$f_2(i)$ 优化了簇内节点间距离的选择，使得簇内节点间距离更近的节点充当簇头；$f_3(i)$ 优化了簇间的距离，使的节点完成数据汇聚融合后，以更低的能耗发送至汇聚节点。各分量通过系数 α、β、γ 进行调整，比如在网络运行初期，节点剩余能量充足，优先考虑距离因素的影响；随着网络的运行出现个别节点能量消耗严重时，需优先考虑能量因素的影响，来避免过早出现节点由于能量耗尽而失效的情况。

五、算法的流程

我们提出的基于虚拟坐标动态自适应分簇协议的流程图如图 3-4 所示，可见该设计充分利用了汇聚节点能力强的特点，根据监测区域内感知节点的剩余能量、位置等信息，动态决定分簇最优数量，运用虚拟极坐标进行分簇，簇头节点的选取考虑节点的剩余能量、节点间的距离以及与汇聚节点的距离等因素，选择最合适的节点担任簇头节点；监测区域内节点的剩余能量、位置等信息通过捎带的方式，减少了能量开支，延长了网络的寿命。在每轮运行中，我们确保监测区域内簇头数量为 k_o 个，且随系统中节点数量的变化而动态变化，使每轮中能量消耗降到最低。

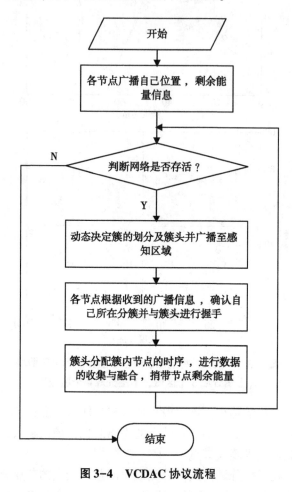

图 3-4　VCDAC 协议流程

第五节　算法性能评估

本节详细介绍 VCDAC 的性能分析与评价结果，并与 SEP、VAP-E 等算法进行比较。

从算法的执行流程与代码我们可以看出，SEP、VAP-E 以及 VCDAC 算法的时间复杂度为 $O(n)$，但 VCDAC 算法在本轮数据收集过程中，节点需"捎带"自身的状态信息，需进行更多的运算。

一、仿真环境

根据我们设定的网络模型，假设信道条件理想，忽略干扰与随机信号的影响等因素，考虑在 100m×100m 的区域内随机部署 100 个传感器节点，汇聚节点位于（50，50）m 处即监测区域中心，如图 3-4 所示，其他实验参数的设置见表 3-1，仿真实验各参数的取值与中取值相同，各参数的具体物理意义在第二章第二节中已给出。

表 3-1　仿真实验参数

参数	值
Network coverage	$(0, 0) \sim (100, 100)\ m$
Sink location	$(50, 50)\ m$
E_0	$0.5J$
E_{elec}	$50nJ/bit$
ε_{fs}	$10pJ/bit/m^2$
ε_{mp}	$0.0013pJ/bit/m^4$
E_{DA}	$5nJ/bit/message$
α	$0.1 \sim 0.8$
β	$0.7 \sim 0.1$
γ	$0.2 \sim 0.1$
Data message size	100 bytes
Data message header	15 bytes

二、仿真结果

下面我们就采用 VCDAC 协议与采用 SEP、VAP-E 协议时的网络性能进行比较，首先我们看一下无线传感器网络中存活节点数量，我们可以定义为网络的生存时间或者生命周期，如图 3-5 所示。从图中我们可看到，我们提出的 VCDAC 协议较 SEP、VAP-E 能明显延长网络的生存周期，首个节点耗尽能量的时间也明显延长，这是因为我们提出的协议在簇头选择时，充分考虑的节点的剩余能量、节点间的距离等因素，并选择最优节点当作簇头。而 SEP 协议在选择簇头时为考虑节点的剩余能量及节点间的距离因素，并且随机选择簇头节点，簇头分布不一定合理，且分簇不均匀；VAP-E 在簇头选择时虽然考虑了节点的剩余能量，但是仍然随机选择簇头，导致簇头节点选择不合理，从而导致能耗过高且不均衡。我们提出的 VCDAC 协议能有效的避免这些缺点，从而能明显延长网络的生命周期。

汇聚节点（基站）接收的数据情况如图 3-6 所示，从图中我们可以看出，VCDAC 能接收较多的数据，这是因为我们提出的协议提高了数据的传输效率，并且选择剩余能量高，簇内间距小，并且距离汇聚节点近的传感器节点作为簇头，有效地节省了能量，减少了数据的传输时延，发送更多的数据至基站。LEACH 在选择簇头节点时未考虑节点的剩余能量，可能选择能量接近耗尽的节点作为簇头，导致数据丢失；VAP-E 在选择簇头时虽然考虑能量，但是未考节点间的距离，且簇头的选择仍然是随机的。

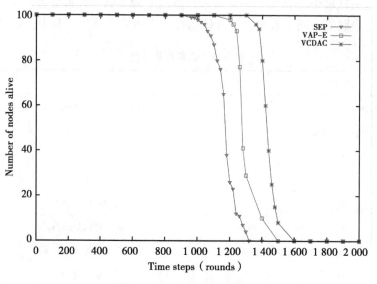

图 3-5　网络中存活节点数

无线传感器网络中剩余的能量如图 3-7 所示，从本图可以看出三种协议的直接节能对比。在网络运行过程中，我们提出的 VCDAC 协议消耗能量较 LEACH、VAP-E 少，更节省能量。这主要是由于我们提出的协议簇分布均匀，节点距离簇头的距离较近；而 LEACH 由于簇的随机性，簇分布不均匀，导致有的簇内节点与簇头的间距过大，从而能耗过高，VAP-E 虽然分簇较均匀，但是簇头选择具有随机性，导致簇内节点与簇头间距过大。我们的协议充分考虑了簇分布的均匀性，在此基础上考虑簇内间距以及簇头与基站间的距离，簇头选择当前最优节点，减少且均衡了能量消耗，从而所耗费的能量更少。

第四章 一种基于小生境粒子群优化的自适应分簇协议

无线传感器网络是一种与应用密切相关的网络，针对不同的应用场景，采用不同的拓扑控制协议。分簇方法作为拓扑管理中一种层次清晰的方法，能延长网络的生存时间，经常被转化为优化问题来求最优解。传统分析优化方法需要大量的计算，计算量随问题规模成指数增长，对于微型传感器来说，少的计算量与资源需求，并能产生可接受结果的优化方法是适宜的，粒子群算法就是一种计算效率高的仿生智能优化方法。我们首先提出了一种无参数小生境粒子群优化算法，并进行验证，然后利用该算法，设计了一种基于小生境粒子群优化的自适应分簇协议（an Adaptive Clustering Protocol using Niching Particle Swarm Optimization，ACP-NPSO）协议，该协议形成的分簇分布均匀，均衡了系统的负载，提高了系统的吞吐量。针对目前很多分簇算法如 LEACH、HEED 等簇头选择的随机性，我们采用粒子群优化算法来优化分簇与簇头的选择。簇头的选择考虑监控区域内节点的具体状态，选出的簇头分布均匀，均衡了网络的能量消耗。本章给出了 ACP-NPSO 的详细设计与仿真结果，实验结果表明，该协议具有良好的效率，能明显延长网络的生存周期，提高系统的吞吐量。

第一节 引 言

Russell Eberhart 和 James Kennedy 于 1995 年分别发表了 "A new optimizer using particle swarm theory" 和 "Particle swarm optimization" 的文章，标志着粒子群优化（Particle Swarm Optimization，PSO）算法诞生，粒子群亦可成为微粒群。粒子群优化方法是一种非常受欢迎的优化方法，具有参数少、算法简单、不需要梯度信息、易于实现等特点，在数据分类、数据聚类、生物系统建模、决策支持、流程规划、模式识别等方面表现出良好的应用前景，成为智能优化领域研究的热门。

基于对简化社会模型的模拟的基础上，提出了粒子群优化算法，许多自然界中的生物具有一定的群体行为，如蚁群、鸟群、鱼群等。人工生命主要对自然界中生物群体行为进行探索，并在电脑上建立群体模型，虽然群体行为通常看起来仅由几条简单的规则，但是群体的行为却异常复杂。生物社会学家 Wilson E O 在 1975 年针对鱼群进行了大量细致的研究，在所发表论文中指出"虽然食物资源不可预知的四处分散，至少从理论上来讲，鱼群中的个体成员受益于其他个体在寻找食物资源过程中与以前发现的经验，并且收益远超过鱼群个体间的竞争所带来的利益上的损失"，这成为粒子群优化算法的基础。Eberhart 和 Kennedy 等对鸟群模型进行修正研究，促使粒子在搜索寻优时飞向解空间，并且在降落在最优处，从而进一步得到了粒子群优化算法。

小生境的思想同样来源于自然界，是一种广泛应用的优化技术，在自然界中，在特定环境下一种生物在进化过程中，通常会与自己相同的物种生活在一起，繁衍生息。我们可把小生境的概念应用到进化计算中，将进化中的每一代个体划分为若干类，在每类中选出一些适应度较大的个体作为这个类的优秀代表组成一个类，在种群中与其他类交互信息等完成任务。我们可把小生境的思想引入到粒子群算法中，提高种群中子群的多样性，快速选出最优解，我们提出了一种无参数小生境粒子群算法，并进行测试验证。

分簇方法可转化成最优化问题来求解，且分簇拓扑管理方法是无线传感器网络的一项重要的技术，对路由、数据融合等提供基础支持。我们充分利用小生境粒子群的优良特性，对无线传感器网络进行最优分簇，提出了一种基于小生境粒子群优化的自适应分簇协议，理论分析和仿真实验证明，该算法能明显延长网络的生存周期，使网络的分簇更加均匀合理，均衡了网络的负载，表现出优良的性能。

第二节 相关工作

在无线传感器网络中，大量传感器节点部署在监测区域内，直接传输等技术由于低效率需尽量避免，LEACH 协议通过随机轮流转换簇头来形成分簇；SEP 针对异构无线传感器网络，在竞选簇头时考虑节点的剩余能量，一定程度上提高了系统的稳定性；BEES 是一轻量级的仿生架构协议，能提高网络的性能。

粒子群在分簇拓扑控制方面的研究较多，S. Guru 等提出了四种不同的 PSO-Clustering，分别是时变惯性权重 PSO（PSO-TVIW）、时变加速系数 PSO（PSO-TVAC）、层次型 PSO（HPSO-TVAC）和主管学习模式 PSO（PSO-SSM），这系列算法在分簇时仅考虑节点间的距离影响，而对节点的剩余能量未考虑；N. M. A. Latiff 等提出了 PSO-C 算法，该算法考虑了节点的剩余能量和节点间的距离影响，但未考虑节点与汇聚点的距离因素；X. Cao 等提出了 MST-PSO，目的是构建一个基于距离的最小生成树，未能很好地均衡能量的消耗。J. Tillet 等提出了采用 PSO 分簇的算法，该算法通过均衡簇内成员的数量来减少能耗。Y. Liang 等用 PSO 来解决分簇，簇头的选择考虑位置与能耗信息。S. X. Yang 等利用 PSO 算法来优化层次式分簇结构，来寻找多极值中的最优值。

我们提出的 ACP-NPSO 协议充分利用了 NPSO 的优点，充分考虑了影响网络的状态，在此基础上进行最优分簇。当然还有一些方法可用来减少、均衡能量的消耗，如分布式采样速率控制，负载均衡，以及一些通信控制调度方法的结合运用。

第三节 系统模型

一、网络模型

我们设定在 $M \times M$ 的方形区域内，随机部署 N 个传感器节点，传感器网络部分结构如图 4-1 所示，传感器节点集合为 $S = \{s_1, s_2, s_3, \cdots, s_n\}$，其中 s_i 代表第 i 个传感

器节点，且 | S | = N。我们对无线传感器网络做如下设定。

汇聚节点远离监测区域。

所有传感器节点没有移动性。

每个节点知道自己的位置信息。

节点周期性的采集数据。

节点电池供电，能量受限，具备一定的初始能量。

节点的发射功率根据通信距离的远近可进行调整，双向通信且链路可靠。

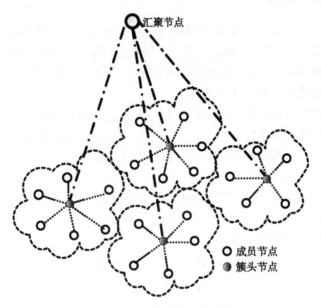

图 4-1　无线传感器网络结构

该网络模型运行 ACP-NPSO 算法，簇头节点负责簇内管理协调，并进行数据的汇聚融合，并发送至远处的汇聚节点（基站），成员节点采集数据，避免了簇头节点能量消耗过快，这种结构设计具有如下优点。

采用小生境粒子群优化分簇，减少能耗，分簇更加合理。

仅簇头节点进行数据融合工作，进一步减少了能耗。

每个节点加入最优分簇中，能量消耗最低。

同样把每个周期定义为一轮，在每一轮中，运行 ACP-NPSO 协议形成分簇，簇头发布簇内每个成员节点的传输时间片；成员节点采集感知信息后，在各自的时间片内发送数据至簇头，其他时间进入休眠状态；簇头节点汇聚成员节点采集的数据，并将汇聚后的数据发送给远处的汇聚节点，汇聚节点通过网关接入 Internet，方便用户的使用。

二、能耗模型

我们采用与上章相同的能量消耗模型，在此不再赘述。

第四节　无参数小生境粒子群优化算法的改进

我们已介绍了粒子群优化算法，下面了解一下小生境技术在粒子群优化中的应用，就小生境粒子群优化存在问题依赖性强，以及参数过多等问题，提出了一种无参数小生境粒子群优化算法，然后运用单、多值测试函数来证明该算法的有效性。

一、粒子群优化算法

粒子群算法可描述如下：在 D 维搜索空间中，由 m 个粒子组成的一粒子群体，群体中的每个粒子以一定的方向和速度飞行，每个粒子在飞行搜索的过程中，考虑群体（邻域）内其他粒子的历史最优点和自身曾经搜索到的历史最优点，在此基础上进行位置、速度等状态的变化，如图 4-2 所示。

第 i 个粒子的位置表示为：$x_i = (x_{i1}, x_{i2}, \cdots, x_{id})$，$1 \leq i \leq m$，$1 \leq d \leq D$；

第 i 个粒子的速度表示为：$v_i = (v_{i1}, v_{i2}, \cdots, v_{id})$，$1 \leq i \leq m$，$1 \leq d \leq D$；

第 i 个粒子经历过的历史最好点表示为：$p_i = (p_{i1}, p_{i2}, \cdots, p_{id})$，$1 \leq i \leq m$，$1 \leq d \leq D$；

群体（邻域内）所有粒子经历过的历史最好点表示为：$p_g = (p_{g1}, p_{g2}, \cdots, p_{gd})$，$1 \leq i \leq m$，$1 \leq d \leq D$；

粒子的速度和位置按照下面公式变化：

$$v_{id}^{k+1} = \omega v_{id}^k + c_1 \xi (p_{id}^k - x_{id}^k) + c_2 \eta (p_{gd}^k - x_{id}^k) \tag{4-1}$$

$$x_{id}^{k+1} = x_{id}^k + v_{id}^{k+1} \tag{4-2}$$

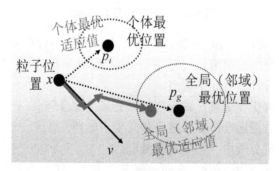

图 4-2　粒子 x 以 v 速度飞行

其中 ω 为惯性权重，决定粒子对当前速度的继承能力，值越小继承越少，选择合适的惯性权重可使粒子具有优良的探索能力（广域搜索能力）和开发能力（局部搜索能力）；c_1 和 c_2 为学习因子或加速系数，通常为正数，使粒子具有向群体中的优秀个体学习和进行自我总结能力，从而促使粒子朝自己历史最优点和群体（邻域）内历史最优点飞行；ξ 和 η 是均匀分布在 [0，1] 内的伪随机数。粒子最大速度 $V_{\max}$ 通常进行限制，粒子的位置有时也进行限制，因为粒子跑出解空间对实际问题没有意义。

粒子群优化的局部版本是把群体内某部分成员看成邻域，粒子群优化的全局版本是把群体内全部粒子均看成是邻域中的成员。小生境指的就是一种邻域的构造方法，目前进行了很多研究，存在问题依赖性以及参数过多等问题。

1. 算法流程（流程图如图 4-3 所示）

(1) 在取值范围内，对每个粒子速度和位置进行随机初始化。

(2) 利用适值函数计算粒子的适应值。

(3) 把每个粒子的适应值同该粒子的历史最优适应值相比较，若好于历史最优值，则将其更新为该粒子的历史最优适应值，同时最优粒子位置也进行更新。

(4) 把每个粒子的适应值同群体（邻域）内粒子的历史最优适应值相比较，若好于历史最优值，则将其更新为群体（邻域）内粒子的历史最优适应值，粒子的历史最好位置也同步更新。

(5) 根据式（4-1）和式（4-2）对粒子的位置和速度进行更新。

(6) 判断是否达到结束条件，若未达到则转（2），否则结束输出。

2. 算法构成要素

(1) 群体大小 m。群体大小 m 是整数类型参数，当 m 较小时，粒子群算法易陷入局部最优；当 m 较大时，虽然陷入局部最优概率减少，但是若过大会导致计算时间大幅增长。当粒子群中的粒子增到一定数量之后，再继续增加对优化求解问题作用不明显，故需选择合适的群体大小。

(2) 惯性权重。探索能力和开发能力的平衡非常关键，决定智能优化算法运行是否成功，粒子群优化算法可通过惯性权重 ω 来进行平衡调节。当惯性权重较小的时候，粒子较少的继承了原来方向上的速度，从而飞的比较近，开发能力较好；当惯性权重较大的时候，粒子继承了较大的原来方向上的速度，在原来方向上飞的较远，探索能力较强。开发能力与探索能力的平衡可通过惯性权重来进行调整。

(3) 学习因子 c_1 和 c_2。学习因子（加速系数）促使粒子向群体中优秀个体进行学习和自我总结学习能力，从而引导粒子向群体（邻域）中最优点飞行。

(4) 最大速度 $V_{\max}$。粒子在进行一次迭代中所能移动的最大距离为最大速度 $V_{\max}$，当 $V_{\max}$ 较小时，粒子具有较强的开发能力，但是容易陷入局部最优；当 $V_{\max}$ 较大时，粒子具有较强的探索能力，但是容易飞过最优解。因此需根据情况选择合适的最大速度。

(5) 邻域拓扑结构。当我们把整个粒子群体内所有粒子看成邻域时，可得粒子群优化的全局版本，这时粒子飞行速度较快，但有时候容易陷入局部最优而无法飞出；当把某些个体看作邻域时，我们得到粒子群优化算法的局部版本，收敛速度慢，但是不易陷入局部最优。我们提出的小生境就是一种构建粒子邻域的方法。

(6) 停止标准。通常采用可以接受的满意解或者最大迭代次数作为停止标准。

(7) 粒子初始化。粒子初始化空间是有问题依赖性的，较好的选择初始化空间，将明显减少搜索时间。

粒子群优化需要调节的参数并不算多，邻域的定义和惯性权重相对比较重要。下面我们就粒子群的邻域讨论一种无参数小生境邻域构造方法。

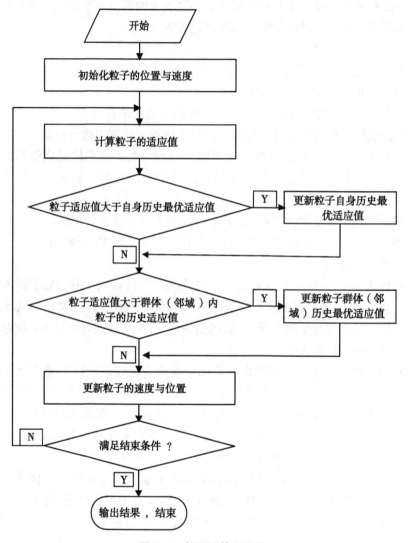

图 4-3 粒子群算法流程

二、小生境技术

小生境受大自然的启示，来自生物学的一个概念。在自然生态系统中，物种个体为了生存竞争必须承担不同的角色，不同的物种依据环境进化成不同小生境（子群体），而不是仅进化成一个群落中毫不相干的个体。换句话说，生物在进化过程中，通常会与自己相同的物种生活在一起，共同繁衍生息。

小生境都是在某一特定的区域中，如北极熊不能生活在热带，大熊猫不能生活在北极等，把里面包含的思想提炼出来，运用到智能优化中去，保持群体内物种的多样性，利于优化算法探索能力与开发能力的平衡，提高算法的效率。

小生境技术将来源于大自然中的小生境概念应用到各种智能优化算法中去，在智能优化算法中，每一代划分为若干子群体，每个子群体中选出一些最优适应值的个体，作为子类的优秀代表组成群，然后在种群中，或者不同的种群之间进行个体的变异、杂交等产生新一代个体群，同时可辅助分享、排挤、选择等机制协助完成任务。

三、无参数小生境粒子群优化算法改进

粒子邻域的组成如何定义，是粒子群算法实现的一个基本问题，通常有两种构成邻域结构的方式，一种是索引号相邻的粒子构成邻域，另一种是位置相邻或相近的粒子组成邻域。

目前众多小生境构造方法存在严重的问题，算法的性能严重依赖于小生境参数的设置，这对于普通用户来说比较困难。在 fitness sharing 中的共享参数 σ_{share}、species conserving GA（SCGA）中的种群距离 σ_s、clearing 中的距离测量 σ_{clear}、speciation-based PSO（SPSO）中的种群半径 r_s 等，均定义了不同的距离量度参数，Shir 等试图采用固定小生境半径、Bird 和 Li 等试图降低小生境半径对结果的敏感性，但是参数依然存在或者引入了新的参数，我们提出了一种采用环形结构的小生境构造方法，该方法不引入任何参数，相比以上小生境的构造方法，该方法具有明显的优势，且性能优良。

Xiaodong Li 在环形拓扑结构的基础上，利用左右邻居节点的构成小生境，如图 4-4 所示，该小生境的构造方法不需要参数，且取得了较好的性能，但是我们认为，仅依靠左右节点序号的邻居信息，在粒子的社会经验的学习与传播方面过于局限，为此我们将左右邻居进行重新定义，定义左邻居为节点左侧半径范围内的任意节点，右邻居为节点右侧半径范围内的任意节点，我们定义的左右邻居节点能增加一定的干扰性，使粒子在飞行时受社会经验吸引更大或者逃离局部最优的可能加大，从而增加求解的质量。

根据以上的描述，算法的实现过程可以用以下的伪代码来表示。

begin
粒子群初始化；
While（not met termination criterion）{
for（i=1；i<=Population Size；i++）
　　　if（$fit(\vec{x_i}) > fit(\vec{p_i})$）
$\vec{x_i} -> \vec{p_i}$；
　　for（i=1；i<=Population Size；i++）
$neighborhoodBest(\vec{p_l}, \vec{p_i}, \vec{p_r}) -> \vec{p_{n,i}}$；
　　for（i=1；i<=Population Size；i++）{
运用公式（4-1）、（4-2）对粒子进行更新
}
}
End

四、算法验证

我们对粒子群优化算法采用我们提出的小生境构造方法的性能进行验证，为了进一

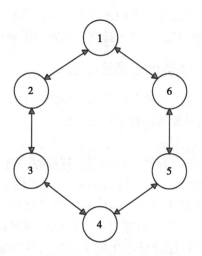

图 4-4　环形拓扑左右邻居节点构成一种小生境

步说明问题，我们验证具有多个相等最大值的函数 $f(x) = \sin^6(5\pi x)$，$0 \leq x \leq 1$，在该范围内，该函数有 5 个极大值点，函数图形如图 4-5 所示。

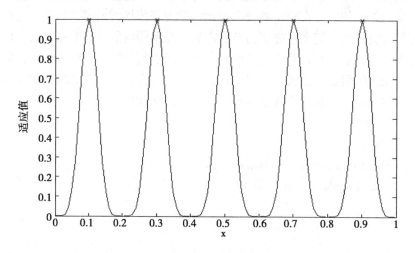

图 4-5　函数 $f(x) = \sin^6(5\pi x)$，$(0 \leq x \leq 1)$

我们将提出的小生境粒子群算法与 Xiaodong Li 的进行比较，就 $f(x)$ 在取值范围内的所有极大值点，我们运行两种算法进行搜索，实验结果如图 4-6 所示。

从图中可以看出，我们提出的小生境构造方法把所有的极大值均搜索到，且分布比较均匀；而原来的小生境方法还有一个极大值未发现，且分布不均匀。造成这种现象的主要原因是我们构造的小生境充分考虑了粒子的探索能量与开发能力，在一定范围内赋予每个粒子不同的探索能量与开发能力，而原来的小生境强度粒子的开发能力，导致在很多情况下收敛，失去了探索能力，找到更多极大值的概率减少所致。

另外我们验证非均匀减少最大值函数在 [0, 1] 内求解最优值的运算迭代情况如

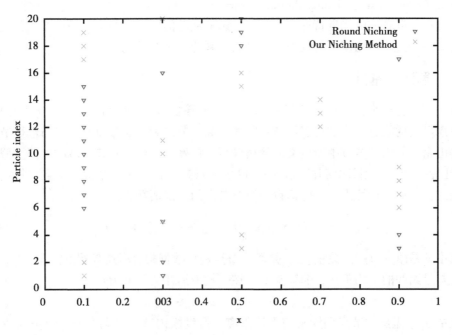

图 4-6 函数 $f(x)=\sin^6(5\pi x)$，$(0\leqslant x\leqslant 1)$

表 4-1 所示，其中 $f(x)=\exp\left(-2\log(2)\cdot\left(\dfrac{x-0.08}{0.854}\right)^2\right)\cdot\sin^2[5\pi(x^{3/4}-0.05)]$。

表 4-1 优化结果

粒子个数	迭代次数（1 000）	
	Round niching	Our niching method
20	825	564
30	784	536
40	718	489

从表 4-1 我们可以看出，我们提出的小生境构造方法在求解最优值时迭代次数较少，能快速找到最优值。这是因为我们提出的小生境构造方法在搜索过程中强调探索能力与开发能力的平衡，使粒子能迅速定位到可能极值点；而原来的小生境构造方法过于强调粒子的开发能力，有时容易陷入局部最优而无法跳出。

从上面单值（多值）函数最优值求解比较中，可以看出，提出的小生境构造方法能更好的平衡粒子的探索能力与开发能力，避免粒子陷入局部最优，同时算法简单，无小生境参数，易于实现，在单值、多值寻优中具有良好的表现。

第五节 一种基于小生境粒子群优化的自适应分簇协议

本节详细设计了一种基于小生境粒子群优化的自适应分簇协议（an Adaptive

Clustering Protocol using Niching Particle Swarm Optimization，ACP-NPSO），我们充分利用小生境粒子群算法的优良特性，充分考虑影响能耗的因素，根据网络中节点的具体情况进行分簇，进一步均衡了能量消耗，延长了系统的生存周期。

一、算法的基础

我们利用基于小生境粒子群优化的自适应分簇协议 ACP-NPSO 进行分簇的时候，充分考虑网络的状态，比如传感器节点间的距离、节点剩余能量以及节点与汇聚节点的距离等因素，如第三节节中所定义的网络模型，汇聚节点位于感知区域的外部，假设第 r 轮选出 k 个簇头，则每个簇内包含一个簇头节点和 $(N/k) - 1$ 个成员节点，根据能量消耗公式，我们可计算出在一轮运行中簇头节点的能量消耗为：

$$E_{CH} = (\frac{N}{k} - 1) l \cdot E_{elec} + \frac{N}{k} l \cdot E_{DA} + l \cdot E_{elec} + l \cdot \varepsilon_{mp} d_{toBS}^4 \qquad (4-3)$$

其中 k 是簇头节点的数量，E_{DA} 是簇头节点进行数据融合时所耗费的能量，d_{toBS} 是簇头节点与基站间的平均距离，非簇头节点在一轮中消耗的能量为：

$$E_{nonCH} = l \cdot E_{elec} + l \cdot \varepsilon_{fs} \cdot d_{toCH}^2 \qquad (4-4)$$

其中 d_{toCH} 是簇内节点到簇头的平均距离，假设簇内节点均匀分布，且分布密度为 $\rho(x, y)$，则有：

$$d_{toCH}^2 = \int_{x=0}^{x=x_{max}} \int_{y=0}^{y=y_{max}} (x^2 + y^2) \rho(x, y) dx dy = \frac{M^2}{2\pi k} \qquad (4-5)$$

所以单个簇内的能量消耗大约为：$E_{cluster} = E_{CH} + \frac{n}{k} E_{nonCH}$

这样在一轮的运行过程中，整个网络的能量消耗为：

$$E_r = l \left(NE_{DA} + 2NE_{elec} + k\varepsilon_{mp} d_{toBS}^4 + N\varepsilon_{fs} \frac{1}{2\pi} \frac{M^2}{k^2} \right) \qquad (4-6)$$

对于最优分簇的数量，我们将 E_r 对 k 求导数并令结果为 0，即 $E'_r = l \left(\varepsilon_{mp} d_{toBS}^4 - \frac{N}{2\pi} \varepsilon_{fs} \frac{M^2}{k^2} \right) = 0$，得：

$$k_o = \frac{\sqrt{N}}{\sqrt{2\pi}} \sqrt{\frac{\varepsilon_{fs}}{\varepsilon_{mp}}} \frac{M}{d_{toBS}^2} \qquad (4-7)$$

可见在我们设定的场景下，簇头数量与节点数量、监测区域大小以及汇聚节点与监测区域的平均距离等有关。

二、ACP-NPSO 的适应值函数

适应值函数的定义与问题域有密切的关系，直接决定求解算法最优解的性能，所以说适应值函数相当重要，我们在认真分析可能影响系统性能因素的基础上，定义了适应值函数 f 如下：

$$f = \alpha f_1 + \beta f_2 + \gamma f_3 \qquad (4-8)$$

$$f_1 = \sum_{k=1}^{k_o} \sum_{\forall n_i \in C_{p,k}} d(n_i, CH_{p,k}) / |C_{p,k}| \qquad (4\text{-}9)$$

$$f_2 = \sum_{i=1}^{N} E(n_i) / \sum_{k=1}^{k_o} E(CH_{p,k}) \qquad (4\text{-}10)$$

$$f_3 = \frac{1}{k_o} \sum_{k=1}^{k_o} d(CH_{p,k}, BS) \qquad (4\text{-}11)$$

$$\alpha + \beta + \gamma = 1, \quad 0 \leq \alpha, \beta, \gamma \leq 1$$

其中 $C_{p,k}$ 是在粒子 p 中属于分簇 C_k 的节点个数，$CH_{p,k}$ 是簇头节点集，即在粒子 p 中包含 k 个候选簇头节点信息，可见 f_1 优化了簇内节点间的距离；$E(n_i)$ 是节点 n_i 的剩余能量，可见 f_2 优化了簇头节点的能量，使具有更多剩余能量的节点更容易当选簇头；f_3 优化了簇头节点与汇聚节点（基站）间的距离。α、β、γ 是适应值函数各分量的权重，权重的大小决定了各分量所占的比例，在算法运行初期，我们重点强调簇间距离的优化，随着算法的运行，有些节点耗尽能量，这是我们加大能量分量的权重，使剩余能量更多的节点有更多的机会当选簇头。从上面定义的适应值函数可以看出，我们选择具有最小函数适应值的节点当选簇头，这样的节点同时优化了簇内距离、簇头与基站的距离以及节点的剩余能量，是本轮运行中最适合的簇头节点。

由于最优簇头节点数 k_o 根据公式（4-7）可得，我们在每轮运行中，选择 k_o 个节点为簇头节点并把网络分簇，ACP-NPSO 是周期性执行的，簇头节点充当簇内控制中心，协调簇内的数据传输，并为簇内成员节点分配 TDMA 时间片，成员节点在自己的时间片内发送数据，其他时间进入休眠状态，进一步减少了系统的能量消耗，提高了系统的性能。

三、ACP-NPSO 协议

由于分簇经常转化成优化问题来解决，而小生境粒子群优化算法相对于传统优化方法来说是一种高效的智能优化方法，我们利用小生境粒子优化的优点来解决无线传感器网络的分簇问题。

我们利用定义的适应值函数式（4-8）来选择最优簇头节点集，并将网络分簇。根据定义的适应值函数，最优簇头具有充足的能量，且能量消耗较小，簇头负责收集、汇聚数据，并发送至远处的基站，我们的仿真实验证明该协议有效可靠，能明显延长网络的生存时间，提高系统的性能。

在如图所示无线传感器网络模型中有 N 个传感器节点，我们利用公式（4-7）计算出最优簇头数量 k_o，然后按照如下步骤运行。

(1) 随机初始化 m 个粒子，每个粒子包含 k_o 个随机簇头信息。
(2) 将每个粒子包含的 k_o 个簇头节点对应至实际传感器节点。
(3) 利用公式（4-8）计算每个粒子的适应值。

针对存活节点 $n_i (i = 1, 2, 3\cdots, N)$，计算 n_i 和所有簇头节点之间的距离，若满足 $d(n_i, CH_{p,k}) = \min_{\forall k=1,2,\cdots,k_o}\{d(n_i, CH_{p,k})\}$，将 n_i 分配到簇头 $CH_{p,k}$ 对应的簇中运用公式（4-8）计算粒子的适应值。

(4) 寻找个体和小生境内粒子最优适应值。个体最优指的是该粒子运行过程中曾经的最优适应值；小生境最优是在我们构造的小生境内粒子的曾经（历史）最优适应值。并对最优适应值进行更新。

(5) 运用公式（4-1）和式（4-2）更新粒子的位置与速度。

(6) 限制粒子位置和速度的更新。

(7) 重复（2）至（6）直到达到终止条件或者达到最大迭代次数，算法的流程图如图4-7所示。

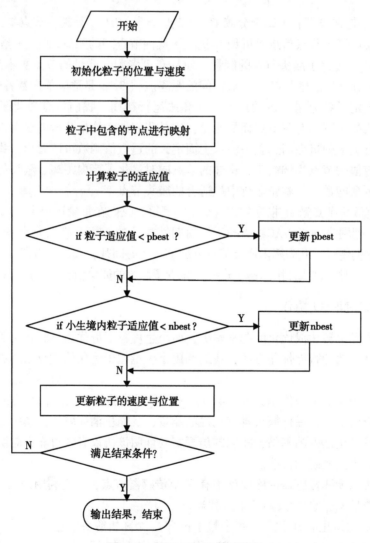

图 4-7　ACP-NPSO 算法流程

我们选择最优适应值粒子对应的簇头集作为簇头，并确定其成员节点，基站将包含簇头 ID 与簇内节点的信息发布整个网络中，成为簇头节点充当本地的控制中心来协调簇内数据的传输。簇头节点分配 TDMA 时间片给各成员节点，各成员节点得到自己的时间片后,系统进入数据传输阶段,每个成员节点仅需在自己的时间片内发送数据,其他时间进入休眠状态,进一步减少了能量消耗。

第六节 算法性能评估

本节我们评估 ACP-NPSO 的性能,对仿真实验结果进行分析,并与 LEACH、PSO-C 等分簇协议进行比较。

一、仿真环境

根据我们设定的网络模型,考虑 100 个节点随机分布在 $100m \times 100m$ 的区域内,汇聚节点远离监测区域,位于 $(50m, 175m)$ 处,如图 4-8 所示,能量消耗模型参数与中相同,小生境粒子群参数如表 4-2 所示,各个参数的物理意义在前面章节中已经介绍。

表 4-2 粒子群优化实验参数

参数	值
m	20
ω	0.9~0.4
c_1	2.5~0.5
c_2	0.5~2.5
α	0.2~0.9
β	0.65~0.05
γ	0.15~0.05

二、仿真结果

我们就采用 ACP-NPSO 协议与采用 LEACH、PSO-C 协议时的网络性能进行比较,首先我们比较一下三种协议所形成的分簇,如图 4-8 所示。从图中我们可以看出 LEACH 所形成的分簇数量过多,这是由于簇头选择的随机性造成的,因为在 LEACH 中每个传感器节点根据一定的概率决定是否成为簇头,具有很强的随机性。PSO-C 所形成的分簇较 LEACH 所形成的分簇有所改进,但与 ACP-NPSO 相比较则存在大小不一的情况。这是因为 ACP-NPSO 充分考虑了网络的状态来决定最优簇头数量,然后根据节点的剩余能量、节点间的距离以及节点与基站的距离来决定最优簇头的选择,因此形成的分簇分布均匀,保持最优簇头数量,降低网络的能量消耗,提高系统的性能。

无线传感器网络中存活节点的数量随时间变化如图 4-9 所示。从图中我们可以看到,我们提出的 ACP-NPSO 能明显延长网络的生存时间,且首个节点耗尽能量的时候也明显延迟,这是由于我们提出的分簇协议根据网络的状态决定最优簇头数量,然后运用小生境粒子群算法进行最优簇头的选择,且簇头的选择充分考虑了节点的剩余能量、节点间距离等因素,使簇头分布更加合理,从而减少能量消耗。而 LEACH 协议每个节点根据一定的概率决定是否成为簇头,具有较强的随机性,且簇头选择未考虑节点剩余能量以及节点间的距离等因素，从而簇头数量具有较强的随机性，且分簇不均匀；PSO-C 协议采用粒子群优化分簇选择，但是簇头数量未根据网络状态动态变化，且未考虑节点与基站间的距离影响，从而导致能量消耗较高。我们提出的 ACP-NPSO 协议能避免以上缺点，从而能明显延长网络的生存时间。

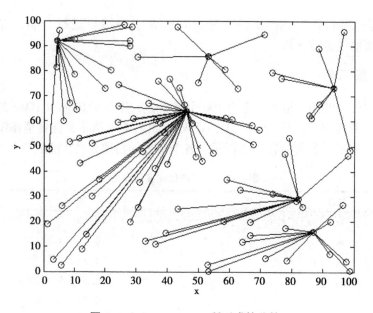

图 4-8（a） LEACH 所形成的分簇

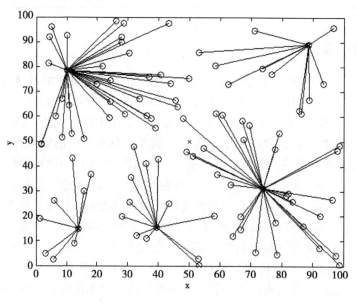

图 4-8（b） PSO-C 所形成的分簇

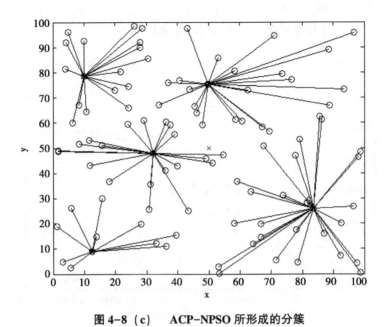

图 4-8 (c)　ACP-NPSO 所形成的分簇

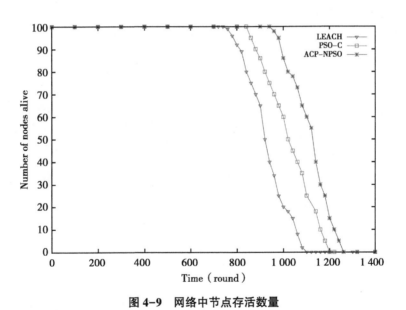

图 4-9　网络中节点存活数量

基站（汇聚节点）接收数据情况如图 4-10 所示。从图中我们可以看出，我们提出的 ACP-NPSO 能接收更多的数据，这是因为我们在分簇时考虑了节点的剩余能量、簇内间距以及与基站的距离等因素，有效地减少了能量消耗，减少了数据的传输时延，提高了数据的传输效率，从而发送更多的数据至基站。而 LEACH 在进行分簇时未考虑节点的剩余能量等因素，且随机选取节点作为簇头，有可能选择能量即将耗尽的节点为簇头，导致数据丢失；PSO-C 在分簇时虽然考虑了剩余能量等因素，但是最优簇头数量不随网络状态变化，且未考虑簇头与基站间的距离的影响。

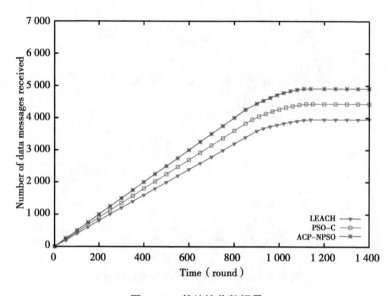

图 4-10　基站接收数据量

第五章 一种基于粒子群蚁群优化的动态分簇路由协议

无线传感器网络具有很强的应用相关性，在家居健康监测、智能医疗、交通监控、海洋探测等领域应用广泛，在不同应用场景中路由协议差别较大，没有一个通用的路由协议。根据路由协议逻辑结构的不同，分为平面和分簇路由协议，同平面路由协议相比，分簇路由协议结构清晰，能高效利用节点资源。本文设计了一种基于粒子群蚁群优化的动态分簇路由协议（a Dynamic Clustering Routing Protocol using Particle Swarm Optimization and Ant Colony Algorithm, DCRP-PSOACA），使得在数据的传输过程中，减少节点的能量消耗，使节点的能量得到高效利用，从而延长了网络的生命周期，实现网络的高效通信。我们充分利用粒子群优化算法优点，先将网络分簇，利用蚁群算法的优点，在汇聚节点和簇头间形成多跳最优路径的选择。本章给出了 DCRP-PSOACA 的详细设计与仿真实验，仿真实验结果表明，该协议能提供稳健的路由路径，减少节点的能耗，从而延长网络的生存时间，实现高效通信。

第一节 引 言

蚁群算法是一种模仿蚂蚁群体行为的智能算法，兴起于 20 世纪 90 年代，该算法将正反馈机制引入，该算法具有较强的鲁棒性、分布式计算能力强，且易于与其他算法结合等优良特性。现在蚁群算法已经渗透到各个领域，在无线传感器网络中也存在广泛应用，由于蚁群算法性能优异，鲁棒性强，前景广阔，成为专家学者研究和关注的焦点与热点。

昆虫行为学家在研究群居性昆虫行为时发现，群体中的昆虫通过自组织的合作，这种合通常比较简单，但是可用来解决复杂问题。1991 年，意大利学者 Dorigo M 等人在法国巴黎召开的第一届欧洲人工生命会议上首次提出了蚁群算法。Dorigo M 和 Bonabeau E 等于 2000 年在《Nature》上发表了蚁群算法的综述，把蚁群算法推向学术研究的前沿。

粒子群优化方法作为一种流行的智能优化方法，具有易于实现、参数少、不需要梯度信息、算法简单等特点，在模式识别、决策支持、生物系统建模、流程规划、数据分类等方面表现出良好的应用前景，成为智能优化领域研究的热门。

粒子群算法的提出基于对简化社会模型的模拟，在自然界中，许多生物具有一定的群体行为，如蚁群、鸟群、鱼群等。人工生命探索的自然界生物的群体行为，并在计算机上构建群体模型，虽然群体行为由几条看似简单的规则进行建模，但是群体的行为却特别复杂。Kennedy 和 Eberhart 修正了鸟群模型，引导粒子飞向解空间，并在最优处降落，在此基础上得到粒子群优化方法。

我们充分利用粒子群优化的优点，先将网络分簇，利用蚁群算法的优点，在汇聚节点和簇头间形成多跳最优路径的选择。通过优化分簇与路由路径来减少节点的能量消耗，均衡网络的负载，提供系统的性能。

第二节 相关工作

在无线传感器网络中，分簇拓扑控制存在广泛的应用，LEACH 协议中簇头的选择具有一定的随机性，且未充分考虑影响系统性能的其他因素，比如传输距离，网络的动态性等。PEGASIS 协议通过构造接近最优的链来减少能量消耗，减少簇重构的开销，但仍需知道邻居节点的能量状态，仍需要动态调整网络的拓扑结构。粒子群在分簇拓扑控制方面的研究较多，N. M. A. Latiff 等提出了 PSO-C 算法，该算法考虑了节点的剩余能量和节点间的距离影响，但未考虑节点与汇聚点的距离因素；X. Cao 等提出了 MST-PSO，目的是构建一个基于距离的最小生成树，未能很好的均衡能量的消耗。J. Tillet 等提出了采用 PSO 分簇的算法，该算法通过均衡簇内成员的数量来减少能耗。

蚁群算法在网络路由中的研究较多，AntNet 较早地将蚁群算法应用到路由优化中，协议中每个节点维护一张路由表和一张网络状态表，路由表记录当前节点到达所有下一跳节点信息素的浓度，路由数据根据信息素的浓度进行下一跳的选择，AntNet 使用了前向蚂蚁和后向蚂蚁。文献对分析了 AntNet 的路由性能，发现与迪杰斯特拉最短路径路由算法性能接近，但是若通信量变化较大时，AntNet 还是性能更好些。文献针对异构无线传感器网络，通过蚁群算法来构建路由路径。文献通过在路径能耗和节点剩余能耗间的权衡，通过蚁群算法来寻找路由路径。ACO-MNCC 通过 ACO 算法发现最大数目的连接覆盖集，来提高异构网络的生命周期。

第三节 系统模型

一、网络模型

我们设定在 $M \times M$ 的方形区域内，随机部署 N 个传感器节点，传感器节点集合为 $S = \{s_1, s_2, s_3, \cdots, s_n\}$，其中 s_i 代表第 i 个传感器节点，且 $|S| = N$，传感器网络部分结构如图 5-1 所示。我们对无线传感器网络做如下设定。

汇聚节点远离监测区域。

所有传感器节点没有移动性。

每个节点知道自己的位置信息。

节点电池供电，能量受限，具备一定的初始能量。

节点的发射功率根据通信距离的远近可进行调整，双向通信且链路可靠。

该网络模型运行 DCRP-PSOACA 算法，首先将网络分簇，簇头节点负责簇内管理

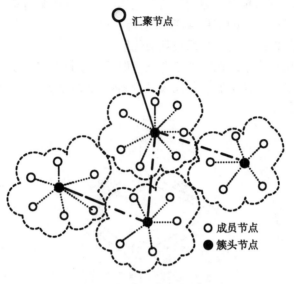

图 5-1　无线传感器网络结构

协调，并进行数据的汇聚融合，然后在簇头节点中运行蚁群算法，选择出一条到汇聚节点的最优路径。簇头节点沿最优路径路由数据，最后一个簇头汇聚数据发送至汇聚节点，这种结构设计具有如下优点。

采用粒子群优化分簇，减少能耗，分簇更加合理。

仅簇头节点进行数据融合工作，进一步减少了能耗。

每个节点加入最优分簇中，能量消耗最低。

采用蚁群算法选择最优路由路径，减少能量消耗。

二、能耗模型

我们采用与上章相同的能量消耗模型，在此不在赘述。

第四节　蚁群算法及性能分析

通过种群中个体间的竞争与合作，可以在群体智能中找到很好的解决方案和优化方法，很多进化搜索方法源于自然界，如模拟退火、遗传算法、禁忌搜索等。本节我们先介绍基本蚁群算法，给出算法的具体流程，并对影响算法的因素与性能进行分析。

一、蚁群算法

在蚁群算法中待求解问题的一种可行解利用人工蚂蚁的行走路线来表示，每只蚂蚁在解空间中独立搜索可行解，每当遇见还没走过的路口时，蚂蚁随机挑选一条继续行走，并释放出信息素，且信息素浓度与路径长度有关，路径越短浓度越大。当后面的蚂蚁碰到这个路口的时候，以相对大的概率选取信息素浓度大的路径，并留下自己的信息素，这样后来的蚂蚁也受影响，从而形成正反馈机制。随着算法的推进，代表最优解路径上的信息素浓度越来越大，导致该路径上蚂蚁汇聚的越来越多，同时为了避免信息素的无线聚集，每条路径上的信息素随时间逐渐挥发减弱，最后整个蚁群通过正反馈机制集中到代表最优解的路径上，从而找到最优解。

单只蚂蚁在寻优过程中的选择能量有限，但是整个蚁群自组织性较高，利用信息素交换路径信息，借助正反馈机制，找出最优路径。图 5-2 是一个利用蚁群算法中的蚂蚁寻找最短路径的例子。

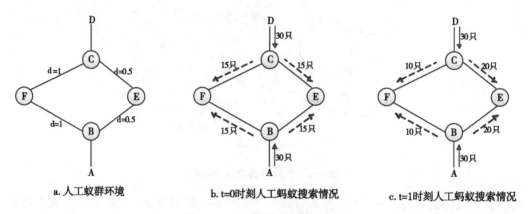

图 5-2　人工蚁群路径搜索实例

如图 5-2a 所示，路径 BEC、BF、CF 的长度为 1 个单位，路径 BEC 的重点为 E。假设每个时间单位内从 A 到 B 的蚂蚁为 30 只，从 D 到 C 的蚂蚁为 30 只，每只蚂蚁在单位时间内移动一个长度单位的距离，前进过程中，蚂蚁在单位时间内留下信息素浓度为 1 个单位，经过一个时间段（t，$t+1$）后，信息素在瞬间完全挥发掉。

如图 5-2b 所示，在时刻 $t=0$，各有 30 只蚂蚁在 B、C 处，因为以前没有留下信息素，每只蚂蚁随机选择路径，各有 15 只蚂蚁在 BF、BE、CF、CE 路径上。

如图 5-2c 所示，在时刻 $t=1$，30 只蚂蚁到达 B 处，蚂蚁发现路径 BF 上信息素浓度为 15，路径 BE 上信息素浓度为 30，这样选择路径 BE 的蚂蚁数量的期望值是选择路径 BF 的 2 倍，从而 10 只蚂蚁选择路径 BF，20 只蚂蚁选择路径 BE。这个过程持续下去，最终所有的人工蚂蚁选择路径 CEB（BEC）。

为便于蚁群算法数学模型的理解，蚁群觅食过程与旅行社问题（Travelling Salesman Problem, TSP）求解相类似，我们以 n 个城市 TSP 问题来分析蚁群算法的求解。

TSP 问题描述：设定 n 个城市为 $C=\{c_1, c_2, c_3, \cdots, c_n\}$，城市间的距离为 d_{ij}（$1 \leqslant i \leqslant n, 1 \leqslant j \leqslant n, i \neq j$）。TSP 问题求解一条回到起点，经过每个城市仅且一次的最短路径。

TSP 问题在求解过程中，假设蚁群中的蚂蚁具有如下智能。

（1）每只蚂蚁在其经过的路径（i, j）上都会留下信息素。

（2）蚂蚁选择城市的概率与当前路径上遗留信息素浓度以及城市间的距离有关。

（3）直到一次周游完成后，蚂蚁才可以访问已经经过的城市。

定义蚁群算法中的变量与常量：城市个数 n；蚂蚁的总数量 m；城市 i 和城市 j 之

间的距离 d_{ij}，$i,j \in (1, n)$；在初始时，每条路径上有相同的信息量，t 时刻在 (i, j) 路径上遗留信息量为 $\tau_{ij}(t)$，并设定 $\tau_{ij}(t) = $ const（const 为常数）。

蚂蚁 $k(k = 1, 2, \cdots, m)$ 在遍历行走时，路径上信息素的浓度决定蚂蚁的转移方向，$p_{ij}^k(t)$ 表示在 t 时刻蚂蚁 k 从城市 i 行走到城市 j 的状态转移概率，可以根据每条路径的的启发信息 η_{ij} 和残留信息素浓度 $\tau_{ij}(t)$ 来计算，如公式（5-1）所示，该公式表示蚂蚁尽量选择信息素浓度大且距离短的路径。

$$p_{ij}^k(t) = \begin{cases} \dfrac{[\tau_{ij}(t)]^\alpha \cdot [\eta_{ij}]^\beta}{\sum\limits_{k \in allowed_k}[\tau_{ik}(t)]^\alpha \cdot [\eta_{ik}]^\beta}, & j \in allowed_k \\ 0, & \text{otherwise} \end{cases} \quad (5\text{-}1)$$

其中 $allowed_k = \{C - tabu_k\}$ 代表蚂蚁 k 下一步可以选择的城市；
$tabu_k(k = 1, 2, \cdots, m)$ 为禁忌表，表中记录蚂蚁 k 已走过的城市；
α 是信息启发式因子，代表蚂蚁运动过程中残留信息量的重要程度；
β 是期望启发式因子，代表期望值的重要程度；
η_{ij} 表示从城市 i 转移到 j 的期望程度，在 TSP 中通常取值 $\eta_{ij}(t) = \dfrac{1}{d_{ij}}$，对于蚂蚁 k，城市间的距离越小，该路径被选择的期望程度就越大。

为避免启发式信息被过多的残留信息素淹没，蚂蚁在每完成一步，就对残留信息素进行更新，$(t + n)$ 时刻在路径 (i, j) 上的信息素可按如下两个公式进行调整：

$$\tau_{ij}(t + 1) = (1 - \rho) \cdot \tau_{ij}(t) + \Delta\tau_{ij} \quad (5\text{-}2)$$

$$\Delta\tau_{ij}(t) = \sum_{k=1}^{m} \Delta\tau_{ij}^k(t) \quad (5\text{-}3)$$

其中 ρ 表示信息素挥发系数，为了避免信息素的无限累积，取值范围 $[0, 1)$，则信息素残留系数为 $1 - \rho$；

$\Delta\tau_{ij}(t)$ 代表路径 (i, j) 在本次循环中信息素增量，初始为 0；

$\Delta\tau_{ij}^k(t)$ 代表第 k 蚂蚁在本次循环中在路径 (i, j) 上留下的信息素，根据不同信息素更新策略，Dorigo M 等提出了三种蚁群算法模型。我们仅了解蚁周模型，$\Delta\tau_{ij}^k(t) = \dfrac{Q}{L_k}$ 当第 k 只蚂蚁在本次循环中经过 (i, j)，其他时候为 0，Q 是常量，代表蚂蚁在经过一条路径时释放信息素总量，L_k 代表第 k 只蚂蚁在本次所走路径的总长度。

二、蚁群算法的具体步骤

如上所述的蚁群算法的流程图如图 5-3 所示，其实现步骤为：

（1）初始化参数。$t = 0$ 时，设定循环最大次数 N_{cmax}，循环当前次数 $N_c = 0$，另路径 (i, j) 上的初始化信息量 $\tau_{ij}(t) = $ const，在初始时刻 $\Delta\tau_{ij}(t) = 0$。

（2）在 n 个城市上随机放 m 只蚂蚁。

（3）$N_c < - N_c + 1$。

（4）禁忌表的索引号设置为 $k = 1$。

(5) $k=k+1$。

(6) 依据状态转移概率公式（5-1）来计算选择城市 j 的概率，$j \in \{C - tabu_k\}$。

(7) 蚂蚁选择最大状态转移概率的城市，把该城市录入禁忌表，并将蚂蚁移动到该城市。

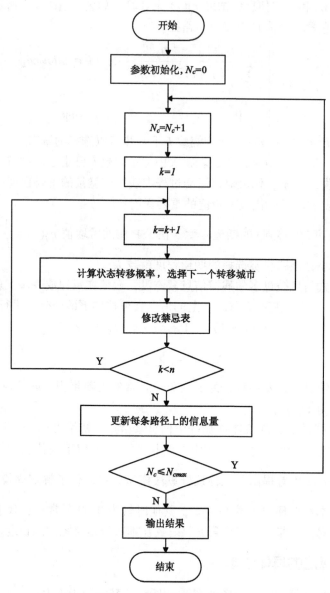

图 5-3 蚁群算法的流程

(8) 若所有城市没有访问完，即 $k<m$，程序转至第 5 步；否则转至第 9 步。

(9) 根据公式（5-2）和（5-3）更新每条路径上的信息素。

(10) 若满足结束条件，循环结束输出计算结果；否则清空禁忌表并跳转至第 3 步。

三、蚁群算法的复杂度

算法的时间复杂度指的是求解该问题的所有一切算法中,具有最小时间复杂性的算法的时间复杂度。

为了便于衡量算法的时间复杂度,我们定义数量级的概率,给定自然数 n 的两个函数 $F(n)$ 和 $G(n)$,当且仅当存在一个正常数 K 和一个 n_0,使得当 $n \geq n_0$ 时,有 $F(n) \leq KG(n)$,则称函数 $F(n)$ 以函数 $G(n)$ 为界,记作 $F(n) = O(G(n))$,或称 $F(n)$ 是 $O(G(n))$。

蚁群算法的问题规模可表示为 n 的函数,其时间复杂度记为 $T(n)$,我们从蚁周模型的实现过程可以看出该算法各个环节的时间复杂度如表 5-1 所示。

表 5-1 蚁群算法各环节的时间复杂度

执行步骤	执行内容	$T(n)$
1	初始化 $t=0$;$N_c=0$;$\tau_{ij}(t)=\mathrm{const}$;$\Delta\tau_{ij}=0$ 将 m 只蚂蚁随机放在 n 个节点上	$O(n^2+m)$
2	设置蚂蚁禁忌表 $s=0$; for $k=1$ to m do 置第 k 只蚂蚁的起始城市到禁忌表 $tabu_k(s)$	$O(m)$

3	每只蚂蚁单独进行求解
	循环计算直到禁忌表满//需循环 $(n-1)$ 次
	$s=s+1$;
	for $k=1$ to m do
	    根据转移概率 $p_{ij}^k(t)$ 选择下一个城市
	    将城市序号 j 加入禁忌表 $tabu_k(s)$
		$O(n^2m)$

4	解的评价与轨迹更新
	for $k=1$ to m do
	    将第 k 只蚂蚁从禁忌表 $tabu_k(n)$ 移到 $tabu_k(1)$
	    计算本次循环中第 k 只蚂蚁经过的路径长度
	    最优路径更新
	    计算每条路径上信息素反馈量 $\Delta\tau_{ij}$
		$O(n^2m)$

5	信息素轨迹浓度的更新 计算各条路径上在下一轮循环开始前的信息素浓度 $t=t+n$;$N_c=N_c+1$	$O(n^2)$

6	判断是否达到终止条件
	如果 $N_c \leq N_{\max}$ 且搜索没有出现停止
	    清空全部禁忌表
	    返回步骤2
	否则
	    打印最优路径
	结束
		$O(nm)$

从以上我们可以看出，蚁群算法的时间复杂度 $T(n)=O(N_c n^2 m)$，若参与搜索的蚂蚁数量与问题规模 n 大致想当时，算法的时间复杂度为 $T(n)=O(N_c n^3)$。对于 TSP 问题，若采用穷举法，所有可能的路径共有 $\frac{(n-1)!}{2}$ 条，若把路径长度比较作为基本操作，则需要 $\frac{(n-1)!}{2}-1$ 次比较。若 24 个城市的所有路径比较可在 1 秒内完成，则 30 个城市的所有路径比较需要 10.8 年，31 个城市的所有路径比较需要 325 年，远远超出了可接受程度。

四、参数选择对蚁群算法的影响

开发能量和探索能力的均衡是影响智能优化算法性能的重要因素，具体到蚁群优化算法来说，探索能力指的是在解空间中，蚁群算法在不同区域中尝试找到一个最优解的能力；开发能力指的是在有希望的区域内，蚁群精确搜寻最优解的能力。蚁群算法参数之间存在关联且参数空间比较大，下面分析参数最蚁群算法的性能影响。

1. 启发函数和信息素

启发函数表示未来信息的载体，信息素表示过去信息的载体，影响算法的收敛性和求解效率。

2. 信息素残留因子

信息素的挥发因子 ρ，信息素残留因子为 $1-\rho$，该参数关系到算法的收敛速度和搜索能力，反映蚂蚁个体间的相互影响力的强弱。蚁群算法的收敛性受信息素残留因子影响较大，在 0.1~0.99 范围内，$1-\rho$ 几乎与迭代次数成正比，这由于 $1-\rho$ 较大时，在搜索路径上残留信息占主导，正反馈较弱，蚂蚁搜索的随机性变强，导致算法收敛速度变慢。若 $1-\rho$ 较小时，正反馈作用突出，搜索的随机性弱，导致收敛速度快，但是容易陷入局部最优。

3. 蚂蚁数目

蚁群算法在多个候选解构成的群体进化过程中搜索最优解，所以蚂蚁的数量影响算法。对问题规模来说，蚂蚁数量越多，能提高蚁群算法的稳定性和全局搜索能力，但随着大量曾被搜索过的路径上信息量趋于均衡，导致信息正反馈减弱，增强随机性，减慢算法的收敛速度。相反若蚂蚁数量相对于问题规模来说过少，则导致从来没被搜索过的路径上的信息量趋于 0，减弱算法的全局搜索性，加快收敛速度，导致算法的稳定性差。通过大量实验仿真表明，当问题规模大约是蚂蚁数目的 1.5 倍时，算法的收敛速度和收敛性均不错。

4. 启发式因子

启发式因子反映在运动寻径过程中，信息量的积累在蚂蚁搜索路径中的影响作用的程度，较大的启发式因子，导致蚂蚁以较大的可能性搜索以前走过路径，搜索的随机性减弱；较小的启发式因子，导致蚂蚁以较小的可能性搜索以前走过的路径，搜索的随机性增强。

5. 期望启发式因子

该参数反映在引导蚁群搜索过程中,启发式信息的相对重要程度,表现为在搜索寻优过程中启发式信息的确定性、先验性因素。较大的启发式因子,选择局部最短路径的可能性加大收敛速度变快,随机性减弱,容易陷入局部最优。

6. 信息素强度

信息素强度是蚂蚁在完成一周循环时所释放的信息素总量。较大的信息素强度,导致在已经遍历的路径上积累信息素加快,增强了蚂蚁的正反馈性,算法的收敛速度得到提高。

第五节 一种基于粒子群蚁群优化的动态分簇路由协议

我们充分利用蚁群和粒子群算法的优良特性,设计了一种基于粒子群蚁群优化的动态分簇路由协议,该协议首先执行分簇算法,运行粒子群优化分簇,然后运行蚁群算法,在簇头节点中形成最优路径,各簇头汇聚后的数据沿着该路径传输,并进行数据的汇聚融合,发送至远处的基站。

一、DCRP-PSOACA 协议适应值

适应值函数的定义与问题域有密切的关系,直接决定求解算法最优解的性能,所以说适应值函数相当重要,在的基础上,我们在认真分析可能影响系统性能因素的基础上,定义了适应值函数 f 如下。

$$f = \alpha f_1 + \beta f_2 \tag{5-4}$$

$$f_1 = \sum_{k=1}^{k_o} \sum_{\forall n_i \in C_{p,k}} d(n_i, CH_{p,k}) / |C_{p,k}| \tag{5-5}$$

$$f_2 = \sum_{i=1}^{N} E(n_i) / \sum_{k=1}^{k_o} E(CH_{p,k}) \tag{5-6}$$

$$\alpha + \beta = 1, \quad 0 \leq \alpha, \beta \leq 1$$

其中 $C_{p,k}$ 是在粒子 p 中属于分簇 C_k 的节点个数,$CH_{p,k}$ 是簇头节点集,即在粒子 p 中包含 k 个候选簇头节点信息,可见 f_1 优化了簇内节点间的距离;$E(n_i)$ 是节点 n_i 的剩余能量,可见 f_2 优化了簇头节点的能量,使具有更多剩余能量的节点更容易当选簇头;α、β 是适应值函数各分量的权重,权重的大小决定了各分量所占的比例,在算法运行初期,我们重点强调簇间距离的优化,随着算法的运行,有些节点耗尽能量,这是我们加大能量分量的权重,使剩余能量更多的节点有更多的机会当选簇头。从上面定义的适应值函数可以看出,我们选择具有最小函数适应值的节点当选簇头,这样的节点同时优化了簇内距离以及节点的剩余能量,是本轮运行中最适合的簇头节点。

二、算法的流程

我们在每轮运行中,运行 DCRP-PSOACA 协议将网络分簇,簇头节点充当簇内控

制中心，协调簇内的数据传输，并为簇内成员节点分配 TDMA 时间片，成员节点在自己的时间片内发送数据，其他时间处于休眠状态，提高了系统的性能，减少了系统的能耗。然后运行蚁群算法，在簇头节点间形成最优路由路径，各簇头汇聚后的数据沿着该路径传输，并进行数据的汇聚融合，发送至远处的基站。算法流程如图 5-4 所示。

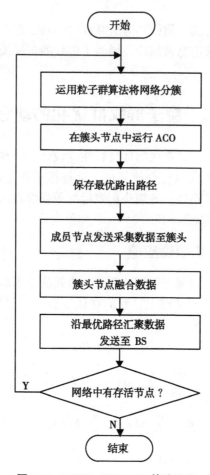

图 5-4　DCRP-PSOACA 算法流程

第六节　算法性能评估

本节我们评估 DCRP-PSOACA 的性能，对仿真实验结果进行分析，并与 PEGASIS、MST-PSO、ACO-MNCC 等协议进行比较。

一、仿真环境

根据我们设定的网络模型，考虑 100 个节点随机分布在 $100m \times 100m$ 的区域内，汇聚节点远离监测区域，位于 $(50, 175)m$ 处。蚁群算法参数如表 5-2 所示，各个参数的物理意义在前面章节中已经介绍。

表 5-2　蚁群算法实验参数

参数	值
m	10
α	1.0
β	5.0
ρ	0.15
σ	5

二、仿真结果

我们就 DCRP-PSOACA 协议的性能与 PEGASIS、MST-PSO、ACO-MNCC 等协议进行比较。首先我们比较一下网络的生存周期，亦即网络中存活节点数量随时间变化情况，如图 5-5 所示。从图中我们可以看出 DCRP-PSOACA 能明显延长网络的生存周期，这是因为 DCRP-PSOACA 采用 PSO 算法来优化分簇，然后采用 ACA 算法寻找合适路径将数据发送至基站。而 PEGASIS、MST-PSO 与 ACO-MNCC 要么通过构造近似最优链来减少能耗，但是仍需动态调整网络结构，带来能耗浪费；或者仅考虑距离的影响，构造最小生成树；或者仅强调通过 ACO 算法发现最大数目的连接覆盖集；均未充分考虑影响能力消耗的多种因素，造成能量浪费。

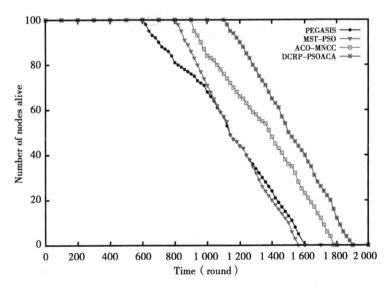

图 5-5　网络中存活节点数量

基站接收数据情况如图 5-6 所示。从图中可以看出我们提出的协议 DCRP-PSOACA 能接收更多的数据至基站，这是因为我们利用 PSO 的优化的优势来进行分簇，然后利用 ACA 算法优化路由路径选择，从而减少能量消耗。而 PEGASIS 通过构造接近最优的链来减少能量消耗，减少簇重构的开销，但需要动态调整网络的拓扑结构，带来能量浪费；MST-PSO 仅构建一个基于距离的最小生成树，未能很好地均衡能量的消耗；ACO-MNCC 通过 ACO 算法发现最大数目的连接覆盖集，来提高异构网络的性能，但是对节点剩余能量、节点间距离未充分考虑。我们提出的 DCRP-PSOACA 协议充分考虑了影响能量消耗的因素，利用 PSO 算法优化分簇，簇头的选择考虑剩余能量、节点间距等因素，然后利用 ACA 算法优化簇间路由，选择最合适的路径把数据发送至基站，减少能量消耗，提高了网络的利用效率。

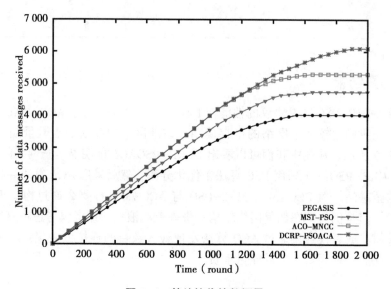

图 5-6　基站接收的数据量

第六章 附 录

缩略词

英文缩写	英文全拼
ACA	Ant Colony Algorithm
CDMA	Code Division Multiple Access
CSMA	Carrier Sense Multiple Access
GEAR	Geographical and Energy-Aware Routing
IoT	the Internet of Things
MST	Minimum Spanning Tree
MWSN	Mobile Wireless Sensor Networks
PSO	Particle Swarm Optimization
NPSO	Niching Particle Swarm Optimization
WoT	the Web of Things
WSN	Wireless Sensor Networks

参考文献

马静, 唐四元, 王涛. 2012. 物联网基础教程 [M]. 北京: 清华大学出版社.

孙利民, 李建中, 陈渝, 等. 2005. 无线传感器网络 [M]. 北京: 清华大学出版社.

王忠宏. 2012. 如何促进物联网从概念转化为创造价值的生产力 [EB/OL]. http://ftp.drcnet.com.cn/DRCNET.Channel.Web/gylt/20120301/index.html.

Abbasi A. A., Younis M. 2007. A survey on clustering algorithms for wireless sensor networks [J]. Computer Communications, 30 (14-15): 2826-2841.

AbdelSalam H. S., Olariu S. 2012. BEES: BioinspirEd backbonE Selection in Wireless Sensor Networks [J]. IEEE Trans. on Parallel and Distrib. Sys., 23 (1): 44-51.

Abdul Latiff N. M., Tsimenidis C. C., Sharif B. S. 2007. Energy-aware clustering for wireless sensor networks using particle swarm optimization [R]. PIMRC. IEEE 18th International Symposium on.

Akyildiz I. F., Su W., Sankarasubramaniam Y., et al. 2002. A Survey on Sensor Networks [J]. IEEE Communications Magazine, 40 (8): 102-114.

Akyildiz I. F., Su W., Sankarasubramaniam Y., et al. 2002. Wireless sensor networks: A survey [J]. Computer Networks, 38 (4): 393-422.

Akyildiz I. F., Su W. 2002. Sankarasubramaniam Y., et al. A Survey on Sensor Networks [J]. IEEE Communications Magazine, 40 (8): 102-114.

Al-Karaki J., Kamal A. 2004. Routing techniques in wireless sensor networks: a survey [R]. IEEE Wireless Communications.

Atzori L., Iera A., Morabito G. 2010. The internet of things: a survey [J]. Computer Networks, 54 (15): 2787-2805.

Bahl P., Padmanabhan V. N. 2000. RADAR: An in-building RF-based user location and tracking system [R]. in Proc. of INFOCOM'2000, Tel Aviv, Israel.

Baronti P., Pillai P., Chook V., et al. 2007. Wireless sensor networks: a survey on the state of the art and the 802.15.4 and ZigBee standards [J]. Computer Communications, 30 (7): 1655-1695.

Battiti R., Tecchiolli G. 1995. Training neural nets with the reactive tabu search [J]. IEEE transactions on neural networks, 6 (5): 1185-2000.

Beasley D., Bull D. R., Martin R. R. 1993. A sequential niche technique for multimodal function optimization [J]. Evol. Comput, 1 (2): 101-125.

Bird S., Li X. 2006. Adaptively choosing niching parameters in a PSO [R]. in Proc. Genet. Evol. Comput. Conf. (GECCO'06), Seattle, WA: ACM.

Blackwell T. 2012. A Study of Collapse in Bare Bones Particle Swarm Optimization [J]. IEEE Trans. Evol. Comput., 16 (3): 354-372.

Bonabeau E., Dorigo M., Theraulaz G. 2000. Inspiration for optimization from social insect behavior [J]. Nature, 406 (6): 39-42.

Boulis A, Ganeriwal S, Srivastava M B. 2003. Aggregation in sensor networks: an energy-accuracy trade-off [J]. Ad hoc networks, 1 (2): 317-331.

Brits A. E. R., van den Bergh F. 2002. A niching particle swarm optimizer [R]. in Proc. 4th Asia-Pacif. Conf. Simul. Evol. Learn (SEAL 2002), Singapore.

Brits A. E. R., van den Bergh F. 2012. A niching particle swarm optimizer [R]. in Proc. Asia-Pacif. Conf. Simul. Evol. Learn.

Bucherer E, Uckelmann D. 2011. Business Models for the Internet of Things [M]. Architecting the Internet of Things. Springer Berlin Heidelberg.

Bui N, Zorzi M. 2011. Health care applications: a solution based on the internet of things [R]. in Proc. the 4th International Symposium on Applied Sciences in Biomedical and Communication Technologies. ACM.

Cao X., Zhang H., Shi J., et al. 2008. Cluster heads election analysis for multi-hop wireless sensor networks based on weighted graph and particle swarm optimization [R]. Conf. Natural Comput.

Castellani A. P., Bui N., Casari P., et al. 2010. Architecture and protocols for the Internet of Things: a case study [R]. In Proc. 31st IEEE Conf. on local computer Net looks.

Castillo-Effer M, Quintela D H, Moreno W, et al. 2004. Wireless sensor networks for flash-flood alerting [R]. in Proc. 5th International Caracas Conference on Devices, Circuits and Systems.

Castro M., Jara A., Skarmeta A. 2012. Architecture for improving terrestrial logistics based on the Web of Things [J]. Sensors, 12 (5): 6538-6575.

Chander R., Elias S., Shivashankar S., et al. 2012. A REST based design for Web of Things in smart environments [R]. in Proceedings of the 2nd IEEE International Conference on Parallel Distributed and Grid Computing (PDGC).

Chen W., Hancke G., Mayes K., et al. 2010. NFC mobile transactions and authentication based on GSM network [R]. in Proc. International Workshop on Near Field Communication.

Chen W. H., Srivastava B. 1994. Simulated annealing procedures for forming machine cells in group technology [J]. European Journal of Operational Research, 75 (1): 100-111.

Cheng P., Zhang F., Chen J., et al. 2013. A Distributed TDMA Scheduling Algorithm

参考文献

for Target Tracking in Ultrasonic Sensor Networks [J]. IEEE Transactionson Industrial Electronics, 60 (9): 3836-3845.

Cunha A. B. da, Silva D. C. da 2012. Behavioral Model of Alkaline Batteries for Wireless Sensor Networks [J]. IEEE Latin America Transactions, 10 (1): 1295-1304.

Davidor Y. 1991. A naturally occurring niche & species phenomenon: The model and first results [R]. in Proc. 4th Int. Conf. Genet. Algorith. , San Mateo, CA: Morgan Kaufmann.

Davidyuk O. , Georgantas N. , Issarny V, et al. 2011. MEDUSA: A middleware for end-user composition of ubiquitous applications [R]. in Handbook of Research on Ambient Intelligence and Smart Environments: Trends and Perspectives, N. -Y. Chong and F. Mastrogiovanni, Eds. Hershey, PA, USA: IGI Global.

Deb B. , Bhatnagar S. , Nath B. 2001. A topology discovery algorithm for sensor network with applications to network management [R]. DCS Technical Report DCS-TR-441, Rutgers University.

Deb K. , Goldberg D. 1989. An investigation of niche and species formation in genetic function optimization [R]. in Proc. 3rd Int. Conf. Genet. Algorith.

Deb K. 1989. Genetic algorithms in multimodal function optimization, the Clearinghouse for Genetic Algorithms, M. S thesis and Rep [R]. Univ. Alabama: Tuscaloosa.

Dhillon S. S. , Vanmieghem P. 2007. Performance analysis of the AntNet Algorithm [J]. Computer Networks, 51 (8): 2104-2125.

Di Caro G. , Dorigo M. 1998. AntNet: distributed stigmergetic control for communications networks [J]. Journal of Artifical Intelligence (9): 317-365.

Dorigo M. , Ganbardella L. M. 1997. Ant colony system: a cooperative learning approach to traveling salesman problem [J]. IEEE Trans. Evolutionary Computation, 1 (1): 53-66.

Dorigo M. , Maniezzo V. , Colorni A. 1996. Ant System: Optimization by a Colony of Cooperating Agents [J]. IEEE Trans. Systems, Man, and Cybernetics-PART B, 26 (1): 29-41.

Drossos N. , Goumopoulos C. , Kameas A. 2007. A conceptual model and the supporting middleware for composing ubiquitous computing applications [J]. J. Ubiquit. Comput. Intell. , 1 (2): 174-186.

Duquennoy S. , Grimaud G. , Vandewalle J. 2009. The Web of Things: Interconnecting Devices with High Usability and Performance. in Proc. of the International Conference on Embedded Software and Systems, ICESS'09 [R]. Hangzhou, Zhejiang, China.

Engelbrecht A. , Masiye B. , Pampara G. 2005. Niching ability of basic particle swarm optimization algorithms [R]. in Proc. IEEE Swarm Intell. Symp.

Estrin D. , Govindan R. , Heidemann J. S. , et al. 1999. Next century challenges: Scalable coordinate in sensor network [R]. in Proc. 5th ACM/IEEE Int'l Conf on Mobile

Computing and Networking.

Fielding R T. 2000. Architectural styles and the design of network-based software architectures [D]. University of California.

Fielding, Roy T., Taylor, et al. 2002. Principled Design of the Modern Web Architecture, ACM Trans [J]. Internet Technology (TOIT), 2 (2): 407-416.

Fleisch E. 2004. Business Impact of Pervasive Technologies: Opportunities and Risks [J]. Human and Ecological Risk Assesment, 10 (5): 817-829.

Fortuna C., Grobelnik M. 2011. Tutorial: The Web of Things [R]. in Proc. of the World Wide Web Conference Hyderabad, India.

Fred G., Said H. 2002. Tabu search and finite convergence [R]. Discrete Applied Mathematics.

Fun T. S., Beng L. Y., Likoh J., et al. 2008. A Lightweight and Private Mobile Payment Protocol by Using Mobile Network Operator [R]. in Proc. International Conference on Computer and Communication Engineering (ICCCE 2008).

Gao T, Greenspan D, Welsh M, et al. 2006. Vital signs monitoring and patient tracking over a wireless network [R]. in Proc. 27th International Conference on Engineering in Medicine and Biology Society.

Gen M., Liu B. R., Ida K. 1996. Evolution program for deterministic and stochastic optimizations [J]. European Journal of Operational Research, 94 (3): 618-625.

Ghose A., Bhaumik C., Das D., et al. 2012. Mobile healthcare infrastructure for home and small clinic [R]. in Proc. 2nd ACM international workshop on Pervasive Wireless Healthcare (MobileHealth'12), ACM, New York.

Goldberg D. E., Richardson J. 1987. Genetic algorithms with sharing for multimodal function optimization [R]. in Proc. 2nd Int. Conf. Genet. Algorith., Cambridge, MA.

Griod L., Estrin D. 2001. Robust range estimation using acoustic and multimodal sensing [R]. in Proc. IEEE/RSJ Int'l Conf Intelligent Robots and System (IROS'01), Maui, Hawaii, USA.

Grønli T M, Ghinea G, Younas M. 2013. A Lightweight Architecture for the Web-of-Things [R]. in Proc. Mobile Web Information Systems (MobiWis), Springer Berlin Heidelberg.

Gubbia J., Buyyab R., Marusica S., et al. 2013. Internet of Things (IoT): A vision, architectural elements, and future directions [J]. Future Generation Computer Systems, 29 (7): 1645-1660.

Guinard D, Trifa V, Pham T, et al. 2009. Towards physical mashups in the web of things [R]. in Proc. 6th IEEE International Conference on Networked Sensing Systems (INSS 2009).

Guinard D, Trifa V. 2009. Towards the web of things: Web mashups for embedded devices [R]. in proc. WWW (International World Wide Web Conferences)

Workshop on Mashups, Enterprise Mashups and Lightweight. Madrid, Spain.

Guinard D., Trifa V., Pham T., et al. 2009. Towards Physical Mashups in the Web of Things [R]. in Proc. 6th International Conference on Networked Sensing Systems (INSS'09), Pittsburgh, USA.

Guinard D., Trifa V., Wilde E. 2010. A Resource Oriented Architecture for the Web of Things, in Proc. of IoT 2010 (IEEE International Conference on the Internet of Things) [R]. Tokyo, Japan.

Guinard, Dominique. 2009. Vlad Trifa, Towards the Web of Things: Web Mashups for Embedded Devices [R]. in Proc. of the International World Wide Web Conferences. Madrid, Spain.

Guru S., Halgamuge S., Fernando S. 2005. Particle swarm optimisers for cluster formation in wireless sensor networks [R]. in Proc. Int. Conf. Intell. Sens., Sens. Netw. Inf. Process.

He T., Huang C., Blum B. M., et al. 2003. Range-free localization schemes for large scale sensor networks [R]. in Proc. 9th Annual Int'l Conf on Mobile Computing and Networking (MobiCom), San Diego, CA.

Heinzelman W. B., Chandrakasan A. P., Balakrishnan H. 2002. An Application-Specific Protocol Architecture for Wireless Microsensor Networks [J]. IEEE Trans. Wireless Communications, 1 (4): 660-670.

Heinzelman W. R., Chandrakasan A., Balakrishnan H. 2000. Energy-efficient communication protocol for wireless microsensor networks [R]. in Proc. 33rd Int'l Conf System Sciences (HICSS'00).

Holland J. H. 1975. Adaptation in Natural and Artificial Systems [M]. Ann Arbor: University of Michigan press, USA.

Hui J. W., Culler D. E. 2008. IP is dead, long live IP for wireless sensor networks [R]. in Proc. SenSys'08: Proceedings of the 6th ACM Conference on Embedded Network Sensor Systems New York, USA: ACM.

Hui J. W., Culler D. E. 2008. IP is dead, long live IP for wireless sensor networks [R]. inProc. 6th ACM Conference on Embedded Network Sensor Systems. New York, NY, USA: ACM.

Ian Smith. 2002. The Internet of Things 2012: New Horizons [R]. Znternet of Things European Research Cluster Platinum, Halifax, U. K.

Intanagonwiwat C, Govindan R, Estrin D. 2000. Directed diffusion: a scalable and robust communication paradigm for sensor networks [R]. in Proc. 6th annual international conference on Mobile computing and networking, ACM.

Jackson D. E., Holcombe M., Ratnieks F. F. W. 2004. Trail geometry gives polarity to ant foraging networks [J]. Nature (432): 907-909.

Jaromczyk J. W., Toussaint G. T. 1992. Relative neighborhood graphs and their relatives [J]. Proceedngs of the IEEE, 80 (9): 1502-1517.

Jiang C., Yuan D., Zhao Y. 2009. Towards Clustering Algorithms in Wireless Sensor Networks-A Survey [R]. in Proc. IEEE WCNC.

Jiang X., Zhang C. 2011. A Web-based IT framework for campus innovation [R]. in Proc. 3rd International Workshop on Education Technology and Computer Science.

Juels A. 2006. RFID security and privacy: a research survey [R]. IEEE Journal on Selected Areas in Communications.

Kanniainen L. 2010. Alternatives for banks to offer secure mobile payments [J]. Int J Bank Mark, 28 (5): 433-444.

Karenos K., Kalogeraki V., Krishnamurthy S. 2008. Cluster-based congestion control for sensor networks [R]. ACM Transactions on Sensor Networks.

Kawadia V., Kumar P. 2003. Power control and clustering in Ad Hoc networks [R]. in Proc. IEEE INFOCOM, San Francisco, CA.

Kennedy J., Eberhart R. 1995. Particle swarm optimization [R]. in Proc. IEEE Int. Conf. NeuralNetw.

Kindberg T., Barton J., Morgan J., et al. 2000. People, places, things: web presence for the real world [R]. Proceding third IEEE Workshop on Mobile Computing Systems and Applications. Los. Alamitos CA, USA.

Kirkpatrick S., Gelatt Jr. C. D., Vecchi M. P. 1983. Optimization by simulated annealing [J]. Science, 220: 671-680.

Kozlov D., Veijalainen J., Ali Y. 2012. Security and privacy threats in IoT architectures [R]. in Proc. 7th International Conference on Body Area Networks (BodyNets'12). ICST Brussels, Belgium.

Kubisch M., Karl H., Wolisz A., et al. 2003. Distributed algorithm for transmission power control in wireless sensor networks [R]. IEEE WCNC'2003, New Orleans, Louisiana.

Kulkarni R. V., Venayagamoorthy G. K.. 2011. Particle Swarm Optimization in Wireless Sensor Networks: A Brief Survey [J]. IEEE Trans. Systems, Man, and Cybernetics, 41 (2): 262-267.

Kumar P., Chaturvedi A., Kulkarni M. 2012. Geographical location based hierarchical routing strategy for wireless sensor networks [R]. in Proc. ICDCS'2012.

Latiff N., Tsimenidis C., Sharif B. 2007. Perfor-mance comparison of optimization algorithms for clustering in wireless sensor networks [R]. in Proc. IEEE Int. Conf. Mobile Ad Hoc Sens. Syst.

Lee J., Cheng W. L. 2012. Fuzzy-Logic-Based Clustering Approach for Wireless Sensor Networks Using Energy Predication [J]. IEEE Sensors Journal, 12 (9): 2891-2897.

Leminen S, Westerlund M, Rajahonka M, et al. 2012. Towards IoT Ecosystems and Business Models, Internet of Things, Smart Spaces, and Next Generation Networking

[R]. Springer Berlin Heidelberg.

Li C. H., Yang S. X., Nguyen T. T. 2012. A Self-Learning Particle Swarm Optimizer for Global Optimization Problems [J]. IEEE Trans. Systems, Man, and Cybernetics, 42 (3): 627-646.

Li J., Balazs M., Parks G. 2002. A species conserving genetic algorithm for multimodal function optimization [J]. Evol. Comput., 10 (3): 207-234.

Li N, Hou J C, Sha L. 2003. Design and analysis of an MST-based topology control algorithm [R]. in 22 Annual Joint Conference of the IEEE Computer and Communications (INFOCOM 2003).

Li N, Hou J C. 2004. Topology control in heterogeneous wireless networks: Problems and solutions [R]. in 23 Annual Joint Conference of the IEEE Computer and Communications Societies (INFOCOM 2004).

Li W., Taheri J., Zomaya A. Y., et al. 2013. Nature – Inspired Computing for Autonomic Wireless Sensor Networks [R]. Large Scale Network-Centric Distributed Systems.

Li X. 2004. Adaptively choosing neighborhood bests using species in a particle swarm optimizer for multimodal function optimization [R]. in Proc. Genet. Evol. Comput. Conf.

Li X. 2010. Niching Without Niching Parameters: Particle Swarm Optimization Using a Ring Topology [J]. IEEE Trans. Evol. Comput., 14 (1): 150-169.

Li Y., Thai M., Wu W. 2008. Topology control for wireless sensor networks, Wireless Sensor Networks and Applications [R]. Heidelberg: Springer.

Liang Y., Yu H., Zeng P. 2006. Optimization of cluster-based routing protocols in wireless sensor network using PSO [R]. Control and Decision.

Lin Y., Zhang J., Chung H. S., et al. 2012. An Ant Colony Optimization Approach for Maximizing the Lifetime of Heterogeneous Wireless Sensor Networks [J]. IEEE Trans. Systems, Man, and Cybernetics-PART C, 42 (3): 408-420.

Lindsey S., Raghavendra C., Sivalingam K. M. 2002. Data Gathering Algorithms in Sensor Networks using Energy Metrics [J]. IEEE Trans. Parallel and Distributed Systems, 13 (9): 924-935.

Lorincz K, Malan D J, Fulford-Jones T R F, et al. 2004. Sensor networks for emergency response: challenges and opportunities [J]. Pervasive Computing, IEEE, 3 (4): 16-23.

Ma D., Ma J., Xu P., et al. 2013. An Adaptive Node Partition Clustering Protocol using Particle Swarm Optimization [R]. in Proc. IEEE ICCA'13.

Manjeshwar A., Agrawal D. P. 2002. APTEEN: A hybrid protocol for efficient routing and comprehensive information retrieval in wireless sensor networks [R]. in Proc. 16th int'l Parallel and Distributed Processing Symp.

Massoth M., Bingel T. 2009. Performance of different mobile payment service concepts compared with a NFC-based solution [R]. in Proc. of 4th International Conference on Internet and Web Applications and Services (ICIW'09).

Mazhelis O, Luoma E, Warma H. 2012. Defining an internet-of-things ecosystem, Internet of Things, Smart Spaces, and Next Generation Networking [R]. Springer Berlin Heidelberg.

Messer A., Kunjithapatham A., Sheshagiri M., et al. 2006. InterPlay: A middleware for seamless device integration and task orchestration in a networked home [R]. in Proc. 4th IEEE Conf. Pervasive Computing and Communications (PERCOM), Washington DC.

Murty R. N., Mainland G., Rose I., et al. 2008. CitySense: an urban-scale wireless sensor network and test bed.

Muruganathan S. D., Ma D. C. F., Bhasin R. I., et al. 2005. A Centralized Energy-Efficient Routing Protocol for Wireless Sensor Networks [R]. IEEE Radio Communications, 43 (3): S8-13.

Niculescu D, Nath B. 2003. Ad hoc positioning system (APS) using AOA [R]. in 22nd Annual Joint Conference of the IEEE Computer and Communications (INFOCOM 2003).

Niculescu D., Nath B. 2003. Ad hoc positioning system (APS) using AOA [R]. Proc. INFOCOM'2003.

Ning H. S., Wang Z. O. 2011. Future Internet of Things architecture: like mankind neural system or social organization framework [R]. IEEE Communications Letters.

Noury N., Herve T., Rialle V., et al. 2000. Monitoring behavior in home using a smart fall sensor and position sensor [R]. in Proc. IEEE-EMBS Special Topic Conf on Microtechnologies in Medicine and Biology, Lyon France.

Okdem S., Karaboga D. 2006. Routing in wireless sensor networks using ant colony optimization [R]. in Proc. 1st NASA/ESA Conf. Adapt. Hardware Syst.

Pahlavan K., Levesque A. 1995. Wireless Information Networks [M]. New York: Wiley, USA.

Parrott D., Li X. 2006. Locating and tracking multiple dynamic optima by a particle swarm model using speciation [J]. IEEE Trans. Evol. Comput., 10 (4): 440-458.

Pautasso C., Wilde E. 2009. Why is the Web Loosely Coupled A Multi-Faceted Metric for Service Design [R]. in Proc. 18th International World Wide Web Conference (WWW'09), Madrid, Spain.

Pautasso C., Wilde E. 2009. Why is the Web Loosely Coupled A Multi-Faceted Metric for Service Design [R]. in Proc. 18th International World Wide Web Conference (WWW'09). Madrid, Spain.

Pautasso C. 2008. BPEL for REST [R]. in Proc. 6th International Conference on Business Process Management (BPM'08).

Petrowski A. 1996. A clearing procedure as a niching method for genetic algorithms [R]. in Proc. 3rd IEEE Int. Conf. Evol. Comput., Nagoya, Japan, May.

Pister K., Hohlt B., Jeong J., et al. 2003. A sensor network infrastructure [EB/OL]. http://www-bsac.eecs.berkeley.edu/projects/ivy.

Pompili D, Melodia T, Akyildiz I F. 2006. Deployment analysis in underwater acoustic wireless sensor networks [R]. in Proc. 1st ACM international workshop on Underwater networks.

Pottie G. J., Kaiser W. J. 2000. Wireless integrated network sensors [J]. Communications of the ACM, 43 (5): 51-58.

Priyantha N B, Balakrishnan H, Demaine E, et al. 2003. Anchor-free distributed localization in sensor networks [R]. in Proc. 1st international conference on Embedded networked sensor systems, ACM.

Priyantha N., Chakraborthy A., Balakrishnan H. 2000. The cricket location-support system [R]. in Proc. Int'l Conf on Mobile Computing and Networking, Boston, MA.

Raghunathan V., Kansai A., Hse J., et al. 2005. Design considerations for solar energy harvesting wireless embedded systems [R]. in Proc. the IPSN.

Raghunathan V., Schurgers C., Sung P. 2002. Energy-Aware Wireless Microsensor Networks [J]. IEEE Signal Processing Magazine, 19 (2): 40-50.

Rahimi M., Shah H., Sukhatme G. S., et al. 2003. Studying the feasibility of energy harvesting in mobile sensor network [R]. in Proc. IEEE ICRA.

Rappaport T. 1996. Wireless Communications: Principles & Practice [M]. Englewood Cliffs, NJ: Prentice-Hall.

Reddy S., Murthy C. R. 2012. Dual-Stage Power Management Algorithms for Energy Harvesting Sensors [J]. IEEE Trans. wireless communications, 11 (4): 1434-1445.

Rentala P., Musunuri R., Gandham S., et al. 2002. Survey on sensor networks [R]. Technical Report, UTDCS-33-02, University of Texas at Dallas.

Richardson L, Ruby S. 2008. RESTful web services [R]. O'Reilly Media, Inc.

Richardson, Leonard, Ruby, et al. 2007. RESTful Web Services [R]. O'Reilly.

Roundy S., Rabaey J. M., Wright P. K. 2004. Energy Scavenging for Wireless Sensor Networks [R]. Springer-Verlag, New York, LLC.

Santi P, Simon J. 2004. Silence is golden with high probability: Maintaining a connected backbone in wireless sensor networks [R]. Wireless Sensor Networks, Springer Berlin Heidelberg.

Savvides A., Han C. C., Srivastava M. B. 2001. Dynamic finge-grained localization in ad-hoc networks of sensors [R]. in Proc. 7th Annual Int'l Conf on Mobile Computing and Networking (MobiCom), Rome, Italy.

Savvides A., Park H., Srivastava M., et al. 2002. The bits and flops of the N-hop multilateration primitive for node localization problem [R]. in Proc. 1st ACM int'l Workshop on Wireless Sensor Networks and Application, Atlanta, GA.

Seema Bandyopadhyay, Edward J. 2004. Coyle, Minimizing communication costs in hierarchically-clustered networks of wireless sensors [R]. Computer Networks.

Shah R. C., Rabaey J. M. 2002. Energy aware routing for low energy ad hoc sensor networks [R]. in Proc. IEEE Wireless Communications and Networking Conference (WCNC'02).

Shi L., Yuan Y., Chen J. 2013. Finite Horizon LQR Control with Limited Controller-System Communication [J]. IEEE Transactions on AutomaticControl, 58 (7): 1835-1841.

Shi Y., Eberhart R. 1998. A modified particle swarm optimization [R]. IEEE Int. Congress on Evolutionary Computation.

Simon G, Maróti M, Lédeczi Á, et al. 2004. Sensor network-based countersniper system [R]. in Proc. 2nd international conference on Embedded networked sensor systems, ACM.

Smaragdakis G., Matta I., Bestavros A. 2004. SEP: A stable election protocol for clustered heterogeneous wireless sensor networks [R]. in 2th International Workshop on Sensor and Actor Network Protocols and Applications.

Smaragdakis G., Matta I., Bestavros A. 2004. SEP: A stable election protocol for clustered heterogeneous wireless sensor networks [R]. in Proc IEEE SANPA'04.

Sohrabi K., Gao J., Ailawadhi V., et al. 2000. Protocol for self-organized of a wireless sensor network [R]. IEEE Personal Communications.

Stirbu V. 2008. Towards a RESTful Plug and Play Experience in the Web of Things [R]. IEEE International Conference on Semantic Computing.

Stirbu V. 2008. Towards a RESTful Plug and Play Experience in the Web of Things [R]. in Proc. IEEE International Conference on Semantic Computing.

Thomas E. 2005. Service-Oriented Architecture: Concepts, Technology, and Design [R]. Prentice Hall PTR, Upper Saddle River, NJ.

Tilak S., Abu-Ghazaleh N., Heinzelman W. 2002. A taxonomy of wireless micro-sensor network models [R]. ACM Mobile Computing and Communications Review.

Tillet J., Rao R., Sahin F. 2002. Cluster-head identification in ad hoc sensor networks using particle swarm optimization [R]. in Proc. IEEE PWC'02.

Toumpis S., Tassiulas T. 2006. Optimal deployment of large wireless sensor networks [J]. IEEE Transactions on Information Theory, 52: 2935-2953.

Toussaint G. 1980. The relative neighborhood graph of a finite planar set [J]. Pattern Recognition, 12 (4): 261-268.

Trifa V., Wieland S., Guinard D., et al. 2009. Design and Implementation of a Gateway for Web-based Interaction and Management of Embedded Devices [R]. in

参考文献

Proc. 2nd International Workshop on Sensor Network Engineering（IWSNE'09），Marina del Rey，USA.

Uckelmann D, Harrison M, Michahelles F. 2011. An architectural approach towards the future internet of things［R］. Architecting the Internet of Things, Springer Berlin Heidelberg.

Valle Y., Venayagamoorthy G. K., Hernandez J. C., et al. 2008. Particle swarm optimization: Basic concepts, variants and applications in power systems［J］. IEEE Trans. Evol. Comput., 12（2）: 171-195.

Voss C. A., Hsuan J. 2009. Service architecture and modularity［J］. Decision Sciences, 40（3）: 541-569.

Wakasa Y., Tanaka K., Akashi T., et al. 2012. Decay rate and l^2 gain analysis for the particle swarm optimization algorithm［J］. Asian Journal of Control, 14（1）: 125-136.

Wan J., Yuan D., Xu X. 2008. A review of cluster formation mechanism for clustering routing protocols［R］. in Proc. 11th IEEE International Conference on Communication Technology, Hangzhou, China.

Wang A., Heinzelman W., Chandrakasan A. 1999. Energy-scalable protocols for battery-operated microsensor networks［R］. in Proc. 1999 IEEE Workshop Signal Processing Systems（SiPS'99）.

Wang R., Liu G., Zheng C. 2007. A Clustering Algorithm based on Virtual Area Partition for Heterogeneous Wireless Sensor Networks［R］. in Proc. IEEE Mechatronics & Automation.

Welbourne E., Battle L., Cole G., et al. 2009. Building the Internet of Things using RFID: The RFID ecosystem experience［J］. IEEE Internet Computing, 13（3）: 48-55.

Werner-Allen G, Lorincz K, Ruiz M, et al. 2006. Deploying a wireless sensor network on an active volcano［J］. Internet Computing, IEEE, 10（2）: 18-25.

W. B. Heinzelman, A. P. Chandrakasan, H. Balakrishnan. 2002. An Application-Specific Protocol Architecture for Wireless Microsensor Networks［J］. IEEE Trans. Wireless Communications, 1（4）: 660-670.

Xu R., Xu J., Wunsch D. C. 2012. A Comparison Study of Validity Indices on Swarm-Intelligence-Based Clustering［J］. IEEE Trans. Systems, Man, and Cybernetics, 42（4）: 1243-1256.

Xu Y, Heidemann J, Estrin D. 2001. Geography-informed energy conservation for ad hoc routing［R］. in Proc. 7th Annual Int'l Conf on Mobile Computing and Networking（MobiCOM）, Rome, Italy.

Yang S. X., Li C. H. 2010. A Clustering Particle Swarm Optimizer for Locating and Tracking Multiple Optima in Dynamic Environments［J］. IEEE Trans. Evol. Comput., 14（6）:

959-974.

Yick J, Mukherjee B, Ghosal D. 2005. Analysis of a prediction-based mobility adaptive tracking algorithm [R]. in Proc. 2nd BroadNets.

Yick J., Mukherjee B., Ghosal D. 2008. Wireless sensor network survey [J]. Computer Networks, 52 (12): 2292-2330.

Younis O., Fahmy S. 2004. HEED: A Hybrid, Energy-Efficient, Distributed Clustering Approach for Ad Hoc Sensor Networks [J]. IEEE Trans. Mobile Computing, 3 (4).

Younis O., Krunz M., Ramasubramanian S. 2006. Node clustering in wireless sensor networks: recent developments and deployment challenges [J]. IEEE Network, 20 (3): 20-25.

Yu J., Chong P. 2005. A survey of clustering schemes for mobile ad hoc networks, IEEE Communications Surveys & Tutorials, 7 (1): 32-48.

Yu Y, Govindan R, Estrin D. 2001. Geographical and energy aware routing: A recursive data dissemination protocol for wireless sensor networks, Technical report ucla/csd-tr-01-0023 [R]. UCLA Computer Science Department.

Yun M., Yuxin B. 2010. Research on the architecture and key technology of Internet of Things (IoT) applied on smart grid [R]. in Proc. Advances in Energy Engineering (ICAEE).

Zhang P, Sadler C M, Lyon S A, et al. 2004. Hardware design experiences in ZebraNet [R]. in Proc. 2nd international conference on Embedded networked sensor systems, ACM.

Zhang Y., He S., Chen J., et al. 2013. Distributed Sampling Rate Control for Rechargeable Sensor Nodes with Limited Battery Capacity [J]. IEEE Transactions on Wireless Communications, 12 (6): 3096-3106.

Zhong Y. P., Huang P. W., Wang B. 2007. Maximum lifetime routing based on ant colony algorithm for wireless sensor networks [R]. in Proc. IET Conf. Wireless, Mobile, Sensor Networks.

第三篇

关键技术在农业上的应用

茶树种质及基因资源发掘与开发利用数据库系统构建

当前农业信息化高度发达，构建茶树优质抗寒种质及基因资源发掘与开发利用数据库系统，对现有茶树资源信息进行搜集、整理和发掘，实现茶树种质及基因资源的信息化发掘与利用，对茶产业的发展以及茶树育种、栽培等工作有重要的指导意义。

我国茶区辽阔，有着丰富的茶树品种资源，茶类多样。随着茶的保健功能与作用机理为大众所接受与熟悉，对茶叶的需求呈逐年递增趋势，茶业的发展前景极其广阔。我国是茶树原产地，拥有丰富的茶树种质资源，其遗传多样性举世瞩目，到目前为止共收集3 300多份茶树种质资源[1]。

随着茶叶科学技术的迅速发展，建立茶树种质及基因资源的规范化管理就显得越来越重要了。在茶树种质资源资料分布广泛，有些资料需要尽快更新。因此在茶树种质及基因资源信息的全面化，性状的详细化，分类的合理化基础上，通过对现有茶树种质资源资料的收集、整理和更新，建设了茶树种质及基因资源数据库系统。该系统的完成有利于整合茶树种质资源，高效利用和保护种质资源，充分发掘和利用资源的价值。该数据库系统设立了各种查询方式及过滤功能，同时录入了茶树种质资源植株和芽叶的图片，使得茶树资料更加丰满全面。对茶树资料进行了合理有效的归纳更新，实现了茶树种质及基因资源的信息化利用与管理。

1 建设目的

建立茶树优质抗寒种质及基因资源发掘与开发利用数据库的目的在于整合汇总我国的各种茶树资料，搜集各种不同生态环境、地理以及遗传类型的茶树资源，同时要避免重复记录同一种材料，保证在便于交流与管理的基础上，方便茶学科研的进行。

在建立数据库系统时，需要对茶树资料进行系统的分类，保证数据库的丰度与完整性。记录数据时，确保采集数据详细准确，并对花朵、牙叶、植株和果实等配备图片或视频资料，便于利用。

2 建设原则

构建茶树优质抗寒种质及基因资源数据库时，在避免同一材料过多的同时，保证最大可能的各类茶树资源（地理、生态、遗传性状等）；既要有质量，又要有数量；要有外来品种，便于研究，便于查询；既要保证基本数据丰富，又要有基因数据等原则。

对茶树种质资源的整理，必须处理好不同地区茶树资源的分类，避免同名异种或者同种异名，使数据库资料保持一致性与完整性。同时加大对茶树种质资源的征集、评价与鉴别，基本数据的采集要详细、准确，既要有生物学特性，又要有生态环境信息，同时配备必要的图片视频等资料，便于数据库建设。构建茶树种质及基因资源数据库的目的是为了便于交流和利用，因而需考虑操作规范、数据可靠、信息量大以及交流的方便性。

3 建设方法

3.1 茶树种质及基因资源的描述

茶树种质及基因资源的描述，可参考陈亮等编著的《茶树种质资源与遗传改良》[2]中制定的茶树种质资源数据标准，设立包括基本数据、特征数据、评价鉴定数据在内的110多个描述项目，增加照片和视频项目，在种质类型项目中，增加株系选项。以全国统一编号作为主键，避免种质资源在数据库中的重复录入，便于数据库内容的规范。

3.2 软件平台技术

本茶树种质与基因资源数据库系统采用J2EE架构，是Java2平台企业版，其核心是一组技术规范与指南，包含各类组件、服务架构和技术层次，遵守相同的标准与规格，有良好的兼容性。数据库系统采用MySQL，是一种关联数据库管理系统，将数据存在不同的表中，增加了速度、提高了灵活性，SQL语言是用于访问数据库最常用的标准化语言。

3.3 采集及录入

在《茶树品种志》[3]与《中国茶树品种志》[4]的基础上，广泛采集各种茶树种质与基因资源资料。查阅了《中国茶叶》《茶叶科学简报》等大量关于茶叶的书籍、期刊与论文，并对资料进行反复核对与整理，保证采集数据的权威性与可靠性。借助网络，搜集最近新育成品种的种质资源等信息，最大可能的保证了信息的全面性与广泛性。

在搜集大量茶树种质及基因资源的基础上，完成数据的规范化、标准化工作后录入数据库系统。在录入过程中，为了避免输入错误以及资料间的细微差别或误差，确保信息的可靠性与准确性，在录入完成后进行多次全面核对工作。对于同一种质资源，若在某个性状上有差异，进行反复验证、观察、推敲，同时咨询相关专家，最大可能的保证材料信息的权威性与准确性。

在《茶树品种志》与《中国茶树品种志》提供的图片资料基础上，借助现代化的途径，更加广泛的采集茶树种质资源的图片及视频信息。获得图片或者视频资料后，使其与茶树种质资源的文字资料进行对应。图片与视频资料的添加，使茶树种质资源的资料更加形象生动，表现更直观。

4 建设意义

种质决定遗传性状，并将遗传信息从父代传递给子代。而种质资料也称基因资源、遗传物质。野生近缘物种、各种地方品种以及新培育的品种都属于种质资源的范畴。茶树种质资源是开发利用、生物技术研究与品种创新的物质基础，具有重要的价值与意义。

我国的茶树资源分布在不同的区域，缺乏统一有效的管理，很难做到资源共享与有效利用，因此建设茶树优质抗寒种质及基因资源发掘与开发利用数据库系统，可提高资源的利用效率，更好的促进新品种的选择培育工作的进行。

5　结束语

茶树优质抗寒种质及基因资源发掘与开发利用数据库的建立，有利于整合茶树资源，便于茶树种质资源的交流与对外宣传，使优质茶叶走向世界；利于利用和保护茶树资源，促进资源转化；利于茶树资源信息化，促进茶叶生产、科研、教育等对茶树资源的需求。

参考文献

[1] 虞富莲，俞永明，李名君，等．茶树优质资源的系统鉴定与综合评价［J］．茶叶科学，1992，12（2）：95-125.

[2] 陈亮，虞富莲，杨亚军．茶树种质资源与遗传改良［M］．北京：中国农业科学技术出版，2006.

[3] 福建省农业科学院茶叶研究所茶树品种志［M］．福州：福建人民出版社，1979.

[4] 白堃元，等．中国茶树品种志［M］．上海：上海科学技术出版社，2001.

茶园水肥一体化技术应用现状与发展前景

茶园广泛分布于中国各地，茶产业作为我国农业产业化的重要组成部分，其产业发展的健康与否直接关系到农业、农村、农民的发展。基于物联网的茶园水肥一体化技术是将施肥与灌溉有机结合的农业新兴技术。我们主要介绍茶园水肥一体化现状、存在的问题及解决方法，针对茶园水肥一体化的发展提出一些建议。

引 言

中国是茶的故乡，千百年来的茶文化深受世界各地的喜爱，是我国丰厚的文化遗产。目前中国茶产业的规模不断扩大，但是分布却很零散，茶叶多采用传统的人工生产模式，茶叶生产规模和自动化程度低导致茶叶的整体经济效益降低。如今互联网时代快速发展融入了生产管理的各行各业，茶产业想要长久持续的发展，就要对传统生产工艺进行智能优化和改造，提高茶叶生产效率，促进茶叶产业转型升级。水肥一体化技术可以满足人们对于建设农业可持续发展的需求，因此探索水肥一体化在茶树种植方面的应用颇为重要。

1 茶园水肥一体化应用现状

我国总体上是一个干旱缺水的国家，农业是中国的主要用水户。茶园水肥一体化技术是将灌溉用水与肥料通过可控制的管道系统混合后形成水肥溶液输送到茶树根部为其提供养分和水分的农业新技术。由于该技术集成了计算机技术、传感技术和无线通信技术等现代技术，因此，在灌溉的过程中，可以通过计算机或移动设备自动监测土壤湿度，并可以收集土壤水分数据。自动感知灌溉需求信息，根据施肥浇水的需要启动泵阀进行自动或半自动灌溉操作。茶树适宜在温暖湿润的环境中生长，影响茶树丰产、优质形成的主要因素有土壤酸碱度、营养状况、光照、气候条件等一些外在自然因素，也包括施肥技术、耕作技术、采摘技术以及综合管理技术等一些人为控制因素。

目前茶叶生产管理基本上依靠传统经验，通过管理者观察茶园的土壤水分情况，进行灌溉控制。由于这种人为干预受到诸多因素的影响，譬如管理者的经验、观察判断的准确度等，如果掌握情况不到位不及时的话，往往会造成施肥用水的过量或不足，施肥不均匀造成茶树根部灼伤或养分吸收不足，从而造成不必要的损失，影响茶树的优产。这也是现在很多茶树种植户面临的问题。

2 茶园水肥一体化优势及存在的问题

2.1 优势

过去常用的漫灌等灌溉方法加上肥料的不科学使用，长此以往会导致土壤压实、盐分积累、盐渍化程度加重，不利于作物连续种植使用土地。基于物联网的水肥一体化技术通过自动控制、连接监测土壤墒情等数据信息可以适时、适量地为茶树供给水分和肥料，具有节水节肥、省时省工、调温调湿、减少病害、增加产量，改善品质，提高经济效益的优势。不合理的水肥一体化系统，不仅不能保证优质增产，还有可能造成费工费水，因此需要一套合理的水肥一体化设备达到以上种种优势。除了合理配套的水肥一体化系统，还需要灌溉方式和水源等技术要点的配合，可以降低水资源以及化肥的过度使用，保护环境[1]，在改善土壤环境的同时，提高茶树的产量和质量，释放更多劳动力，将茶园引向精细农业、智慧农业的方向。

2.2 存在的问题

2.2.1 农户的认可度不够高

智能水肥一体机、智能水肥一体化设备在一些示范基地和种植大棚中得到一定的应用，就目前来看，农户还是不太认可，因循守旧，受到一些老观念、老经验的束缚，考虑到水肥机的成本，只看到眼前利益，所以还是采用传统种植茶树的方式。

2.2.2 水肥一体化集成技术模式没有完全成熟

不同灌溉方式、灌水量、施肥量、水肥配比是水肥一体化技术模式的重点，技术参数的把握水平有待提高，茶园水肥一体化技术模式的针对性和实用性有待提高。

3 茶园水肥一体化实施方案

3.1 灌溉方式的选择

我国南方和北方气候条件差异较大，茶树生长受地域影响较大。相关专家针对各地不同差异提出了茶园水肥一体化应用的不同方案，孙海伟等[2]着力研究了北方茶园专用水肥喷、滴灌一体化技术，提出了适用于北方茶园的水肥一体化模式。北方茶园多在山地丘陵地区，水资源稀缺、土壤条件差，不适合茶树的生长，对茶叶的优产造成一定的影响。因此，有很多温室茶园，将茶树生长所需要的光照、温湿度等配置好，以此来适应茶树的生长需求。

灌溉方式的选择对于茶树的生长来说尤为重要。喷灌和滴灌是大多数茶园的灌溉方式。喷灌灌溉面积大，因为喷灌可以控制喷水量和均匀性，避免了地面径流和深层渗漏损失的发生[3]，使水的利用率大大提高，还可以省力增产。但如果遇到较强的风力阻碍，就会降低喷灌的均匀性，影响喷灌效果。水肥一体化滴灌技术是根据传感器判断茶树对水的需求，通过低压管道系统与安装在毛管上的灌水器，控制水肥合理配置的灌溉方式，慢慢均匀地将水和肥料滴入茶根区土壤中。滴灌不会破坏土壤结构，有助于维持土壤中的水、肥料、气体和热量等指标。不会造成地表径流和深层渗漏，是一种经济而科学的灌溉方法，可以准确地将水输送到作物，并且损失低，以满足作物的生长需求，增加产量，省力节水。

3.2 灌溉制度的确定

作物需水量是农业用水的重要组成部分,作物水分田间消耗主要有叶面蒸腾、根间蒸发和深层渗漏,对于茶树来说,其生育对温湿度的要求较高,需水量较大,因此水的分布要与茶树生长的各阶段相适应。茶树需水量与当地气候密切相关,随着茶树树冠覆盖度的增加而提高,也与土壤水分有很大关系,但茶园土壤水分过多同样会对茶树造成不利影响。水肥一体化控制系统可结合彭曼—蒙特斯方程判断作物需水量,提出基于茶树的蒸发蒸腾量模型来更好地控制需水量,这是节水的关键。

3.3 水源

茶园实施水肥一体化的水源要求清洁、无污染,pH值呈中性,杂质少,含盐量低,不堵塞管道,如果不满足以上条件,需要对水源进行过滤。

4 前景展望

4.1 依托物联网大数据,打造智慧茶园

互联网、物联网、大数据等新一代网络技术的到来改变了人们的生产方式、生活方式、思维方式,传统产业面临着前所未有的挑战,为了更好地适应"互联网+"的发展潮流,包括茶产业在内的一些传统农业要形成智能化的生产管理模式,有效促进茶业的农业现代化。

4.2 建设生态茶园,扩大茶园面积

任何行业的发展都要注意保持生态环境,应在整合茶园,扩大茶园面积的基础上应用水肥一体化。

4.3 普及现代化茶园管理办法

在茶园中应用农业物联网智能系统,可以充分利用现有的节水设备,优化调度问题,提高节水节能效率,降低灌溉成本,提高灌溉质量,使灌溉施肥变得更加优化、科学、便捷。

4.4 因地制宜,选择适合本地的灌溉施肥种植方式

不同地区对种植茶树的要求不同,要创新适合本地的茶树水肥一体化模式,因地制宜,可在各类展会、宣传中向农民推广水肥一体化。

参考文献

[1] 苏火贵,郑靖雅,吴月德,等. 茶园水肥一体化技术应用及发展前景 [J]. 广东茶业,2015 (1): 38-40.

[2] 孙海伟,张虹,高红,等. 北方茶园专用水肥喷、滴灌一体化技术 [J]. 山东林业科技,2017,47 (5): 79-81.

[3] 张光旭,王捷,王宪,等. 茶园水肥一体化技术应用及发展前景探析 [J]. 南方农业,2018,12 (24): 50-51.

滴灌在樱桃种植区的应用

樱桃是一种经济价值很高的农作物，樱桃树不同生长时期对水分的供给要求均十分严格，适时适量的灌溉对增强樱桃树的根系活性和提高樱桃品质有着极大的作用。而滴灌技术是目前比较流行的节水灌溉方法，我们就滴灌技术在樱桃种植项目区的应用进行分析。

1 樱桃种植项目区基本情况

1.1 樱桃种植项目区地理位置情况

樱桃种植项目区位于山东省青岛市平度市，属于胶东半岛西部，地处北纬36°28′15″~37°02′46″、东经119°31′30″~120°19′13″，属暖温带东亚半湿润季风区大陆性气候。樱桃种植项目区年均气温为11.9℃，年均无霜期为195.5d，年均日照时间为2 700h，平均降水量为680mm。棕壤、褐土、潮土、盐土和砂姜黑土为樱桃种植项目区的主要土壤类型[1]。

1.2 樱桃种植项目区水资源基本情况

青岛市缺水是一个正常状态。根据水利局的统计，2018年青岛市全年降水量740.1mm，平均比2017年同期增加60.0mm，水库蓄水量为2.7亿 m^3，比2017年增加6 000多万 m^3，但水库蓄水量比历年同期减少2 800万 m^3 [2]。在现代青岛市的历史上，至少出现2次严重缺水现象，一次是1945—1948年，另一次是1981年。1981年，青岛市遭受了历史上罕见的严重干旱，除了少量的积水和水库存储容量外，其他水库大坝的水资源均已用尽。

2 樱桃的种植特点

樱桃 [Cerasus pseudocerasus（Lindl.）G. Don]，属于乔木，高2~6m。樱桃原产于加勒比海的热带美洲西印度群岛，所以也被称为西印度樱桃。其适合在热带和亚热带地区种植，是世界公认的"天然维生素C之王"和"生命之果"。樱桃在土壤营养匮乏的情况下具有耐受性，对土壤的要求不高，在弱酸弱碱性的土壤中可以正常生长，而且对温度的要求不高，在-20℃下仍可以生存。不同树龄的樱桃树与同时期不同生长阶段的樱桃树具有不同的需水量。幼树期，由于树体小，树枝和树叶数量少，蒸发量相应也小，所以对水分需求量不大；进入结果期后，枝叶数量变大，水的消耗量增加，因此对水的需求相应增加。樱桃喜欢阳光照射，抗旱能力强，耐水性较差。若在阴凉地种植，则不利于其生长；若在山区坡地种植，其根系耐水性较差，成活率不足50%，有的甚至在30%以下。

3 樱桃种植项目区应用滴灌技术的必要性

3.1 樱桃树的用水特点

樱桃树根系很浅，主根发育不充分，主要是由侧根斜发育生长。樱桃树的叶子比较大，水分不断蒸腾，若水分缺失时间过长，叶子就会萎缩。但是，如果樱桃种植项目区土壤中的含水量超过25%，樱桃根系就会因缺氧而使其活性降低，进而导致树木死亡。

3.2 水肥利用效率提高

利用滴灌技术，仅会灌溉作物根部的土壤，其他区域的土壤中水分含量很少。这不仅提高了水分利用率，还控制和抑制了杂草的生长。采用滴灌技术时，地表面不会出现积水，可以准确掌握灌溉的深度和面积。

3.3 基本不影响环境温湿度

滴灌技术使用的水资源很少，而且不是一次性灌注，因此对地温的直接影响很小。此外，滴灌是在作物根区附近进行浇水，地面蒸发的水量很小，大大减少了土壤蒸发的热量。因此，滴灌的地温一般高于传统的地面灌溉，因此作物生长快、成熟早。滴灌让水分和肥料直达作物的根部，可以使得其他地面和土壤水分减少，同时可以减少作物生长环境的湿度，从而抑制杂草和其他有害害虫的生长发育。

3.4 提高樱桃的品质

滴灌可以有效地为樱桃提供水分和肥料，为樱桃生长提供最合适的环境，从而可以大大提高樱桃的品质和产量，为农户带来更高的经济效益。

3.5 适应能力较强

滴灌滴头可在大水压环境下工作，滴头排水速度均匀，适用于樱桃园种植项目区的环境，也适用于各种土壤类型。

3.6 减少水费和人工费用支出

由于滴灌是在管道中进行的，可以减少水分的蒸发，并且不会减少由于树叶吸收的水量，也不会减少由于其他的土地面积吸取的水量，不会因为地面径流和土壤深层的渗透，使水量减少。总之，与喷灌、漫灌相比，滴灌可以减少用水量35%~75%。由于一次铺设后，可连续使用滴灌，只需在灌溉时打开开关，大大节省了人工费用，对降低用水费用和人工成本有很大的作用。

4 樱桃种植项目区滴灌系统的设计方案

4.1 确定灌溉系统工作参数

根据樱桃树生长期的需水特征，确定不同时期合理的灌水量。例如，在开花前浇水满足新叶舒展和花朵开放的水分需求；落花10~15d，在其硬核期进行充足浇水；采收前10~15d，进行一次充足的浇水，可有效增产25%左右。

4.2 水源的选择

樱桃种植项目区内有一口人工钻孔的地下水井，水质清洁无毒，pH值在5.5~8.5。另外，每种微量元素的含量低于农田灌溉水质基本控制项的标准值。

4.3 滴灌系统的布置

根据樱桃园种植项目区的情况，结合水源位置确定滴灌系统的首部枢纽的位置。选择合适的灌水器并使用双行平行的毛管，以方便农户管理，并减少资金投入。

4.4 施用肥的选择

为了保证樱桃品质良好，必须增加钾肥的施用量，并补充钙、硼、锌等微量元素，减少氮肥用量。在滴灌系统中，应选择易溶解、溶解快、杂质少的肥料，并根据樱桃树的生长需求合理搭配肥料。

5 结束语

在水资源短缺的背景下，平度市必须调整农业产业结构，发展高效节水农业。今后应向农户大力推广滴灌系统，不局限于一种作物，应开发多种作物的滴灌体系。

参考文献

[1] 郑欣.218省道建设工程项目建设方案选择与评价研究［D］.哈尔滨：哈尔滨工程大学，2014.

[2] 赵兴书．全年降水量740.1毫米青岛水利局长：缺水是常态［DB/OL］.（2018-12-14）［2019-03-25］.http：//m.sohu.com/a/281883285_ 178557.

电子元器件在智能家居领域中的应用发展

智能家居为人们提供舒适、安全、宜居、智能的工作生活环境，必将有长足的发展。概述了智能家居的市场现状，以及其中的元器件应用。基于智能家居中常用的元器件，展望了新型元器件的需求与发展方向。

随着生活水平的不断提高和社会的发展进步，人们对于家居环境提出了更高的要求。目前的家居生活工作环境中存在能量消耗过高、安全防卫措施落后、智能化程度低以及管理混乱等问题，进入云计算 4G 时代，家居生活工作环境正朝着舒适、健康、安全、便利、宜居的方向发展，这迫切需要家居智能化来保障并实现。

智能家居（Smart Home）以家居生活工作环境为平台，利用各种物联网技术，将节能照明、安全防范、网络家电等子系统有机整合为一体，为人们提供舒适、安全、宜居、智能的家庭生活环境。智能家居系统结构如图 1 所示。与传统家居系统比较，智能家居提供舒适宜居的工作生活环境，高效安全利用能源，优化人们的生活工作方式，把家居环境变成能动、智慧的工作生活工具，从而实现节能、低碳、环保[1-3]。

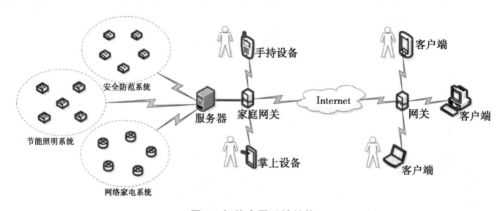

图 1　智能家居系统结构

1　智能家居市场现状

智能家居在我国经过市场培育期后，从 2012 年开始，市场发展提速，随着云计算、4G 技术应用的推广普及，物联网各相关技术的发展，以及手机平板等智能终端的普及与价格迅速下降，智能家居步入快速发展的通道。最新数据表明，2013 年全世界智能家居的市场规模近 7.5 万亿元，且市场规模在高速发展，我国的智能家居市场在 2020 年将会达到近 4 万亿元的规模。随着各种物联网技术应用的发展与推广，智能家居市场需求呈现上升态势，目前已进入理性发展时期，脱离了以前概念混乱与纷争的阶段，智能家居市场发展态势良好。我国涉及智能家居的企业越来越多，技术水平越来越高，同时不少国外智能家居企业进入中国市场，使智能家居市场竞争更加激烈。

中国已有不少企业涉及智能家居市场。清华同方的"e-Home"数字家园计划，该计划的核心是家居环境管理，重点是控制智能化和服务信息化，促进了智能家居的发展。青岛海尔推出了系列智能家居产品"U-Home"，以智能家电系统为载体，通过无线网络，实现各种家电产品、数字媒体以及智能家居系统的互联与管理，从功能上来分有 U-safe、U-play、U-shopping 与 U-service 等一系列覆盖安防、娱乐、服务等功能的家庭数字化解决方案。之前的上海世界博览会上展示的"让城市生活更美好"的关于智能家居的应用场景更加多样化。

从以上智能家居市场情况来看，随着各种物联网技术的飞速发展，科技人才将大量涌入智能家居行业，智能家居系统将更加智能化、平民化、自动化，更加低碳环保，从而进入千家万户的家庭生活。

2 智能家居中元器件应用

物联网、无线通信、NFC、MEMS 传感器、3D 手势控制、智能照明、医疗保健等高端技术可用于智能家居，比如智能手机、平板电脑等设备搭载可识别多种 3D 手势动作的 SMD 超声波传感器，在不用接触面板的情况下，也能识别多种手势动作，包括面板水平与垂直方向的动作。利用智能手机 NFC 功能，结合 NFC 新技术，成功实现小型化、智能化的同时，解决了金属外壳磁场耦合难题。

采用 ZigBee 技术，设计基于 Web 的网关来构建智能家居系统，实现在平板电脑、智能手机上调控、设置家居中的各种设备，同时可把无线通信技术、传感器技术等进行整合，实时监控智能家居中各家电情况，并进行远程控制。比如在电视机上，可安装红外线人感传感器，用来感应人的到来，根据情况自动调整空调风速，自动开启电视，自动开启关闭照明等设备，提供人性化的服务；还可利用 ZigBee、WiFi 产品构建智能家居系统，避免遥控器和布线的麻烦，在床上安装 MEMS 加速度传感器，用来感知人体的微小机械振动，只要躺在床上就能准确测出呼吸率、心率等一系列数据，并给出健康建议。

云计算在智能家居中必将得到广泛应用，云计算的海量存储能力与高速计算处理功能将使家居响应速度更快，更加安全便捷。随着 4G 牌照的发放，必将为智能家居行业创造巨大的机会，使智能化的概念大力推广，更加深入人心，进一步提高各类智能家电的数据传输速率与安全性，进一步提升用户的体验感。云计算与 4G 相关的元器件，比如 MEMS 振荡器、固态硬盘所涉及的元器件、ARM 芯片等必将在智能家居中得到广泛应用。

3 元器件需求与发展

智能家居为我们提供舒适、惬意、轻松、安全的工作生活环境，这些服务与功能的实现离不开智能家居环境中布置的各种传感器，常用的有下列几种。

温度传感器。温度测量仪器仪表的核心元器件，能感受温度并转化成数字或模拟信号输出，种类繁多。按传感器材料与电子元器件的特性可分为热电偶和热电阻两种类型，按测量方式来分，可分为接触式与非接触式两类。近几年随着物联网技术应用的推广，温度传感器件作为传感器中重要的一类，占据了市场 40% 以上的需求量。常用的温度计有电阻式、双金属、压力式、玻璃液体、温差电偶和热敏电阻等，在智能家居以及工农商业等部门广泛应用。

湿度传感器。湿敏元器件是最常用的湿度传感器件，主要有电阻式和电容式两类。湿敏电阻是把感湿材料制作的一层膜覆盖在基片上，电阻值与电阻率随着空气中水蒸气吸附在膜上而发生改变，利用该特征来测量湿度。常用的品牌有：施耐德、三菱、宝力马、西门子、松下、日本神荣、美国 Honeywell 和德国的 HLP 等。

火焰传感器。通过红外线接收管来检测光线的亮度，将火焰亮度转换为电平信号，该器件对火焰非常敏感。不同燃烧物的火焰波长与辐射强度有差别，具有离散光谱气体辐射和连续光谱固体辐射，对应火焰温度的近红外波长域和紫红外光域具有很大的辐射强度，可分为远红外火焰和紫外火焰传感器。

气体传感器。对气体样品（如天然气、沼气）进行杂质去除，制冷干燥处理，将某种气体体积分数转换成对应电信号。气体传感元器件可分为：红外线、半导体、热导式与电化学等方式。气体传感器在二氧化碳、有毒、易爆、易燃等气体检测领域应用极其广泛，广泛应用于智能家居等物联网环境中，成为环境检测的必备产品与有力助手。

传感器的发展，需要不断开发新型敏感材料、创新制作工艺。比如对于 Fe_2O_3、SiO_2 等比较成熟的半导体材料，还可采用化学修饰改性的办法，对敏感膜材料改性处理、进行掺杂以及材料的表面修饰等，并对成膜制作工艺优化与改进，以实现新的性能新的应用。研究新型敏感膜材料，对于资源丰富的混合型或复合型高分子敏感材料、半导体敏感材料等新型材料，应着重改善新材料的稳定性、提供灵敏度以及兼容性。

元器件的创新与智能化是智能家居建设的基础。采取微结构化设计与先进的加工制作技术，选取石英、陶瓷、硅等晶体材料，采用某些新效应和传统原理方法，研制新型元器件。比如微生物元器件、仿生元器件、石英谐振式元器件等。随着新工艺与新材料技术的运用，元器件更加微型化，功能多样化，性能更加完善，并且价格更低，稳定性提高。纳米材料以及超材料的研制与广泛应用使元器件的集成化与智能化成为可能，可结合智能技术、仿生技术、信号处理技术、微电子技术等多学科交叉来提高元器件的智能化，多功能全数字智能化元器件将是元器件发展的重要方向。

4 结束语

智能家居是与人们日常工作生活密切相关的智能应用，有可能会成为智慧城市最大的亮点，市场发展潜力巨大。随着物联网等概念的落地，智慧深度的不断切入，智能技术向纵深发展，应用会更加广泛，势必带动智能型传感器、甚至整个电子元器件产业链的大发展。

参考文献

[1] 马乐,燕炜,姜思羽,等.基于物联网体系结构的智能家居系统设计[J].北京师范大学学报,2013,49(5):458-461.

[2] 严萍,张兴敢,柏业超,等.基于物联网技术的智能家居系统[J].南京大学学报,2012,48(1):26-32.

[3] 俞文俊,凌志浩.一种物联网智能家居系统的研究[J].自动化仪表,2011,32(8):56-59.

基于"互联网+"的农业信息传播应用现状与发展趋势

进入21世纪后,随着经济的发展和移动互联网的兴起,新的商业革命也在悄悄地茁壮成长,电子商务和互网+思维已经渐渐融入国人的日常生活中。农村市场的潜在巨大发展潜力使得互联网+农业展现出蓬勃发展的势头。"互联网+"模式主导的农业信息传播方式已成为我国农业发展的现实需求,为其发展和运用迎来的前所未有的契机。本文从"互联网+"、农业传播、网络传播、农业网络传播存在的问题、基于"互联网+"的农业传播应用近况与发展趋向及建议进行分析。

在我国广大的农村地区,大多数村民过着分散且封闭的生活,报纸和杂志等媒体对农村的影响缓慢随着广播电视和网络以及一些新媒体技术的普及,农民也通过现代化手段获取了大量信息,和过去相比有了明显的改善。但农民被动的受众地位,没有从根本上改变,"知识沟"问题依然严重。利用传统的通信手段,想从根本上解决这些问题是非常困难的。互联网+农业的诞生,为农业信息传播提供了一个新的有效途径。

1 "互联网+"和"互联网+农业信息传播"

"互联网+"是指通过操纵互联网平台、互联网讯息或通讯技术等互联网产业和传统产业相融合,在新的范围内产生的一种新的模式。通俗来说,"互联网+"就是互联网与各个传统行业利用信息通讯技术和互联网平台进行的深度融合。例如:常用微信和QQ就是互联网+通信而得到的即时通信;而淘宝和京东就是互联网+零售得到的电子商务。解决什么是首先要解决一个重要问题。于农业传播和网络传播,2005年1月赵晓春出版的《农业传播学》一书中提到:农业传播是从信息内容的角度对传播进行的划分[1]。如果说传播的本质是信息的传递的话,那么农业传播可以界定为社会领域内的涉农信息的传递。其中涉农信息又包括农业生产信息、农村农民信息、有关农业生产资料生产与流动的信息、农业科技信息。网络传播包含在传播学当中,网络传播是以计算机通信网络为基础,进行信息传递、交流和使用,从而达到其社会文化传播目的的传播形式。它是当代信息革命的产物。网络信息来源的广度和开放性,确保了网络传播可以在第一时间、第一速度获取信息资料,并迅速且及时地传递给观众[2]。

综上所述,所谓"互联网+农业信息传播"就是农业信息网络传播,即是指通过互联网的农业信息传播形式[3]数字形式的农业信息网络存储在网络云端,通过高速互联网通信,利用PC客户端或可移动互联网终端设备进行阅读农业相关内容[4]。

2 农业网络传播存在的问题

在我国，无论是在农业信息资源开发还是在利用上都依然和发达国家有很大差距，必须理性客观的看待这些问题，才能为农业信息网络传播提出有力建议。农业网络传播中主要存在以下几个方面的问题：第一，政府宏观调控作用在农业信息网络传播建设中得到没有充分运用；第二，社会方面涉及有关农业信息的网站所发布信息的适用性低，质量差；第三，农业信息收集渠道有问题；第四，农村区域基层咨询信息服务滞后性；第五，缺乏有效整合信息资源开发；第六，教育培养方面，缺乏高素质人才[5]。

3 "互联网+"农业信息传播应用近况

我国的农业计算机网络年龄虽小但是成长迅速。1993年我国开始筹备建立农业信息网，现在所有的农业院校、地方农科院都有自己的网站。目前，随着我国农业信息网络的建设与发展，从根本上改变传统信息的获取方式，使信息的传播速度和利用效率显著提高。从网上，农民能够轻松获得他们需要的科学技术信息和市场信息，正应了那句话"不出门便知天下事"。农民朋友可以随时随地收取相关专家建议，获取的渠道也变得丰富起来[6]。

3.1 农业论坛社区 App

"互联网+农业信息传播"农业论坛社区包括一些手机 App 软件或网站，农民朋友只需要上网或登录手机 App 即可有自己的一个 ID，可以在论坛上发表自己的问题、回答他人问题、表明自己观点，是一种互联网服务。农民在遇到问题时，可以通过网络社区把自己的问题放在网上，其他 App 用户看到后，只要有自己的答案，就可以在他的问题下面回答。论坛会定期邀请农业方面专家进行网络现场直播，解答问题和谈论交流。使用手机 App 的好处众所周知，方便快捷地使用网络、随时随地想看就看。

3.2 农业网络电子阅览室

现如今，几乎每个村都有自己的电子阅览室，农民朋友可以通过电脑上网获取自己想了解的农业知识，并且电子阅览室的电脑都会装有农业类的专用软件方便用户更好的搜索、查询、使用。

3.3 远程教育

由于农村农民知识文化水平普遍不高。专业成效的大规模培训是一项重要任务。中国农村远程教育网的创办，主要为了加强教育和农民的组织管理，协调农民科技教育培训资源，使农村教育培训按部就班高效进行。目前，全国已有统一域名互联网网站的省级农广校60个。

3.4 农业资讯微信公众号

现如今，几乎每一个人都有微信号并且每天都在使用，对于农民来说通过手机来获取信息无疑是最快捷、最及时的方法。微信公众号可以定期推送农业相关资讯，并同时获得农民的反馈和建议。大大的节省了人力物力财力，微信推广方式相对于传统的媒介推广更方便快捷可行性高。另外，但在保障日常工作的同时也要加强微信公众号的监管，不要成为不能够定时推送信息的摆设账号，而没有起到应当发挥的实际作用。

4 "互联网+"农业信息传播发展趋势和建议

全球已从工业时代逐步走向信息时代。信息化、网络化浪潮正在席卷全球。如今由于"互联网+"传统行业热潮的兴起,现代信息技术的一部分通过网络广泛应用到农业领域,网络信息技术表现出不可替代的重要性。一定要把握时机迎接挑战,争取在农业信息化研究成长进程中占据有利位置。目前,我国传统农业正处在重要的转型期,也是把握机会加快农业信息化发展的关键时期。所以,一定要抓住机遇快速发展[7]。

4.1 将"三农问题"作为发展重点

应将"三农问题"作为发展的重中之重,是因为我国是一个农业大国,全国上上下下、老老少少、男男女女有四分之三都是农村户口。国家想要安定、改革和发展必须要把"农业、农村、农民"作为依靠和基础。因此,任何时候都不能忽视和放松"三农"工作。解决农民的问题是国家长远发展的重要任务。相关部门理应出台相关扶持政策,这样既可以在社会上得到广泛关注又可以发挥其主导作用[8]。

4.2 加强网站管理,完善奖励机制

网络监管部门要加强各种农业类网站的管理并且提出奖励机制,为农民朋友满意的、优秀的网站和相关网络产品颁发奖品和荣誉。这样既可以起到积极鼓励作用也可以增加人们对"互联网+农业"的认识,从而配合活动顺利开展。

4.3 增加基层服务人员的监管

调查农业信息和收集农业信息的渠道必须严格核实,以保证信息的真实性与有效性。对于一些虚假、错误、片面信息充分利用网络的即时性高效性,第一时间及时向农民朋友通知,以免造成不必要的恶劣影响和财产利益损失。

4.4 坚持与党的基本路线相结合,理论与实践相结合

安排专业农业信息人士下乡到基层调研农民朋友最关心、最根本、最直接的利益问题。只有明白问题的所在,精准对症下药才能从真正意义上解决问题。因为实践才是检验真理的唯一标准,要不断在实践中检验真理、发展真理。

4.5 充分运用主流的新媒体传播方式

充分利用"互联网+"的大数据时代背景,适当运用当下主流的新媒体传播方式,要求各个组织,各个部门积极配合,整合信息资源开发利用。我国是一个农业大国,虽然有一些传统生产经验,但在农业科学技术方面,发达国家领先很多。必须把农业信息技术作为重点,克服困难,大力开展电子商务在农业上的实践。

4.6 加大资金投入,加强高素质人才培养

加大对农业科研院校以及科研部门的投人,改善农民知识水平加强高素质人才培养,对为农业传播方面有杰出贡献的人才设立特殊奖项,加强教学硬件软件设施建设,提高农业网络传播效果,要更重视对农村受众文化素质的培养,只有农村整体文化水平提高,才能从根本上提高传播效果和提升农民辨别不良信息的能力[9]。

5 结束语

中国农业网络传播有其与众不同的优势,但也不能忽略其劣势。由于信息社会的光

速发展,在"互联网+"热潮的巨大影响力下,农业网络传播也逐步显现出前所未有的美好光明前景。就农业而言,网络传播为农业发展带来了全新的机遇,促使利用新的传播方法加快推动了农业信息化的进程。同时,这对我国高等农业院校的教育也提出了更高的要求,教师科研人员队伍也要不断加强自身建设。这是一个关于农业的构想,互联网+的热潮无疑为实现这个构想创造了机遇。现在作为农业信息化研究领域的一员,应努力从实践中取得经验勤于调研多做总结,为解决三农问题不遗余力的贡献出自己最大的力量。

参考文献

[1] 赵晓春,董成双,徐鹏民. 农业传播学 [M]. 北京:中国传媒大学出版社,2005.

[2] 赵璞. 农业网络平台在农业生产中的传播功效 [J]. 黑龙江农业科学,2011 (7):138-140.

[3] 胡凌. 网络安全、隐私与互联网的未来 [J]. 中外法学,2012 (2):379-394.

[4] 曲桐凤. 网络传播与农业信息化 [J]. 商场现代化,2007 (12):151-152.

[5] 傅颢文. 沈北新区农业信息化发展对策研究 [D]. 北京:中国农业科学院,2013.

[6] 张琪. 信息网络化对农业技术推广的影响及其对策研究 [J]. 现代经济信息,2015 (22):317.

[7] 赵玉姝. 农户分化背景下农业技术推广机制优化研究 [D]. 青岛:中国海洋大学,2014.

[8] 王勇. 河南省农业信息化水平评价与提升对策研究 [D]. 郑州:河南农业大学,2013.

[9] 王世伟. 论信息安全、网络安全、网络空间安全 [J]. 中国图书馆学报,2015 (5):72-84.

基于彭曼-蒙特斯公式的温室茶树腾发量计算模型研究

作物需水量是确定灌溉用水定额的基础，其关键参数是作物的蒸腾蒸发量（腾发量）。以 Penman-Monteith 方程为基础，借鉴 P-M 温室修正式的计算方法，提出了基于常规气象数据和茶树生长发育指标的温室茶树蒸腾蒸发模型 ET_0（Tea）。在试验期间，选取茶园温室 2016 年 3 月 10 日至 4 月 10 日的气象数据，利用 P-M 温室修正式和温室茶树蒸腾蒸发模型 ET_0（Tea）对作物腾发量进行逐日统计，并用水量平衡原理验证，结果表明 ET_0（Tea）与实测值的变化趋势较为一致，误差相对较小，且在晴天条件下比阴天效果更好。我们提出的 ET_0（Tea）计算精度较高，在理论和实践上均具有较好的可行性，可作为北方温室茶树灌溉决策的重要依据。

茶树原产于我国西南部湿润多雨的原始森林中，在长期的生长发育进化过程中，茶树形成了喜温、喜湿、耐阴的生活习性[1]。适宜茶树生长的年降水量约为 1 500mm，生长期间的月降水量最好在 100mm 以上。20 世纪 60 年代我国开始实施"南茶北引"工程，但北方地区降水量相对较少，冬春季节室外温度较低，茶树需温室栽培。由于茶树在生长期间需水量较大，合理确定茶树灌溉用水定额，不仅可以节约水资源，还可以为北方温室茶树灌溉管理提供一定的参考依据[2,3]。

作物需水量是确定灌溉用水定额的基础，其关键参数是作物的蒸腾蒸发量（腾发量）。对于温室作物的腾发量，目前国内还没有形成一套标准的计算公式，主要是借助一些气象学的方法进行估算。由于温室小气候环境与露天环境的水热运移模式有很大不同，Boulard[4]、Demrati[5]等提出利用温室能量平衡和 Penman-Monteith 方程（P-M 方程），推导出基于室内气象数据的温室作物蒸腾量计算模型；王健、陈新明等[6,7]从温室内总辐射和风速因子入手，提出适于参考作物腾发量的 P-M 温室修正式；刘浩等[8]建立了包含气象数据、叶面积指数和冠层高度为主要参数的日光温室番茄蒸腾量估算模型。但是针对茶树这一经济作物，其在温室内的蒸腾蒸发模型的研究目前还相对较少。

我们提出了适于北方地区温室茶树腾发量的计算模型 ET_0（Tea），并用水量平衡原理进行验证。本研究旨在为当地温室茶树生产与实践提供指导，同时为设施农业灌溉技术提供一定的理论依据与技术支持。

1 作物腾发量的计算方法

1.1 Penman-Monteith 方程

在 1998 年，联合国粮农组织 FAO（Food and Agriculture Organization）推荐将 P-M 方程作为计算 ET_0 的首选方法[9]。其方程为：

$$ET_0 = \frac{0.408\Delta(R_n - G) + \gamma \dfrac{900u_2(e_a - e_d)}{T + 273}}{\Delta + \gamma(1 + 0.34u_2)} \tag{1}$$

式中：ET_0 为参考作物蒸腾蒸发量，mm/d；Δ 为饱和水汽压曲线斜率，kPa/℃；R_n 和 G 分别为地表净辐射通量和土壤热通量，MJ/（m²·d）；γ 为干湿表常数 0.064 6 kPa/℃；u_2 为 2m 高度处风速，m/s；e_a 和 e_d 分别为饱和水汽压和实际水汽压，kPa；T 为 2m 高度处的平均温度，℃。

式（1）为组合方程，可分为两部分，前一部分为辐射项（ET_{rad}），后一部分为空气动力学项（ET_{aero}）。

1.2 P-M 温室修正式

由式（1）可知，当温室内风速 $u_2 = 0$ 时，空气动力学项（ET_{aero}）为 0，而实际上此时蒸发和热量输送仍然是存在的，这显然与水汽扩散理论相矛盾。根据相关研究[10,11]，王健、陈新明等[6,7]对 P-M 方程中与风速有关的空气动力学项进行修正，并以 FAO（1998）给出的关于 ET_0 的定义[9]，假设作物高度 $h_c = 0.12$m，推导出适于温室 ET_0 计算的修正式：

$$ET_0(P-M\,修正式) = \frac{0.408\Delta(R_n - G) + \gamma \dfrac{1\,713(e_a - e_d)}{T + 273}}{\Delta + 1.64\gamma} \tag{2}$$

1.3 温室茶树蒸腾蒸发模型 ET_0（Tea）

在 P-M 公式和 P-M 温室修正式中假设作物高度是 $h_c = 0.12$m，而实际作物高度是一个时间变量，所以在计算不同作物腾发量时不能忽略作物实际高度这一参数。为了提高计算温室实际作物腾发量的精度，本文选取温室茶树作为实际作物进行试验，在公式中引入作物高度 h_t 这一参数。结合 P-M 方程和 P-M 温室修正式，推导出适于温室茶树的 ET_0（Tea）计算方法：

$$ET_0(Tea) = \frac{0.408\Delta(R_n - G) + \gamma \dfrac{39\,611(e_a - e_d)}{(T + 273)\left[\ln\left(\dfrac{Z - 0.64h_t}{0.13h_t}\right)\right]^2}}{\Delta + \gamma + \dfrac{15\gamma}{\left[\ln\left(\dfrac{Z - 0.64h_t}{0.13h_t}\right)\right]^2}} \tag{3}$$

式中：h_t 为茶树冠层高度，m；其他参数意义同前。

根据汪小旵等[12]的方法计算 e_a、e_d、Δ：

$$e_a = 0.610\,7\,exp[17.4T/(239 + T)] \tag{4}$$

$$\Delta = 4\,158.6e_a T/(T+239)^2 \tag{5}$$

$$RH = \frac{e_d}{e_a}100\% \tag{6}$$

式中：RH 为温室内空气相对湿度，%。

根据强小嫚[13]的研究可知土壤热通量（G）和地表净辐射通量（R_n）具有很好的线性关系。

$$白天：G = 0.1R_n \tag{7}$$

$$夜晚：G = 0.5R_n \tag{8}$$

1.4 作物系数 K_c 的确定

根据FAO推荐，在充分供水条件下作物需水量计算公式如下：

$$ET_c = K_c ET_0 \tag{9}$$

式中：ET_c 为作物需水量（充分供水条件下作物腾发量，mm）；K_c 为作物系数。FAO-56作物系数表中列出了不同作物 K_{cini}、K_{cmid}、K_{cend} 的典型值，Allen指出 K_c 的变化主要由其具体的作物特征来决定，受气象条件的影响有限[9]。作物系数最重要的影响因素是作物自身的生理生态指标，例如作物种类、品种、生育阶段以及冠层状况等[14]。而近年来，我国学者[15,16]研究发现，温室作物系数比大田环境下有变小的趋势，所以在本文中选取茶树生长阶段作物系数的较低值（K_c = 0.95）。

2 试验设计

2.1 试验概况

试验于2015年11月15日至2016年4月15日在青岛农业大学试验基地茶园温室内进行。试验基地地处青岛市城阳区，东经120°23′，北纬36°19′，海拔54.88m；属暖温带季风大陆性气候，年平均气温12.6℃，1月最低，月平均气温-2℃，8月最高，月平均气温25.7℃；降水主要集中在夏季，年平均降水量700mm左右。

试验温室内的茶树品种为"黄山种"[17]，种植数量2 000余株，高度范围在0.25~0.45m，计算中取其平均值0.35m。此茶园温室为南北走向，长38m，宽12m，顶高2.5m，覆盖无滴聚乙烯薄膜。试验地土壤类型为沙质壤土，地下水埋深大于5m。在每行茶树上方0.5m处均安置一条喷雾灌溉带，喷头间距1.5m；下方两侧各安置一条滴灌带，滴孔间距0.2m，灌水量由数字水表计量。温室内安装有2台小型自动气象站，高度2m，编号分别为0713（S）、9711（N）。

2.2 试验方法

茶园温室内部环境参数，采用山东省计算机中心研制的智慧农业小型气象站0713（S）和9711（N）自动采集，采集的项目包括温室内南北端的温度 T（℃）、相对湿度 RH（%）及土壤含水量 W（%）（近地表层0~20cm）、光照度 L 和土壤热通量 G（W/m²）。茶树新梢一年中有三次生长和休止，全年可以发生4~6轮新梢[18]，试验选取两轮春梢生长期数据进行分析，具体时间为2016年3月10日至4月10日每天的10:00—17:00（7h），系统设定每1h自动采集1组数据，进行自动存储，可通过计算机或手机平台读出。

2.3 利用实测值验证温室茶树蒸腾蒸发模型

为了验证温室茶树蒸腾蒸发模型的可靠性及模拟精度，本试验采用水量平衡法计算茶树需水量，其结果与P-M温室修正式、ET_0（Tea）所求得的作物腾发量进行对比分析。根据潘永安等[19]对温室作物实际耗水量的分析，其水量平衡方程为：

$$ET_{ca} = I + ASW \tag{10}$$

式中：ET_{ca} 为计算时段内的实测腾发量，mm；I 为计算时段内的灌水量，mm；ASW 为土壤含水量的变化量，mm。

3 结果与分析

在本次试验期间内以茶树发生轮次为周期及阴天和晴天作为典型日期，对试验结果进行分析与评价。

3.1 利用 ET_0（Tea）计算作物腾发量的逐日变化

利用 P-M 温室修正式和温室茶树蒸腾蒸发模型 ET_0（Tea）对茶园温室 2016 年 3 月 10 日至 4 月 10 日的作物腾发量进行逐日统计，其变化过程如图 1 所示。

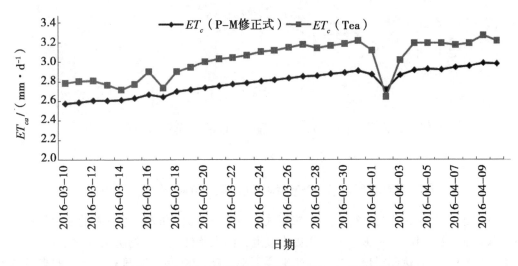

图 1 ET_c（P-M 修正式）及 ET_c（Tea）的逐日变化

从图 1 可以看出，2 种方法计算出的 ET_c 值具有相同的变化趋势，且总体上看，3 月至 4 月茶树腾发量随气温的变暖呈现升高趋势；P-M 温室修正式的计算值几乎均小于温室茶树蒸腾蒸发模型的计算值，这是由于 P-M 温室修正式中参考作物为苜蓿，在计算式中假设植株高度 $h_c = 0.12\mathrm{m}$，远低于"黄山种"茶树，且总叶数量和总叶面积同样小于"黄山种"茶树；在 3 月 13 日、3 月 17 日和 4 月 2 日的蒸腾蒸发量均出现不同程度的下降，推测其原因是这三天的天气状况均是阴天，且 3 月 17 日伴有雾出现，4 月 2 日阴转小雨，太阳辐射遭到不同程度的削弱。

3.2 利用 ET_0（Tea）计算作物腾发量的日变化

为了检验温室茶树蒸腾蒸发模型的计算精度，选取试验时间段内具有代表性的 2 天，3 月 13 日（阴天）和 4 月 5 日（晴天）进行分析，由于清晨温室内湿气较大，作物腾发基本从 9∶00 左右开始，所以选择当天 10∶00—17∶00（7h）的实测数据绘制图 2，利用 P-M 温室修正式和温室茶树蒸腾蒸发模型 ET_0（Tea）对作物 ET_c 进行计算，两种方法的计算结果与实测值进行对比分析（图 3、图 4）。

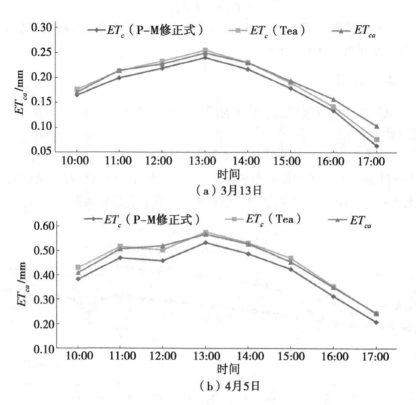

图2 阴天和晴天 ET_c（P-M 修正式）、ET_c（Tea）及 ET_{ca} 的变化

从图 2（a）可以看出 ET_c（P-M 修正式）、ET_c（Tea）和 ET_{ca} 这三者的变化趋势基本一致，而 ET_c（Tea）和 ET_{ca} 的数值更为接近，阴天条件下，作物腾发从 10：00 一直增大，13：00 左右达到峰值，之后开始平缓下降，16：00 后下降显著；但总的来说试验时间段内蒸腾蒸发量的变化幅度不是很大。从图 2（b）可以看出，晴天条件下 ET_c（P-M）、ET_c（Tea）和 ET_{ca} 这三者的变化趋势也基本一致，但试验时间段内腾发量明显高于阴天，ET_c（Tea）和 ET_c 重合度也更好，峰值出现在 13：00 左右。

图 3（a）和图 3（b），分别为 3 月 13 日（阴天）由 P-M 温室修正式求得的 ET_c（P-M 修正式）和温室茶树蒸腾蒸发模型求得的 ET_c（Tea）与实测值 ET_{ca} 的比较，其结果表明温室茶树蒸腾蒸发模型与实测值的变化趋势较为一致，相对误差较小，计算精度较高。

由图 4（a）和图 4（b）可以看出，在晴天条件下，整体变化趋势与阴天一致，温室茶树蒸腾蒸发模型和 P-M 温室修正式计算所得结果与实测值的相关性均都高于阴天；其中，温室茶树蒸腾蒸发模型与实测值相关性仍高于 P-M 温室修正式。

就当地温室茶园而言，在不同天气条件下，利用温室茶树蒸腾蒸发模型计算的作物腾发量与实测值相近，P-M 温室修正式的计算结果则偏小。造成偏差的原因主要是不同植物间生长发育状况不同，茶树的平均高度、总叶面积均大于参考作物苜蓿，所以蒸腾蒸发量较大。温室茶树蒸腾蒸发模型 ET_0（Tea）能够较准确的计算出温室茶树的腾

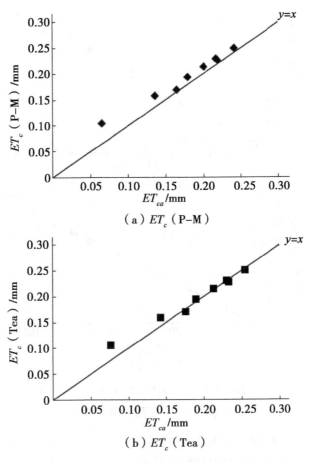

图3 阴天（3月13日）ET_c（P-M修正式）与ET_{ca}、ET_c（Tea）与ET_{ca}的关系曲线

发量，且晴天比阴天的应用效果更好。

4 结论与讨论

作物蒸腾蒸发量的大小受作物本身生理过程和环境条件（气象条件和土壤水分状况）的综合影响，此外还受农业栽培技术、作物品种及生长发育状况、灌溉排水措施等因素的影响[20,22]。同时，温室环境与露天环境特征有很大差异，P-M方程不宜直接应用于温室环境下作物蒸腾蒸发量的计算。本文以P-M方程为基础，结合前人的研究论证，借鉴P-M温室修正式的计算方法，提出了基于常规气象数据和茶树生长发育指标的温室茶树蒸腾蒸发模型ET_0（Tea），在理论上和实践上均具有可行性，可作为北方温室茶树灌溉决策的重要依据。

温室内微气候环境极为复杂，对于没有通风系统的简易温室大棚，假定风速为零，计算结果相对理想[7,19]；但对于有通风设备的温室而言，温室茶树蒸腾蒸发模型计算结果可能存在一定的误差。在验证ET_0（Tea）时，我们使用了水量平衡法，如果出现深

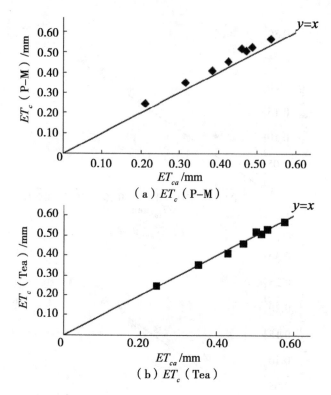

图4 晴天（4月5日）ET_c（P-M 修正式）与 ET_{ca}、ET_c（Tea）与 ET_{ca} 的关系曲线

层渗透或径流量较大的情况，该方法的应用会受到很大的限制，在以后的研究中还有待进一步通过作物试验确定这些误差的影响因子。因而，今后应加强作物腾发物理机制和生理机制在温室环境下的深入研究，或在原有模型的基础上校正相关参数，使模型模拟结果更接近真值。

参考文献

[1] 陈学林. 现代茶叶产业技术 [M]. 北京：中国农业大学出版社，2014.

[2] Dnyaneshwar N W, Manish G. Intelligent Drip Irrigation System [J]. International Journal of Engineering Sciences and Research Technology, 2014, 3 (5): 170-174.

[3] Yang Y T, Shang S H, Jiang L. Remote Sensing Temporal and Spatial Patterns of Evapotranspiration and the Responses to Water Management in a Large Irrigation District of North China [J]. Agricultural and Forest Meteorology, 2012, 164: 112-122.

[4] Boulard T, Wang S. Greenhouse crop transpiration simulation from external climate conditions [J]. Agricultural and Forest Meteorology, 2000, 100 (1): 25-34.

[5] Demrati H, Boulard T, Fatnassi H, et al. Microclimate and transpiration of a greenhouse banana crop [J]. Biosystems Engineering, 2007, 98 (1): 66-78.

[6] 王健, 蔡焕杰, 李红星, 等. 日光温室作物蒸发蒸腾量的计算方法研究及其评价 [J]. 灌溉排水学报, 2006, 25 (6): 11-14.

[7] 陈新明, 蔡焕杰, 李红星, 等. 温室大棚内作物蒸发蒸腾量计算 [J]. 应用生态学报, 2007, 18 (2): 317-321.

[8] 刘浩, 段爱旺, 孙景生, 等. 基于 Penman-Monteith 方程的日光温室番茄蒸腾量估算模型 [J]. 农业工程学报, 2011, 27 (9): 208-213.

[9] Allen R G, Pereira L S, Raes D S, et al. Crop evapotranspiration: Guidelines for Computing Crop Water Requirements [M]. Rome: FAO Irrigation and Drainage Paper No. 56, 1998.

[10] 左大康, 谢贤群. 农田气象研究 [M]. 北京: 气象出版社, 1991.

[11] 水利部国际合作司, 水利部农村水利司, 等. 美国国家灌溉工程手册 [M]. 北京: 中国水利水电出版社, 1998.

[12] 汪小旵, 丁为民, 罗卫红, 等. 长江中下游地区夏季温室黄瓜冠层温度模拟与分析研究 [J]. 农业工程学报, 2007, 23 (4): 196-199.

[13] 强小嫚. ET_0 计算公式适用性评价及作物生理指标与蒸发蒸腾量关系的研究 [D]. 陕西杨凌: 西北农林科技大学, 2008.

[14] 叶澜涛, 彭世彰, 张瑞美, 等. 温室灌溉西瓜作物系数的计算及模拟 [J]. 水利水电科技进展, 2007, 27 (5): 23-25.

[15] 胡永翔, 张莹, 李援农. 陕西关中地区温室枣树作物系数试验研究 [J]. 中国农村水利水电, 2015, (10): 19-22.

[16] 邱让建, 杜太生, 陈任强. 应用双作物系数模型估算温室番茄耗水量 [J]. 2015, 46 (6): 678-686.

[17] 王玉, 张金霞, 丁兆堂. "黄山种" 自然杂交后代叶片解剖结构变异特性研究 [J]. 中国农学通报, 2012, 28 (10): 209-212.

[18] 骆耀平. 茶树栽培学 [M]. 北京: 中国农业出版社, 2008.

[19] 潘永安, 范兴科. 气象资料缺测时 Penman-Monteith 温室修正式的应用 [J]. 西北农林科技大学学报, 2015, 43 (1): 117-124.

[20] 段爱旺. 一种可以直接测定蒸腾速率的仪器——茎流计 [J]. 排水灌溉, 1995, 4 (3): 44-47.

[21] 张祎, 汪小旵, 李聪, 等. 基于作物蒸散量模型的智能化滴灌控制系统研究 [J]. 节水灌溉, 2011, (12): 33-36.

[22] Gao G L, Zhang X Y, Yu T F, et al. Comparison of three evapotranspiration models with eddy covariance measurements for a Populus euphratica Oliv. forest in an arid region of northwestern China [J]. Journal of Arid Land, 2016, 8 (1): 146-156.

基于物联网的水肥精准配比调控技术

首先分析了当前我国农业发展，阐述了水肥一体化技术和农业物联网技术，研究了物联网在农业灌溉中的应用。

传统农业怎么改造升级为智慧农业，这是我国作为农业大国可持续发展的一项长期战略性任务，也是当前我国新旧动能转换的重要课题，更是乡村振兴战略提出后，中国农业与农村面对的机遇与挑战。水肥一体化技术最简单地来讲就是将可溶性肥料与水混合后然后进行施肥。智能水肥一体化技术就是在普通水肥一体化技术基础上以物联网为平台进行水肥精确配比，然后在进行施肥。主要是将互联网技术、电子信息通信技术、自动控制技术进行有机结合，再加入一些专门专家以及当地"土专家"的知识库来辅助，使它能够进行科学的、准确的、有效的进行水肥灌溉。这会大大减少水资源以及肥料的浪费，同时也会减少人工肥用的投入，并且有利于生态环境的保护和农业可持续发展。

1 当前我国农业发展的分析

1.1 农业用水资源极度短缺

我国农田灌溉水有效利用系数远低于 0.7~0.8 的世界先进水平；单位用水的粮食产量不足 $1.2kg/m^3$，而世界先进水平为 $2kg/m^3$ 左右[1]。因此我国水资源总体上极度短缺，进而农业用水就更加的匮乏。当前我国农业的灌溉方式主要还是以传统的农业灌溉为主，这种粗犷的灌溉方式不仅浪费大量的水资源而且还难以满足农作物本身的生长需要，达不到增量增产的效果。当前我国水资源面临的形势十分严峻，水资源短缺、水污染严重、水生态环境恶化等问题日益突出，在农业灌溉方面存在灌溉设备老化，灌溉方式粗放，节水等一系列技术尚未普及，同时人们节水意识也不强，这已经成为制约社会可持续发展的主要瓶颈。因此，改进农业灌溉方式，发展现代农业就显得极其具有战略意义。

1.2 农业施肥基本情况

化肥对于农作物的生长起着不可或缺的作用，是极其重要的农业生产原料，对农业的发展有着至关重要的作用，但是由于施肥的结构以及方法的不合理造成了大量的肥料浪费，利用效率低下。在化肥的使用量上我国也远超其他发达国家，根据中商情报网数据显示：我国单位耕地化肥消费量远超世界平均水平（图1），那么很明显就可以看出我国单位耕地化肥消费量远远超于世界的平均水平。所以当前需要研究解决精确高效的施肥技术，掌握科学的灌溉施肥技术，基于物联网云平台的方式，改变以往的不精确施肥方式，降低肥料的浪费量[2]。

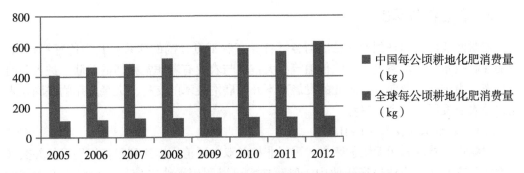

图 1 我国单位耕地化肥消费量远超世界平均水平

2 水肥一体化技术

传统的田间施肥就是通过人工手撒化肥，这种方法的特点是简单、低效、浪费、易污染。简易的水肥一体化是将可溶性的固体肥料注入低压灌水管路，利用文丘里管按照滴灌施肥规律进行施肥，这种方法的缺点是需要人工配肥，泵打，需要亲临现场进行管理操作，比较耗时以及消耗人力资源。随着时代的发展，节水灌溉技术使用的范围正在逐渐地扩大，其中喷灌、微灌面积和低压管灌面积也逐年扩大（表1）。智能的水肥一体化技术就是将灌溉和施肥结合于一体的新型精确进行施肥与灌溉的技术。该技术借助压力灌溉系统，将可溶性固体肥料或液体肥料配对而成的肥液与灌溉水一起，均匀、准确地输送到作物根部土壤，并可按照作物生长需求，进行全生育期水分和养分定量、定时、按比例供应[3]。就目前来说随着物联网技术的不断发展，将其不断融入水肥一体化技术中，会有效解决传统灌溉施肥技术中的缺点，所以发展智慧灌溉施肥技术会是农业发展的新方向。

表 1 近年来中国农业灌溉情况

年份	全国耕地灌溉面积 （×10³hm²）	节水灌溉工程面积 （×10³hm²）	喷、微灌面积 （×10³hm²）	低压管灌面积 （×10³hm²）
2007	57 782	23 489	3 853	5 574
2008	58 472	24 436	4 071	5 873
2009	59 261	25 755	4 596	6 249
2010	60 348	27 317	5 141	6 680
2011	61 682	29 179	5 796	7 130
2012	62 491	31 217	6 600	7 526
2013	63 473	27 109	6 847	7 424
2014	64 540	29 109	7 843	8 271

数据来源：全国水利发展统计公报

3 农业物联网技术

物联网技术从出现到现在已经与各行各业有着不可分割的联系，其中传感器技术、智能控制技术、定位系统、无线传输等相关技术与农业有着非常密切的关联。进入新时代以来，由传统农业向智慧农业的发展离不开物联网技术的支持，大量运用物联网技术这也是农业发展的一个大趋势。

3.1 射频识别技术（RFID）

射频识别技术利用无线射频方式进行非接触双向通信，以达到目标识别目的并交换数据[4]。在日常的农业生产作业中，射频技术可以利用无线电信号进行对所测物体的识别而不必使物体与物体进行直接的接触，并且也无须进行光学接触。利用这些优势，即使是在雨天、雪天等一些恶劣天气，射频技术也可以不受影响能够稳定的进行工作，这些优势特点正是农业生产工作中所需要的，而且这项技术能够减少大量的人工投入，从而实现更高的经济效益。

3.2 定位以及监控技术

随着定位以及监控技术的应用，它们逐渐融入了人们的日常生活中，给我们的生活带来了极大的便利。目前，我国北斗卫星定位系统技术也是越来越成熟，它不仅可以为我们在任何地方提供位置信息、定位服务，而且还具有很高时效性和准确性。卫星系统不仅提供定位系统还可以提供监控服务，通过卫星监控加上在田间管理区安装摄像监控系统就能及时的查看作物的生产状况信息。

3.3 传感器网络技术

传感器是指能够感受被测物体的某些数据，并按照一定的规律转变成可识别的信号的装置。传感器作为农业物联网及其重要的一部分，主要是在需要测量数据的地方安装许多的微型传感器装置，通过连续感知来不断地收集信号并进行反馈。一般在农业生产中常用的传感器主要包括温湿度传感器、光敏传感器以及压力传感器等，通过对这些数据的收集能够准确地去发现各种因素对农作物生长的影响，进而去这些相关数据进行相应的调节。

3.4 智能控制技术

智能控制就像是物联网系统中的大脑中枢，所有被采集的信息都要反馈到这里，它能够将海量的数据进行分析并最终做出最终的决策。以后越来越多的前沿研究成果也会不断的应用到农业物联网中，这会使农业物联网本身拥有更强大的处理能力。

4 物联网在农业灌溉中的应用

4.1 农业物联网现状分析

农业物联网技术就是一种在农业生产方面实现智能、高效、集约的技术手段。在一些农业信息化较为发达的国家，它们在农业物联网方面起步早而且相关的技术已经非常的成熟，能够按照农作物的种类、生长情况与周围的环境等因素对水肥进行有效的控制。我国农业物联网技术方面虽然不是很成熟但随着近些年国家政策的支持，中国在农业物联网方面也取得一定的研究成果，在用于检测土壤水分和温度的传感技术、精准控制等技术取得了重大的突破。因此,物联网技术将会更多的应用于农业中,逐步实现现代农业。

4.2 "农业物联网的水肥施肥机"+"云平台管理系统"的管理模式

"农业物联网的水肥施肥机"+"云平台管理系统"的管理模式是在水肥一体化技术的基础结合计算机控制技术、农艺技术、电子通信技术、物联网云平台的有机结合,并加入当地"土专家"以及专业专家知识库管理配肥,来实现科学、高效、精准、有效的灌溉参数,使整个系统能够进行精确水肥配比。该技术系统路线图如图2所示,它可以根据监测的土壤的温湿度、农作物相应的生长需肥规律,来设置相应的水肥灌溉计划实施,在施肥时会按照数据库中的专家知识库设定配方、灌溉过程参数自动控制灌溉量、吸肥量、肥液浓度、pH值等水肥过程中重要的参数,来完成对灌溉、施肥的定时、定量控制,充分提高肥料利用率并节约用水,提高农作物的生长品质、改善提高土壤环境。该技术系统具有视频监控的作用,在农业管理区中安装放置无死角的全方位红外球形摄像机,可非常清楚直接的实时查看种植区域作物生长情况、设备远程控制执行情况等。在移动管理方面,该技术系统已可以与手机端、平板电脑端、PC电脑端进行无缝连接,这样就方便管理人员通过手机等移动终端设备随时随地查看系统信息,对相关设备进行远程操作。在无线田间气象站模块,在田间生产区放置检测气象设备将其与云平台相连能够实时将数据传送至云端。可对其进行远程设置数据存储和发送时间间隔,无须现场操作并且使用太阳能发电,可持续在田间地头工作。整套技术系统模式能够使在农业生产中最重要的水肥因素得到智能化管理,而且还有强大数据收集能力,为发展智

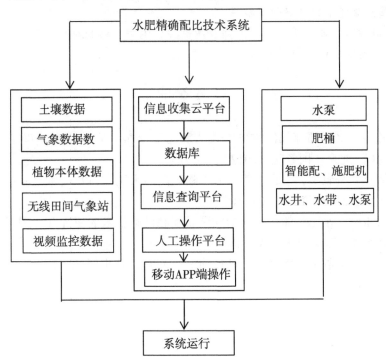

图2 农业物联网的水肥施肥机+云平台管理系统

慧农业积累更多的珍惜的前沿数据。

5 系统技术创新点

5.1 智能配、施肥机

智能配肥机、施肥机能够根据物联网云平台计算的数据进行精确配肥与精准定时灌溉，管理操作的面积大，自动化的设置，水肥资源能够得到充分的利用；可以人为的设定灌溉程序，不再需要人工去值守操作。

5.2 物联网云平台

基于物联网和农业大数据的智能服务平台，通过传感器检测的农作物数据可以实时传送至云平台，充分利用物联网的优势结合专家知识库，控制智能配肥机和施肥机实现整个过程中的精确配比与施肥。

5.3 植物本体传感器

这是整个技术系统的一大亮点，它不仅能够向我们展示检测的农作物外部生长情况数据，而且还能够检测植物的本身数据，如果实膨大、茎秆直径的微小变化、叶片表面温度等，并通过云平台的智能分析将结果以预警的方式发送至使用者，假设植物在生长过程中遇到生长缓慢的情况，它就会给用户发送一个预警，这样用户就可以及时的采取相应的措施。当植物缺水时，系统也会提醒用户，并且可以科学的推荐用水量。这样农作物品质就可以在一定程度上得到保障，而且也预防一些病虫害的侵扰。对于专业的农业工作人员来说，也可以获得更多的与植物密切相关的信息用来研究。

5.4 推广智能水肥精确配比技术

可以促进我国农业可持续发展加强农业资源利用率，降低农民生产成本，加快我国由传统农业向智慧农业发展的脚步。

参考文献

[1] 陈雷. 水利部部长陈雷向全国人大常委会报告农田水利建设工作情况报告[R]. 北京：国务院，2012.

[2] 周杰, 杨景文, 马良, 等. 基于物联网的水肥一体化技术[J]. 资源节约与环保, 2018（10）：106-107.

[3] 吴玉发. 水肥一体化自动精准灌溉施肥设施技术的研究和实现[J]. 现代农业装备，2013（4）：46-48.

[4] 马建. 物联网技术概论[M]. 北京：机械工业出版社，2011.

农村经济地理信息系统的构建

根据国家农村经济动态监测业务要求，构建基于电子地图的农村经济地理信息系统。该系统丰富了农村经济监测点的内容，使监测工作形象直观，实现为领导决策服务，为基层工作服务的目标。

1 建设目标

农村经济动态监测地理信息系统是青岛市发改委农村经济动态监测点项目的延伸，目的是加强市发改委与县发改局监测点间的信息联系，同时，在国家发改委农经司与市发改委、县发改局、镇联系点之间建立起高效的农经动态监测体系，完善农经信息采集渠道，为国家发改委农经司掌握第一手资料服务，提高信息上报的质量和时效性，为"金宏"工程提供及时、准确、全面的农村经济信息源支持，为农村经济宏观调控服务，并且通过网络功能的扩展，为农民提供服务[1]。下面以青岛所辖胶州市为例进行说明，具体目标包括：

（1）构建连接胶州市13个镇的胶州市农经监测网络。做好青岛市农经动态监测主干网络的延伸工作，在胶州市发改局设立监测点，建立与胶州市13个镇高效、快捷的农经动态信息监测网络，使胶州市农经监测纳入国家、市、县、镇四级农村经济动态监测体系。

（2）通过制度建设和业务培训等措施，培养和建立一支掌握信息技术，会采编、懂业务的农经动态信息监测队伍，确保国家监测任务的完成。

（3）开发基于网上电子地图的胶州市农村经济动态监测地理信息系统[2]。通过文字、数据、图片、视频、音频分层地理标注，形象直观展示胶州市农经资源信息，包括：农业概况、种植业、畜牧业、渔业、林业、农业产业化、龙头企业、农经合作组织、农经项目、品牌农业、新农村建设、农村城镇化建设、农经项目绩效评价等，为领导科学决策提供服务。

2 建设原则

该项目建设以国家电子政务建设方针以及"金宏"工程建设规划方案为指导，根据国家《农村经济动态监测点建设项目管理办法》组织实施。

（1）集中统一原则。监测点项目建设与已建成的网络工程及其他信息系统建设相结合，充分利用已有的信息资源和网络平台条件，不搞重复建设，使有限资源发挥最大效益。监测点项目建设要按照区域特色和产业布局进行选点，避免雷同和重复，以使反映的农村经济信息更具代表性。

(2) 实用性原则。项目建设以业务需求为导向，处理好实用性、先进性和超前性的关系，不搞超标准建设，严格控制建设成本。

(3) 开放性原则。在项目建设和应用过程中，采用符合国际标准的信息技术和产品，以便于实现资源共享和对系统进行灵活方便的升级。

(4) 资源共享原则。系统建设中，在建设布局和运行机制上保证与国家开展的其他信息系统的互联互通与资源共享，以最大限度地发挥系统的社会效益和经济效益。

(5) 安全可靠原则。要处理好系统开放性和安全性的关系，严格按照国家有关信息安全的规定和标准建设、管理信息系统，建立健全信息安全保障机制，确保数据传输、存储和发布过程中的安全性和稳定性。

3 建设内容

3.1 构建高效、便捷的农经动态监测网络体系

胶州市农村经济动态监测网络建设将依托 Internet 互联网络，利用青岛市信息中心现有的网络环境和网络通道，以青岛市农村经济动态监测平台为支撑，延伸构建胶州市农经动态监测网络。实现胶州市发改局与市发改委的联系；实现与胶州市 13 个镇的联通，并实现与国家发展改革委农经司的互通。胶州市农村经济动态监测地理信息系统网络拓扑图如图 1 所示。

为构建高效、快捷的网络，保证农村经济动态监测工作的正常有效进行，将重点购置监测点用于信息监测、现场采集、监测数据处理和数据上报等设备。

3.2 建立科学、规范的农经动态监测业务体系

胶州市农村经济动态监测点的主要业务内容是定期（月度、季度、半年、全年）上报农村经济动态形势。同时，对"三农"热点问题进行跟踪监测和系统研究，及时上报有关信息：一是跟踪、监测、分析和评估农村经济动态运行情况；二是跟踪、监测、分析和评估国家各项政策在农村的执行情况及基层干部和群众对国家有关政策的意见和建议等。三是搜集、整理和分析"三农"重大问题和热点问题信息，并根据需要对部分问题进行系统、深入研究。由胶州市监测点负责将上述信息报市发展改革委，再由市发改委经过审核后向国家发改委农经司上报，并接受考核及信息反馈。农村经济动态监测系统业务流程如图 2 所示。

3.3 软件平台技术

本农村经济动态监测地理信息系统采用 ASP.NET，为服务器端应用程序开发的热门工具。由于 ASP.NET 基于通用语言编译运行程序，故适应性强，可运行在几乎所有 Web 应用软件开发平台上，一般分为两种开发语言，VB.NET 和 C#。数据库系统采用 MS SQL 数据库平台，即微软的 SQL Server 数据库服务器，是一数据库管理系统，用于建立、使用与维护数据库，提供从服务器端到用户端的完善解决方案[3]。

3.4 农村经济地理信息系统

开创全国农经动态监测模式，根据国家农村经济动态监测主要业务要求，建立基于胶州市电子地图的胶州市农村经济动态监测地理信息数据库系统[4]。

地理信息数据库系统主要管理资源包括：

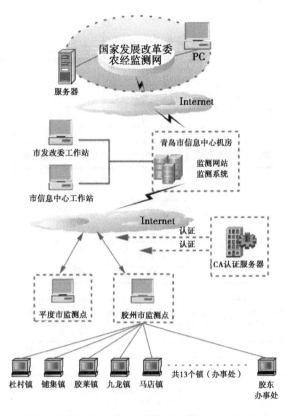

图 1 农村经济动态监测地理信息系统网络拓扑图

(1) 农经概况。

(2) 农业指标（种植业、畜牧业、渔业、林业等）。

(3) 农民人均纯收入。

(4) 农业产业化（国家级、省级龙头企业、规模以上加工出口示范企业等）。

(5) 农业合作组织。

(6) 新农村建设。

(7) 农村城镇化建设。

(8) 农业重点项目（千亿斤粮食工程等项目的建设情况及后评价）。

(9) 农村特色产业（品牌农业、休闲农业、生态农业）。

(10) 农产品价格（专业镇村、专业市场）。

(11) 监测点信息等。

该系统基于 GIS 系统与 FLASH 技术，运行于独立服务器，ASP.NET 环境，使用 SQL Server 数据库。通过基于胶州市电子交通图的地理信息数据库系统提供的分类图层、区域显示、位置标注、比例调整、信息管理、用户管理等功能，实现胶州市农村经济资源各项数据、文字、图片、视频等信息在线标注、展示；实现资源分类浏览、综合检索查询；实现监测点信息的动态管理等。系统可以根据客户需求单独制作某种类别的显示。

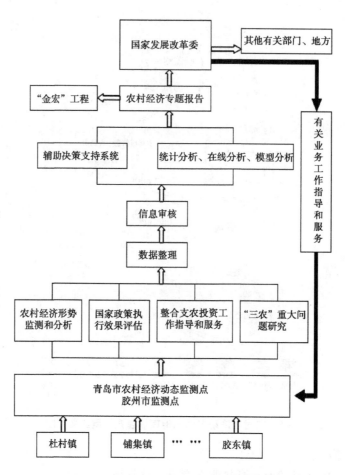

图2 农村经济动态监测系统业务流程

(1) 分类图层。地理信息数据库系统可以按照不同的资源分类标准,提供不同的分类显示的图层。使用者选择某个类别后,地图只显示该分类的指标信息,其余的自动隐藏。

(2) 区域显示。系统适合单独的行政区域的位置标注,县市、镇、街道办事处、龙头企业地址、示范区地址等。

(3) 位置标注。使用者登陆标注系统后,可以选择任意区域进行标注,并在标注时选择标注的指标、项目的属性,添加指标或项目的相关信息。

(4) 信息管理。用户登录系统后可以对所标注的位置进行详细的描述,如项目的文字介绍、数据指标、图片信息、视频信息、音频信息等。

(5) 比例调整。用户可以通过鼠标进行调整视觉比例,以查看某项目的详细位置。

(6) 综合查询。提供农经信息查询服务。用户可以使用关键字搜索,查询结果将以列表形式显示,点击某项搜索结果则进入该指标或项目的详细介绍地理页面及地理位置提示。默认情况下,定位第一项查询结果,并在地图上提示。

4 安全保障

根据国家信息安全有关规定，建立健全信息安全保障机制，确保数据传输、存储和发布过程中的安全性和稳定性。

（1）系统部署。系统分别部署在不同服务器上，数据库服务器实施定时备份，系统服务器加装反篡改系统，有效防止网络攻击和入侵，确保系统运行稳定、安全，可靠。

（2）安全认证。除提供普通用户名和密码登录外，系统还采用了青岛市电子政务数字一证通证书，实现数字证书登录，安全可靠。胶州市发改局可根据监测点用户数量，一次性委托定制数字证书，实现系统安全统一证书管理。

（3）安全统计。用户从登录系统到退出系统，中间进行的所有操作的内容、上网时间及 IP 地址都将详细记录在数据库里，最大限度的保障了系统的安全。

5 结束语

农村经济动态监测地理信息数据库系统的建设，将丰富胶州市农村经济经监测点内容，使监测工作形象直观。为登录该系统的用户，提供胶州市农经资源信息，从而实现为领导决策服务，为胶州市农村经济发展服务，为基层工作服务的目标。

参考文献

[1] 肖燕，孙壮. 山东省区域经济发展状况 GIS 评价[J]. 测绘科学，2012，37（5）：145-147.

[2] 曾庆泳，陈忠暖. 基于 GIS 空间分析法的广东省经济发展区域差异[J]. 经济地理，2007，27（4）：558-561.

[3] 姚晓军，孙美平，刘普幸，等. 基于 GIS 的甘肃省土地利用程度高程模型构建[J]. 干旱区资源与环境. 2012，26（5）：94-100.

[4] Oñte-Valdivieso F, Bosque Sendra J. Application of GIS and remote sensing techniques in generation of land use scenarios for hydrological modeling[J]. Journal of Hydrology，2010，395（3-4）：256-263.

青岛农业科技传播服务平台的构建

在分析农业网络信息的特点及服务方式的基础上，采用 J2EE 多层架构技术，构建了青岛农业科技传播服务平台。

1 引 言

网络环境下农业信息服务方式多样化，服务手段网络化，扩大了传统信息服务的范围，构造一个多元的服务体系，实现网络优势向服务优势的转换，构建农业科技服务平台是一个有效的途径。

2 农业网络信息的特点

2.1 内容丰富

网上农业信息涉及文献资料、动态商情、法律法规、成果通报、专利及技术转让、科研教育等多个方面，几乎对等反映了印刷型信息，并超出了印刷型信息的范围。除文本型信息外，还有大量的图像、声音等多媒体信息，丰富多样。

2.2 形态多样化

农业信息以多种形态存在于网上，数据库与动态信息、文本形态与多媒体形态、文摘题录与全文、浏览信息与下载信息、免费信息与有偿信息等，可满足不同层次用户的不同需求。

2.3 无时空障碍

网络是开放的媒介，用户可以在任何时间地点利用本地或异地的网上资源，不受信息机构开放时间和所在地域的局限。

2.4 开放性的信息发布和获取

因特网是迄今为止限制最少的交流通道，无论是政府部门、专业机构、企业、社会团体，还是个人，无论是官方或民间组织均可以在网上发布或获取信息。

3 农业网络信息服务

广义的信息服务，主要包括信息技术、信息报导、信息咨询、信息交流等服务方式，这些都可以延伸到网络中。网络作为一个由先进技术支持的开放环境，它给农业信息服务带来了服务内容、服务手段、服务方式的创新。

3.1 农业信息推送服务

农业信息推送服务是用户将自己选定的信息需求通过网络提交给信息系统或工作人员，信息工作人员建立用户需求档案，将收集到的用户所需信息按用户指定的时间自动传送给用户。

3.2 网上农业信息发布

网络基础构架的不断完善，使得新的服务方式不断出现，信息发布是互联网的基本服务之一。网上发布信息覆盖面广，发布及时，到达迅速，便于用户及时了解掌握动态变化，适时调整生产、经营计划。

3.3 互动式服务

互动式信息服务是利用网络进行信息交流、咨询的服务。利用这种服务，用户一般应该是某一网站的合法用户，互动服务有以下几种主要形式：①用户信箱，用户通过主页上的用户信箱或用户意见栏，反映自己的意见，管理员可在网上公开答复或通过电子邮件答复；②网上论坛，注册用户可在论坛栏发表论见或与其他用户讨论交换意见；③电子化期刊，网上电子期刊不仅供给用户阅览下载，还可从网上向期刊编辑部提交论文，编者与作者进行网上沟通，修改论文，读者也可与作者、编辑交流，形成三者之间的互动。

3.4 呼叫服务

呼叫服务是集电话、传真、计算机等通讯和办公设备于一体的一对一信息服务，以人工坐席或自动语音方式直接回答用户咨询问题。呼叫服务通常 24 小时值班，若不能 24 小时值班，应开通自动语音系统，以便回答简单的咨询问题。若不能即时回答用户请求，信息人员应在最短时间内利用馆藏或网络为用户查询资料并送达用户。

3.5 中介服务

网上信息的无限与无序造成了使用上的极大不便，因此，信息中介代理服务显得很有必要。网上农业信息中介服务有代检代查。主要是文献信息代查，按代查问题的内容可分为：①一般性代查，用户需要的文献内容较少，涉及面较窄，多为一次性查询；②课题代查。主要为研究人员的课题研究提供服务，所需文献要求具有一定的连续性、系统性、准确性和深度；③租赁式代理服务，对于昂贵的检索系统，超出了农业信息机构的购买力，可租赁数据库，为用户代检信息。

3.6 个人网页

因特网上不乏个人网页，通常提供本人、所属部门等专业信息，以及相关专业网站的链接，这里所指个人网页的网主在所属领域一般具有一定成就，或学者或专家或图书情报工作人员。

4 平台关键技术及结构模型

4.1 平台关键技术

SUN 公司推出的 J2EE 作为一个标准中间件体系结构日趋成熟，其优点在于平台无关、可重用、可移植、易于动态扩展、易于维护、运行稳定和高效安全等。J2EE 作为 SUN 公司定义的一个开放的、基于标准的平台，它提供了一个多层次的分布式应用模型和一系列开发技术规范，用以开发、部署和管理 N 层结构。其主要组件技术有：JDBC、EJB、JSP/Servlet 以及 JMS 等，各组件由容器进行管理，容器间通过相关的协议进行通讯，实现组件间的相互调用，构成一个成熟、完整的分布式应用程序开发模型。

中间层服务器由两部分组成：显示逻辑和业务逻辑部分。其中显示逻辑部分实际上是一个组件结构的 Web 服务器，由 Web 容器为 Servlet 组件提供运行环境，Servlet 组件用于调用业务逻辑部分的 Bean 组件或直接实现生成动态网页所需的处理过程。业务逻辑部分主要由 EJB 技术支持，整个 EJB 体系是一个分布式组件结构，在 RMI（远程方法调用）的基础上，EJB 容器为 Bean 组件提供了一致的分布式运行环境，所以分布于多个 EJB 容器中的组件可以互相操作，共同完成复杂的处理过程；并且 EJB 容器提供了事务管理、安全控制和数据库连接池等功能，构成一个完整的系统框架，使得 Bean 组件的开发更加便捷高效。Bean 组件通过 JDBC 访问各种后台数据库获取信息，并完成对数据的处理。由于广泛采用了组件技术，整个系统具有非常好的可扩展性。

4.1.1 JDBC

JDBC（Java 数据库连接）是一个纯 Java 的 API，屏蔽了各类数据库本身的差异，为 Java 应用程序提供了一致的数据库访问接口。

4.1.2 EJB

EJB 技术是一个分布式组件结构的应用程序开发模型，提供了相对独立的分布式应用开发平台，应用程序各个部分由多个 EJB 组件构成，组件可以运行于不同机器中不同平台的不同 EJB 容器中，组件之间通过 RMI 提供的分布式调用机制建立联系，这样多个分布的组件构成一个互操作网络，共同构成大型的应用程序。虽然 EJB 是一个独立的体系，不依赖于 J2EE 的其他部分，但 EJB 应用可以被 Servlet、Applet（所有的 Java 应用程序）通过 RMI 远程调用。

EJB 系统可以分为两部分：EJB 容器和 EJB 组件。容器为组件提供了运行环境，管理组件运行时的对象，并实现了 Connection Pool、Thread Pool 等机制管理系统资源，容器还提供事务管理、安全管理和 JNDI（Java 命名和发现接口）服务等底层的系统级服务。EJB 组件实际上是遵守 EJB 开发规范的 Java 类，所以理论上说，Bean 组件可以实现任何 Java 程序可以实现的功能。

4.1.3 Servlet/JSP

Servlet/JSP 主要用于显示逻辑的实现。Applet 技术，它是一种嵌入到 HTML 页面中的小应用程序，可以随 HTML 页面一同下载到客户端，并在客户端的虚拟机中运行（客户端需要安装有 JVM）。JSP 是一种动态网页开发技术，经过编译嵌入到 JSP 页面中的 Java 程序段会转化为 Servlet，由 Servlet 处理来自客户端浏览器中 JSP 网页的请求，并返回处理结果，也可以独立开发 Servlet，以扩展 Web 服务器的功能。

4.2 平台结构模型

平台的结构模型如图 1 所示，可分为三部分：客户层、中间层及信息系统层。

客户层组件主要用来处理用户接口，与用户交用。该层提供简洁的人机交互界面，完成数据的输入/输出。同时客户层也包含了一定的安全机制，用户可根据授权范围，控制数据和保密信息。

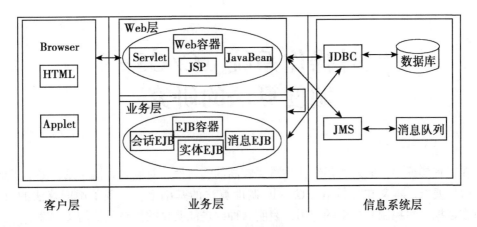

图 1 平台结构模型

中间层通常由 Web 层和业务层构成，其组件部署了应用程序服务器，将客户端和后台数据库联系在一起，Web 层主要使用 JSP 和 Servlet 与客户交互，通过调用 JavaBean 或 EJB 处理的逻辑事务结果产生用户界面表示。业务层经常被称作 EJB 层，主要向商业逻辑组件提供分布式计算的所有访问，既可以将客户数据进行处理，发送到信息系统层存储，也可以从存储中检索数据，送回客户程序。

信息系统层运行信息系统软件，主要由数据库管理系统（DBMS）、Java 消息服务（JMS）等构成，负责数据信息的存储、访问及传递，保证了系统的安全性及事务完整性。

5 结束语

为农民更加方便快捷的提供更多的农业知识和服务，是农业科技传播服务的主要内容之一。本文利用现有的网络技术，采用 J2EE 架构，成功构建了青岛农业科技传播服务平台，通过在实际中的应用，取得了良好的社会效益。

参考文献

［1］ Hougland D, Tavistock H（美）著；马朝晖译. JSP 核心技术［M］. 北京：机械工业出版社，2001.
［2］ Armstrong E, Bodof S. The J2EE 1.4 Tutorial［R］. Sun Microsystems, Inc., 2003.
［3］ 赵强，乔新亮. J2EE 应用开发［M］. 北京：电子工业出版社，2003.
［4］ 唐研，王志诚，李景岭. 网络化信息服务体系的构建与实施［J］. 农业图书情报学刊，2005（2）：28-30.

实时操作系统 μC/OS-Ⅱ 调度算法的研究

在工程实践中，嵌入式系统往往需要较高的实时性，对嵌入式操作系统提出了更高的实时性要求。本文在分析 μC/OS-Ⅱ 调度算法的基础上，实现了调度算法的改进，通过在数据采集控制系统中的应用，表明这种改进能明显的提高系统的实时性。

1 引言

传统的嵌入式系统中广泛采用单任务机制，由此带来系统的实时性、安全性较差，编程复杂等问题，导致系统频繁复位以致无法达到设计目标。本文引入嵌入式操作系统 μC/OS-Ⅱ 是一个多任务的实时内核，主要提供任务管理功能。在实时系统中的多个任务，必须决定这些任务的优先级顺序，任务调度算法需要动态为就绪任务的优先级排序，为了满足对实时性要求越来越高的需要，同时避免频繁改变就绪任务的优先级，在分析 μC/OS-Ⅱ 源代码的基础上，对其调度算法进行改进。

2 μC/OS-Ⅱ 概述

μC/OS-Ⅱ 是一种可移植、可固化、可裁剪及可剥夺的多任务实时内核（RTOS），适用于各种微控制器和微处理器。所有代码用 ANSI C 语言编写，具有良好的可移植性，已被移植到多种处理器架构中，在某些实时性要求严格的领域中得到广泛应用。

2.1 工作原理

μC/OS-Ⅱ 的工作核心原理是：近似地让最高优先级的就绪任务处于运行状态。首先，初始化 MCU，再进行操作系统初始化，主要完成任务控制块 TCB 初始化，TCB 优先级表初始化，TCB 链表初始化，事件控制块（ECB）链表初始化，空任务的创建等等；然后就可以开始创建新任务，并可在新创建的任务中再创建其他新任务；最后调用 OSStart() 函数启动多任务调度。在多任务调度开始后，启动时钟节拍源开始计时，此节拍源给系统提供周期性的时钟中断信号，实现延时和超时确认。

2.2 任务管理

μC/OS-Ⅱ 提供了对 64 个任务的管理，除了系统内核本身所保留的 8 个任务外，用户的应用程序最多可以有 56 个任务。由于 μC/OS-Ⅱ 是一个基于优先级的（不支持时间片轮转调度）实时操作系统，因此每个任务的优先级必须不同。μC/OS-Ⅱ 把任务的优先级当作任务的标识来使用，如果优先级相同，任务将无法区分，所以它只能说是多任务，不能说是多进程，至少不是我们所熟悉的那种多进程。系统中的每个任务都处于以下 5 种状态之一，这 5 种状态是休眠态、就绪态、运行态、等待态（等待某一事件发生）和被中断态。

2.3 任务调度

操作系统在下面的情况下进行任务调度：中断（系统占用的时间片中断OSTimeTick（）、用户使用的中断）和调用 API 函数（用户主动调用）。一种是当时钟中断来临时，系统把当前正在执行的任务挂起，保护现场，进行中断处理，判断有无任务延时到期，若没有别的任务进入就绪态，则恢复现场继续执行原任务。另一种调度方式是任务级的调度即调用 API 函数（由用户主动调用），是通过发软中断命令或依靠处理器在任务执行中调度。当没有任何任务进入就绪态时，就去执行空任务。

3 调度算法的改进

3.1 实时系统的调度策略

在操作系统的多任务调度算法的设计上，要根据系统的具体需求来确定调度策略，实时调度策略可以按不同的方法分为：静态/动态；基于优先级/不基于优先级；抢占式/非抢占式；单处理器/多处理器。其中：静态是指在任务的整个生命期内优先级保持不变，任务的优先级是当系统建立任务时确定的。动态是指在任务的生命期内，随时确定或改变它的优先级别，以适应系统工作环境和条件的变化。

μC/OS-Ⅱ 系统采用静态优先级分配策略，由用户来为每个任务指定优先级。虽然任务的优先级可通过 OSTaskChangePrio（）函数改变，但函数功能简单，仅以用户指定的新优先级来替换任务当前的优先级。随着实时嵌入式技术的发展，对嵌入式系统的实时性要求越来越高，多样化的调度方法已成为一种趋势。本文讨论动态优先级调度中的最优算法截止期最早优先算法的改进及在 μC/OS-Ⅱ 中的实现。

3.2 调度算法的改进

截止期最早优先算法是动态优先级调度算法中的最优算法，但其并不考虑任务的截止期是否已超时，在系统中调用超时的任务会造成系统时间的浪费，在时序要求严格的系统中还可能造成严重的后果。可达截止期最早优先算法是对截止期最早优先算法的一种改进，实质是使截止期最早的任务优先级最高，同时要保证在调度时超过截止期的任务不予调度。系统首先要根据系统的当前时间、任务的调用时间来更新任务的截止时间。如果计算结果小于零，说明任务超时，就要删除这个任务。如果任务没有超时则与截止期优先算法中任务的处理方法相同。在本文所述可达截止期最早优先算法改进中，需在 μC/OS-Ⅱ 系统中增加表 1 所示的项目。

表 1 可达截止期优先算法改进中添加的项目

项目	功能
INT8U deadline	在 OS_TCB 增加的数据项，用来记录任务的截止期。在任务调度时当前系统时间 OSTime 和 cstime 的差更新 deadline
INT8U cstime	在 OS_TCB 增加的数据项，记录任务首次调用时的系统时间

(续表)

项目	功能
INT8U csRec [64] [3]	记录每个优先级任务的调用时间、截止时间、优先级附加值（记录任务优先级的变化）。任务创建或销毁时要根据其截止期分配合理的优先级并插入此数组，在任务发生调用时用数组来更新其相应项的值

本文所述可达截止期最早优先算法要判断任务的截止期是否超时，系统不会调度超时的任务，需要为 μC/OS-Ⅱ 系统编写 OSTaskPrioCreate () 函数，用来在创建或删除一个任务的时候处理任务优先级调整的工作，以及对任务截止期是否超时的判断。该模块流程如图 1 所示。

在对就绪任务优先级进行调整的时候，该模块首先在数组中对任务的优先级完成调整并记录任务优先级的调整情况。在执行此函数后，就绪任务队列中的任务的优先级可能会改变，因此还需要在 μC/OS-Ⅱ 系统中添加 prio_adjust () 函数，该函数应用 μC/OS-Ⅱ 系统原有的 OSTaskChangePrio () 函数来对就绪任务更新，代码如下所示。

```
vid prio_ adjust () {
int i;
for (i=O; i<LOWEST_ PRIO-8; i++) {
if (csRec [i] [0] <0) {
OSTaskChangePrio (i, i+csRec [0]);
}
}
}
```

为防止多个任务同时调用 OSTaskPrioCreate () 函数造成混乱，这段代码应按临界资源来处理，需要在调用前关中断，调用后再开中断。

4 应用及评价

4.1 系统结构

在数控机床 IOC 模块中，系统要求在 15ms 内完成对五个位置的传感器和用户键盘数据的实时采集、处理及显示，且对于采集到的不同数据，要求系统根据任务的紧迫程度，做出优先级不同的实时响应。

系统的结构图如图 2 所示，由外向内分为三层，硬件电路层、任务层、操作系统层。

硬件电路层：本系统主要包括以下几个部分：IOC 模块、用户操作、单片机控制模块，大致功能如下：IOC 模块主要完成传感器数据的实时采集；用户模块主要完成用户的操作；单片机控制模块用于控制数据的接受、处理、发送、短消息的收发等功能。

任务层设计：系统任务层并行存在 8 个任务，每个任务均由以下三部分组成：应用程序、任务堆栈以及任务控制块，主要完成任务优先权的动态设置以及任务状态的

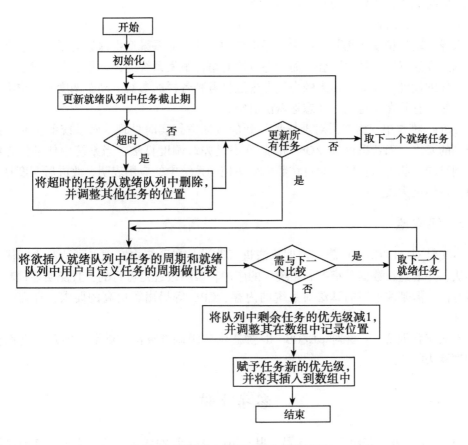

图1 可达截止期最早优先算法流程

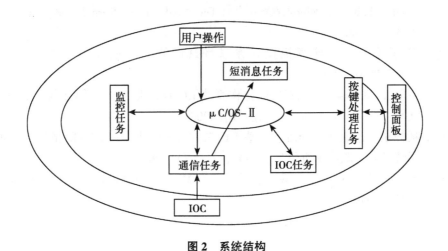

图2 系统结构

转换。

软件层的设计主要是先将 μC/OS-Ⅱ 移植到单片机上，本系统采用 MCS51 系列单片机，同时完成各个任务的具体编程实现。

4.2 算法评估

选择用动态调度还是静态调度是很重要的，这会对系统产生深远的影响。静态调度对时间触发系统的设计很适合，而动态调度对事件触发系统的设计很适合。静态调度必需事先仔细设计，并要花很大的力气考虑选择各种各样的参数。动态调度不要求事先做多少工作，它是在执行期间动态地做出决定。

在 IOC 模块中，由于需对现场采集到的数据进行实时处理，对系统的实时性提出了很高的要求，而若采用 μC/OS-Ⅱ 的静态优先级调度算法，当系统中任务优先级变化时则显的无能为力；同时通过在数据采集控制系统中的应用表明，改进后调度算法的实时性得到极大改善。

5 结束语

本文针对 μC/OS-Ⅱ 静态调度算法进行改进，在系统中实现了可达截止期优先调度算法，并通过在数据采集控制系统中的应用，表明改进是可行的，且能明显提高系统的实时性。但是改进后的算法对系统的内存、CPU 等提出了更高的要求，存在一定的局限性。

本文作者创新点：针对 μC/OS-Ⅱ 静态优先级调度算法的缺陷，实现了动态优先级调度算法的改进。

参考文献

[1] 王海峰,张海丽,张玉林.基于 μC/OS-Ⅱ 的微流体芯片嵌入式实时系统构建 [J].微计算机信息, 2006, 5 (2): 48-50.

[2] Jean J. Labrosse. MicroC/OS-Ⅱ: The Real-Time Kernel [M]. USA: CMP Books, 2002.

[3] 汤子瀛,等.计算机操作系统 [M].西安：西安电子科技大学出版社, 2000.

[4] Jean J. Labrosse. 嵌入式实时操作系统 μC/OS-Ⅱ [M].第 2 版.邵贝贝译.北京：北京航空航天大学出版社, 2003.

[5] 臧怀泉,范亚伟,李海生.基于山东省 C/OS-Ⅱ 与 MSP430 的手持数据采集系统 [J].微计算机信息, 2005, 21 (2): 66-67.

[6] Andrew S. Tanenbaum. 现代操作系统 [M].陈向群等译.北京：机械工业出版社, 1999.

实时操作系统 μC/OS-Ⅱ 的改进与应用研究

传统的嵌入式系统设计大多采用单任务顺序机制，应用程序是一个无限的大循环，所有的事件都按顺序执行，与时间相关性较强的事件靠定时中断来保证，由此带来系统的稳定性、实时性较差；尤其当系统功能较复杂，且对实时性要求较严格时，这种单任务机制的弱点暴露无遗。本文引入的嵌入式操作系统 μC/OS-Ⅱ 是一个多任务的实时内核，主要提供任务管理功能。在实时系统中的多个任务，必须决定这些任务的优先级顺序，任务调度算法需要动态为就绪任务的优先级排序。为了满足对实时性要求越来越高的需要，同时避免频繁改变就绪任务的优先级，在分析 μC/OS-Ⅱ 源代码的基础上，对其调度算法进行改进。

1 μC/OS-Ⅱ 概述

μC/OS-Ⅱ 是一个完整的，可移植、可固化、可裁剪的占先式实时多任务内核；支持 56 个用户任务，支持信号量、邮箱、消息队列等常用的进程间通信机制；适用于各种微控制器和微处理器；所有代码用 ANSI C 语言编写，程序的可读性强，具有良好的可移植性，已被移植到多种处理器架构中，在某些实时性要求严格的领域中得到广泛应用。

1.1 工作原理

μC/OS-Ⅱ 的核心工作原理是：近似地让最高优先级的就绪任务处于运行状态。首先初始化 MCU，再进行操作系统初始化，主要完成任务控制块 TCB 初始化，TCB 优先级表初始化，TCB 链表初始化，事件控制块（ECB）链表初始化，空任务的创建等。然后，开始创建新任务，并可在新创建的任务中再创建其他新任务。最后，调用 OSstart() 函数启动多任务调度。在多任务调度开始后，启动时钟节拍源开始计时，此节拍源给系统提供周期性的时钟中断信号，实现延时和超时确认。

1.2 任务调度

操作系统在下面的情况下进行任务调度：中断（系统占用的时间片中断 OSTimeTick()、用户使用的中断）和调用 API 函数（用户主动调用）。一种是当时钟中断来临时，系统把当前正在执行的任务挂起，保护现场，进行中断处理，判断有无任务延时到期；若没有别的任务进入就绪态，则恢复现场继续执行原任务。另一种调度方式是任务级的调度，即调用 API 函数（由用户主动调用），是通过发软中断命令或依靠处理器在任务执行中调度。当没有任何任务进入就绪态时，就去执行空任务。

2 调度算法的改进

2.1 实时系统的调度策略

在操作系统的多任务调度算法的设计上,要根据系统的具体需求来确定调度策略。实时调度策略按不同的方法可以分为:静态/动态,基于优先级/不基于优先级,抢占式/非抢占式,单处理器/多处理器。其中,静态是指在任务的整个生命期内优先级保持不变,任务的优先级是在系统建立任务时确定的;动态是指在任务的生命期内,随时确定或改变它优先级别,以适应系统工作环境和条件的变化。

μC/OS-II系统采用的是静态优先级分配策略,由用户来为每个任务指定优先级。虽然任务的优先级可通过OSTaskChangePrio()函数改变,但函数功能简单,仅以用户指定的新优先级来替换任务当前的优先级。随着实时嵌入式技术的发展,对嵌入式系统的实时性要求越来越高,多样化的调度方法已成为一种趋势。本文讨论动态优先级调度中的最优算法截止期最早优先算法的改进及其在μC/OS-II中的实现。

2.2 调度算法的改进

截止期最早优先算法是动态优先级调度算法中的最优算法。在截止期最早优先算法中,系统按任务的截止期给每个任务分配优先级。任务的截止期越早其优先级越高,反之亦然。为此,在本文所述截止期最早优先算法的改进中,需在μC/OS-II系统中增加表1所列的项目。

表1 截止期最早优先算法改进中添加的项目

项目	功能
INT8U deadline	在 OS_TCB 增加的数据项,用来记录任务的截止期。在任务调度时,用当前系统时间 OSTime 和 cstime 的差更新 deadline
INT8U csRec [64] [3]	记录每个优先级任务的调用时间、截止时间和优先级附加值(记录任务优先级的变化)。任务创建或销毁时,要根据其截止期分配合理的优先级并插入此数组,在任务发生调用时用数组来更新其相应项的值

在截止期最早优先算法中,需要用户为任务指定其截止期。在本改进中,将 OSTaskCreate() 和 OSTaskCreateExt() 中的参数 INT8U Prio 改为 INT8U deadline,并在函数内定义局部变量 INT8U Prio 来记录分配给任务的优先级。该算法改进也要在系统中增加 OSTaskPrioCreate() 函数,函数优先级分配的方法是按任务的截止期分配。该模块流程如图1所示。

在对就绪任务优先级进行调整时,该模块首先在数组中对任务的优先级完成调整并记录任务优先级的调整情况。在执行此函数后,就绪任务队列中任务的优先级可能会改变,因此还需要在μC/OS-II系统中添加 prio_adjust() 函数。该函数应用μC/OS-II系统原有的函数 OSTaskChangePrio() 来更新就绪任务,代码如下:

```
vid prio_ adjust () {
  int i;
  for (i = 0 ; i < LOWEST_ PRIO-8 ; i + +) {
```

```
    if ( csRec [ i ] [ 0 ] < 0 ) {
        OSTaskChangePrio ( i , i + csRec [ 0 ] ) ;
    }
  }
}
```

为防止多个任务同时调用 OSTaskPrioCreate ( ) 函数造成混乱，这段代码应按临界资源来处理，需要在调用前关中断，调用后再开中断。

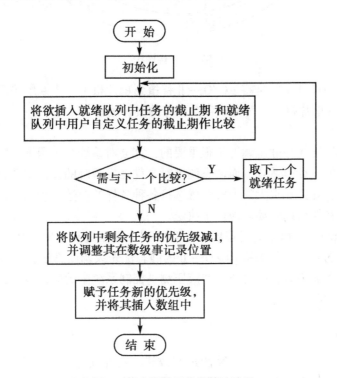

图 1　截止期最早优先算法流程

3　应用及评价

3.1　系统结构

在液压测量控制 HPMC 模块中，系统要求在 18ms 内完成对 7 个位置的传感器和用户键盘数据的实时采集、处理及显示；且对于采集到的不同测量数据，要求系统根据任务的紧迫程度，作出优先级不同的实时响应。

系统的结构如图 2 所示。由外向内分为 3 层：硬件电路层、任务层和操作系统层。

硬件电路层主要包括 HPMC 模块、用户操作、单片机控制模块。大致功能如下：HPMC 模块主要完成传感器数据的实时采集；用户模块主要完成用户的操作；单片机控制模块用于控制数据的接收、处理、发送、短消息的收发等。

任务层并行存在 10 个任务，每个任务均由以下 3 部分组成：应用程序、任务堆栈以及任务控制块，主要完成任务优先权的动态设置以及任务状态的转换。

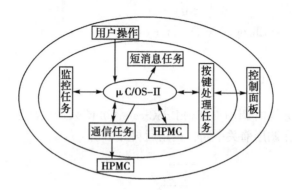

图 2 系统结构

操作系统层的设计主要是将 μC/OS-Ⅱ 移植到单片机上。本系统采用 Atmel 公司的 MCS-51 系列兼容单片机，同时完成各个任务的具体编程。

3.2 算法评估

选择用动态调度还是静态调度是很重要的，这会对系统产生深远的影响。静态调度对时间触发系统的设计很适合，而动态调度对事件触发系统的设计很适合。静态调度必须事先仔细设计，并要花很大的力气考虑选择各种各样的参数；动态调度不要求事先作多少工作，而是在执行期间动态地作出决定。

在 HPMC 模块中，由于需对现场采集到的测量数据进行实时处理，故对系统的实时性提出了很高的要求。若采用 μC/OS-Ⅱ 的静态优先级调度算法，当系统中任务优先级变化时则显得无能为力；同时通过在液压测量控制系统中的应用表明，改进后系统的实时性得到了极大改善。

4 结束语

本文针对 μC/OS-Ⅱ 静态调度算法进行改进，在系统中实现了截止期优先调度算法。通过在液压测量控制系统中的应用，表明这种改进能明显提高系统的实时性；但是改进后的算法对系统的内存、CPU 等提出了更高的要求，存在一定的局限性。

参考文献

[1] Labrosse Jean J. MicroC/ OS-Ⅱ：The Real-Time Kernel [C]. USA：CMP Books，2002.

[2] 汤子瀛，等. 计算机操作系统 [M]. 西安：西安电子科技大学出版社，2000.

[3] Labrosse Jean J. 嵌入式实时操作系统 μC/OS-Ⅱ [M]. 第 2 版. 邵贝贝译. 北京：北京航空航天大学出版社，2003.

数字茶园实验基地建设实践与探索

在分析数字农业的现状及基本组成的基础上,利用现有的数字农业平台,进行数字茶园实验基地建设,进行了有意义的探索。

1 引言

利用现代计算机技术技术,全过程的采用智能控制、自动化作业,实现对数据的采集、记录和分析,同时进行信息服务系统的建设,建设了数字茶园实验基地,进行了有意义的探索与实践。

2 数字农业

2.1 数字农业的涵义

"数字农业"(Digital Agriculture)确切的表达应该是数字化农业。目前,数字农业较为公认的定义是用数字化技术,按人类需要的目标,对农业所涉及的对象和全过程进行数字化和可视化的表达、设计、控制和管理。

数字化农业是农业信息化的核心和具体表现形式,数字化农业技术是对农业所涉及的对象和全过程进行数字化和可视化的表达、设计、控制和管理的现代化农业高新技术。主要包括农业要素(生物要素、环境要素、技术要素、社会经济要素)的数字信息化、农业过程的数字信息化(数字化实施、数字化设计)和农业生产管理的数字信息化。涵盖了信息农业、精准农业、虚拟农业等。

2.2 国外数字农业的发展

进入20世纪90年代以来,美国、日本、西欧等发达国家在农业信息化方面发展很快。以美国、德国和日本为代表的发达国家在完成了农业工业化和农业机械化后已经进入了农业信息化时代,其中,美国的农业信息化强度已高于工业81.6%。这些发达国家相继实现了以数字化农业为基础的农业技术信息服务网络系统,通过公众电话网、专用通讯网和无线寻呼网,对全国各地的农业技术信息进行收集、处理、贮存和传递,将大型数据库系统、Internet系统、气象信息系统、土地资源系统、高效农业生产管理系统联结起来,各地用户可从网上快速获得有关信息,并随时进行信息交换,使农业生产率得到很大的提高。

2.3 我国数字农业的发展

我国面对农业发展的新阶段和中国加入WTO的新形势,也非常重视"数字化农业"。"八五"以来,国家863计划、国家科技攻关计划和国家星火计划均对农业信息化相关内容进行了大力支持。从1990年开始,863计划智能计算机主题连续支持"农业智能应用系统"的研究与应用，已研制出棉花、水稻、玉米、小麦、大豆等多种作物的生育全程调控和农事管理专家系统，以及畜禽疾病防治、果树生产管理专家系统，推出了多个具有较高水平的农业专家系统开发平台，开发出高产型、经济型、优质型的实用农业专家系统200多个，在全国22个示范区应用，取得了显著的经济与社会效益。

3 数字农业的基本组成

数字农业由农业基础信息数据库、农业信息实时（准实时）获取系统、数字网络传输系统、中央处理系统、数字化农业机械等几部分组成。

3.1 农业基础信息数据库

提供农田、种质资源、气候条件、社会经济条件等与农业生产有关的农业基础信息数据。主要包括以气候、土壤为内容的农田信息；以农作物品种、栽培技术、病虫害防治等为内容的农情信息；以种子、农药、农机、农膜等为内容的商情信息等。

3.2 农业信息实时获取系统

完成农业生产的实时监控与数据库更新。包括地下和地面气象、植被、土壤状况等信息的数字化实时采集器，以及机载、星载传感器等遥感、遥测设备。

3.3 数字网络传输系统

数字网络传输系统的主要功能是实现信息的收集与指令的分发。

3.4 中央处理系统

以 GIS（地理信息系统）和农学模型与专家系统为支撑，对收集到的各种信息进行分析处理，做出相应决策，发出控制指令，指导数字化农业机械的作业。

3.5 数字化农业机械

包括数字化播种设备、数字化水肥调控设备、数字化收割设备等数字化农业机械设备。在数字化网络与 GPS（全球定位系统）、GIS 支持下，执行处理分析系统发出的指令，直接或间接通过信息实时获取系统，返回处理结果。

4 数字茶园实验基地建设

在数字茶园试验基地建设时，主要采用中央控制系统、土壤参数检测系统、灌溉的数字化控制、无线控制系统；同时进行网络服务平台的建设，扩大了信息服务的范围，构造一个多元的服务体系，实现网络优势向服务优势的转换。

4.1 中央控制系统

数字农业平台是以桁架驱动，实行厘米级精确位置控制，提供通用作业机架，全过程采用智能控制、自动化作业及数据采集、记录和分析的农业作业平台。数字农业平台由控制系统控制桁架，拖动农机具在直线导轨上行走来完成作业，充分发挥了计算机技术、自动化控制技术、数据存储技术及人工智能专家系统的优势，使机械作业精度提高到厘米级或单株作物，实现了虚拟环境下的自动化作业。

根据数字农业研究平台的要求，控制系统主要完成：作业模块控制参数设定，以此来控制桁架的精确定位和前进速度；监测各功能模块的运行情况；实时显示相关功能参数（例如速度、位移等）及环境和土壤参数测量；种、水、肥的测量；故障报警等。

4.2 土壤参数检测系统

土壤、气候等农业基础信息数据的获取是数字农业平台研究的基础，能够对土壤参数实现精准测量便显得尤为重要。为此，根据数字农业平台的要求，设计了一套完整的土壤综合参数监测系统，对土壤水分、养分、温度等重要参数进行精确快速地监测，以指导农业生产。

土壤参数监测系统主要由控制模块、田间数据采集模块、数据处理及存储模块几部分组成。根据数字农业研究平台的要求，系统主要完成以下几个方面功能：①根据数字农业平台工作机的要求，控制移动探针在指定位置进行监测。②田间数据采集模块主要完成土壤参数信息的采集，单片机系统的一个模拟信号输入通道与可移动土壤水分、温度、养分传感器探针相连，测量土壤水分、养分、温度，然后与上位机相连，进行处理。③数据处理模块主要把单片机传过来的数据进行一系列处理，最终生成水分分布图、养分分布图和温度变化图并保存。

4.3 农业灌溉的数字化控制

农业灌溉的数字化控制就是运用数字逻辑电路中的组合逻辑电路和时序逻辑电路来控制农业灌溉，则可以精确设置灌溉时间，并用 LED 计数显示来监控农业灌溉的进程，以实现农业灌溉的精确化、合理化。

4.4 无线控制系统

数字农业试验平台应能按照预定的路线在田间行走，自动进行田间信息采集、精确控制变量投入。为了实现自动作业的精确控制，设计了基于无线局域网的数字农业试验平台无线控制系统。该系统以无线方式实现网络互连，进行数据传输，使操作人员在远程及时了解现场试验平台的工作状态，实现了试验过程的计算机无线监控与信息反馈，使农业生产突破了空间的限制，进而实现了广阔范围内的自动控制。

系统采用无线局域网技术，将安装在数字农业试验平台上的网络服务器与中心监测室内的数据存储服务器连接起来。网络服务器进行数据采集与处理，并实现数据实时传送数据；存储服务器实时接收网络服务器上传的数据，实现对数字农业试验平台的运行参数进行实时控制、故障报警和 Web 数据库实时更新，同时接收 Internet 网上用户要求，向他们提供平台运行状态的实时信息和历史信息。系统由以下 5 部分组成：①数字农业试验平台。由网络服务器、PLC、数据采集卡、各种传感器及控制阀等组成。利用网络服务器采集与处理数据，并通过 MagicLan 无线局域网与网络工作站通信。②数据存储服务器。配备高性能处理器，作为系统对 Internet 窗口提供数据存储服务。③有线工作站。主机通过 LAN 或 Internet 与数据存储服务器连接，在线远程控制数字农业试验平台。④无线工作站。主机通过 WLAN 与数据存储服务器连接，在线远程控制数字农业试验平台。⑤无线网络系统。采用无线局域网与星型网络混合的拓扑结构通过无线接入点（AP）与网络服务器相连接，WLAN 采用 802.11 b 协议传输速率为 11Mb/s。

4.5 网络传播服务平台

网络环境下农业信息服务方式多样化，服务手段网络化，构造一个多元的服务体系，实现网络优势向服务优势的转换，构建茶叶网络传播服务平台，为广大用户提供更多的知识和服务。

5　结束语

对于数字农业，已经进行了很多研究，在此研究的基础上，进行数字茶园实验基地建设，进行了有意义的探索，积累了丰富的经验，为我国数字农业建设提供重要的启示。

参考文献

［1］马德新，徐鹏民，赵晓春．青岛农业科技传播服务平台的构建［J］．农业网络信息，2007（1）：55-57.

［2］赵晓春．农业网络传播［M］．北京：中国传媒大学出版社，2005.

［3］张健，张帅兵，等．数字农业平台土壤参数监测系统设计［J］．农机化研究，2006（3）：114-115，122.

［4］梅振东，黄建武．农业灌溉的数字化控制［J］．现代电子技术，2007（7）：132-133，136.

数字化农业平台科技传播服务系统的构建

在分析农业网络信息的特点及服务方式的基础上，采用 J2EE 多层架构技术，构建了青岛农业科技传播服务平台。

在网络环境下实现农业信息服务方式多样化，服务手段网络化，扩大了传统信息服务的范围。构造一个多元的服务体系，实现网络优势向服务优势的转换，构建农业科技传播服务平台是一种有效的途径。

1 农业网络信息的特点

1.1 内容丰富

网上农业信息涉及文献资料、动态商情、法律法规、成果通报、专利及技术转让、科研教育等多个方面。除文本型信息外，还有大量的图像、声音等多媒体信息，内容丰富多样。

1.2 形态多样化

农业信息以多种形态存在于网上，数据库与动态信息、文本形态与多媒体形态、文摘题录与全文、浏览信息与下载信息、免费信息与有偿信息等，可满足不同用户的需求。

1.3 无时空障碍

网络是开放的媒介，用户可以在任何时间、地点享用本地或异地的网上资源。

1.4 开放性的信息发布和获取

无论是政府部门、专业机构、企业、社会团体，还是个人均可在网上发布或获取信息。

2 农业网络信息服务

广义的信息服务，主要包括信息技术、信息报道、信息咨询、信息交流等服务方式。网络作为一个先进技术支持的开放环境，它对农业信息服务进行了服务内容、服务手段、服务方式的创新。

2.1 农业信息推送服务

农业信息推送服务是用户将自己选定的信息需求通过网络提交给信息系统或工作人员，信息工作人员建立用户需求档案，将收集到的用户所需信息按用户指定的时间自动传送给用户。目前有 2 种主要的推送技术：一种是自动拉取技术，即最终用户要求发送方按照预先约定的时间自动提交所制定的新信息；另一种是事件驱动技术，即以规则为基础，由推送管理方判断预先设置的规则是否发生，若发生则将相关信息或内容提交给最终用户。

2.2 网上农业信息发布

信息发布是互联网的基本服务之一。网上发布信息覆盖面广，发布及时，到达迅速，便于用户及时了解动态变化，适时调整生产、经营计划。

2.3 网络信息资源导航

利用网上现有搜索引擎，把与某一主题相关的站点进行集中，然后把这些资源分布情况提供给用户，指引用户检索，即建立网上信息资源导航库或指引库。信息资源导航库或指引库是对网络资源搜索并有序化组织的信息产品。

2.4 网上数据库资源建设

网上数据库资源建设包括商业性网络数据库的采集、网上免费数据库网址的收集和特色数据库的开发。根据信息服务对象的需求，利用自身的技术力量，制定特色数据库建设的结构框架，对信息资源和网上信息资源进行筛选、加工、整合，建成有特色、有实用价值及商业价值的数据库群，最终提供网上查询服务，实现信息的增值。

2.5 互动式服务

互动式信息服务是利用网络进行信息交流、咨询的服务。互动服务有以下3种主要形式：①用户信箱。用户通过主页上的用户信箱或用户意见栏，反映自己的意见，管理员在网上公开答复或通过电子邮件答复。②网上论坛。注册用户可在论坛栏发表意见或与其他用户讨论交换意见。③电子化期刊。网上电子期刊不仅供给用户阅览下载，还可从网上向期刊编辑部提交论文，编者与作者进行网上沟通，修改论文，读者也可与作者、编辑交流，形成3者之间的互动。

2.6 呼叫服务

呼叫服务是集电话、传真、计算机等通讯和办公设备于一体的信息服务，以人工坐席或自动语音的方式直接回答用户咨询问题。呼叫服务通常1天24小时值班，若不能1天24小时值班，应开通自动语音系统，以便回答简单的咨询问题。若不能即时回答用户请求，信息人员应在最短时间内利用网络为用户查询资料并送达用户。

2.7 中介服务

网上信息的无限与无序造成了使用上的极大不便，因此，信息中介代理服务显得十分必要。网上农业信息中介服务主要是代检代查，尤其是文献信息代查。按代查问题的内容可分为：①一般性代查。用户需要的文献内容较少，涉及面较窄，多为一次性查询。②课题代查。主要为研究人员的课题研究提供服务，所需文献要求具有一定的连续性、系统性、准确性。③租赁式代理服务。对于昂贵的检索系统，超出了农业信息机构的购买力，可租赁数据库，为用户代检信息。

2.8 个人网页

因特网上的个人网页通常提供本人、所属部门等专业信息，以及相关专业网站的链接，这里所指个人网页的网主要是在所属领域具有一定成就的人。

3 平台关键技术及结构模型

3.1 平台关键技术

SUN 公司推出的 J2EE 作为一个标准中间件体系结构日趋成熟，其优点是可重用、可移植、易于动态扩展，易于维护、运行稳定和高效安全等。J2EE 作为 SUN 公司定义的一个开放的、基于标准的平台，它提供了一个多层次的分布式应用模型用以开发、部署和管理 N 层结构。其主要组件技术有：JDBC、EJB、JSP/Servlet 以及 JMS 等，各组件由容器进行管理，容器间通过相关的协议进行通讯，实现组件间的相互调用，构成一个成熟、完整的分布式应用程序开发模型。中间层服务器由 2 部分组成：显示逻辑部分和业务逻辑部分。其中显示逻辑部分实际上是一个组件结构的 Web 服务器，由 Web 容器为 Servlet 组件提供运行环境，Servlet 组件用于调用业务逻辑部分的 Bean 组件或直接实现生成动态网页所需的处理过程。业务逻辑部分主要由 EJB 技术支持，整个 EJB 体系是一个分布式组件结构。在 RMI（远程方法调用）的基础上，EJB 容器为 Bean 组件提供了一致的分布式运行环境；并且 EJB 容器提供了事务管理、安全控制和数据库连接池等功能，构成一个完整的系统框架，使 Bean 组件的开发更加便捷高效。Bean 组件通过 JDBC 访问各种后台数据库获取信息，并完成对数据的处理，使整个系统具有很好的可扩展性。

3.1.1 EJB

EJB 技术是一个分布式组件结构的应用程序开发模型，提供了相对独立的分布式应用开发平台。应用程序各个部分由多个 EJB 组件构成，组件可以运行于不同机器中不同平台的不同 EJB 容器中，组件之间通过 RMI 提供的分布式调用机制建立联系，这样多个分布的组件构成一个互操作网络，共同构成大型的应用程序。虽然 EJB 是一个独立的体系，不依赖于 J2EE 的其他部分，但 EJB 应用可以被 Servlet，Applet（所有的 Java 应用程序）通过 RMI 远程调用。EJB 系统可以分为 2 部分：EJB 容器和 EJB 组件。容器为组件提供了运行环境和运行对象，并实现了 Connection Pool，Thread Pool 等机制管理系统资源；容器还提供事务管理、安全管理和 JNDI（Java 命名和发现接口）服务等底层的系统级服务。EJB 组件实际上是遵守 EJB 开发规范的 Java 类，因此，Bean 组件可以实现任何 Java 程序可以实现的功能。

3.1.2 Servlet/JSP

Servlet/JSP 主要用于显示逻辑的实现。Applet 技术是一种嵌入到 HTML 页面中的小应用程序，可以随 HTML 页面一起下载到客户端，并在客户端的虚拟机中运行（客户端需要安装有 JVM）。JSP 是一种动态网页开发技术，经过编译嵌入到 JSP 页面中的 Java 程序段会转化为 Servlet，由 Servlet 处理来自客户端浏览器中 JSP 网页的请求，并返回处理结果，也可以独立开发 Servlet 来扩展 Web 服务器的功能。

3.1.3 JDBC

JDBC（Java 数据库连接）是一个纯 Java 的 API，屏蔽了各类数据库本身的差异，为 Java 应用程序提供了一致的数据库访问接口。

3.2 平台架构模型

平台的结构模型如图 1 所示,可分为 3 部分:客户层,中间层及信息系统层。客户层一般由浏览器组成,用户可通过客户层从服务器上访问 Web 层的静态 HTML 页面或由 JSP 或 Servlet 生成的动态页面而获取所需的信息。

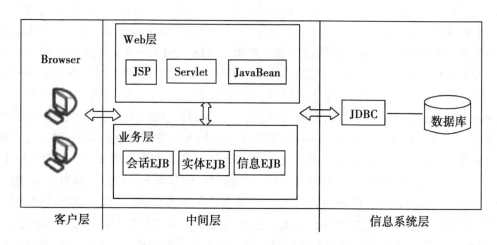

图 1　平台架构模型

中间层通常由 Web 层和业务层构成:①Web 层可以由 JSP 页面、Servlet 或 Applet 组成,通过处理客户层的用户输入请求,并把请求消息发送给运行在业务层上的 EJB 进行处理。JSP 是一种动态 Web 应用技术标准,由传统 HTMLWeb 页的文件中加入 Java 程序片和 JSP 标记所构成,可以在服务器端根据客户请求产生动态页面。Servlet 是服务器端的 Java 小程序,通过 Applet API 响应客户端的请求并进行处理,可以产生动态的 HTML 页面。Web 层和业务层通常被封装在一个应用服务器中。②业务层又被称为 EJB(Enterprise Java Bean)层。EJB 定义了一个面向对象的可重用服务器端分布组件标准,用来封装业务逻辑。EJB 是 J2EE 体系结构的核心,在功能上分为会话 Bean(Session Bean)和实体 Bean(Entity Bean)。Session Bean 执行事务逻辑、算法、规则和工作流程,是具有处理事务过程逻辑的可重用组件。Entity Bean 将底层数据以对象的形式映射到内存中,供其他组件使用。这种将事务逻辑与底层数据分离的方法,增强了应用系统的可移植性和可扩展性。

信息层运行信息系统软件,主要由数据库管理系统(DBMS),JDBC,Java 消息服务(JMS)等构成,负责数据信息的存储、访问及传递,保证了系统的安全性及事务完整性。

4　结束语

为农民更加方便快捷地提供更多的农业服务和知识,是农业科技传播服务的主要内容之一。笔者利用现有的网络技术,采用 J2EE 架构,成功构建了青岛农业科技传播服务平台,取得了良好的社会效益。

参考文献

[1] 王秀东,王永春. 加快我国网络农业建设的对策措施团 [J]. 农业经济,2001(8):17-18.

[2] 姜丽华,王文生. 网络农业信息资源的开发与利用 [J]. 农业网络信息,2005(11):82-83.

[3] 郭建新. 论农业网络信息资源开发与利用 [J]. 现代情报,2002(1):30,65.

[4] ARMSTRONG E, BODOF S. The J2EE 1.5 tutorial [R]. Copyright, Sun Microsystems, 2004.

水肥一体化技术的发展现状与应用对策研究

中国是一个非常缺水的国家,农业灌溉用水约占我国全部供水量的63%。受技术落后等因素的限制,农业用水浪费严重。而水肥一体化技术具备节约水肥资源、节省人力物力、提高水肥利用率等优点,有利于农业发展。本文总结水肥一体化技术的发展过程,分析水肥一体化技术在应用过程中存在的问题,并提出有效对策,以期提升其应用效果。

1 我国农业用水与水肥一体化技术背景

1.1 我国农业生产用水现状

农业是我国的重要产业,农业的发展至关重要。结合我国实际情况,实现农业可持续发展、农作物增产,需要将科学灌溉、高效节约地使用水资源和科学施肥等有效结合起来。

我国是一个严重缺水的国家。虽然拥有全球6%的水资源,位居世界第四位,但人均拥有量仅为2 300m³,是世界平均水平的1/4,在全球位列121位[1],仅高于埃及、阿曼、阿联酋、佛得角、布隆迪、沙特阿拉伯、巴巴多斯、阿尔及利亚、约旦、科威特、科比亚、马耳他、巴林和也门等国家。

据报告,我国的农业灌溉施肥和喷洒农药等用水约为3 900亿 m³,约为我国用水量的70%。在农业领域中,灌溉用水约占全部用水的90%以上,占全国所有用水量的63%左右。长期以来,因我国农业灌溉技术落后,水土管理方法不科学,农业用水浪费较多,大大降低了水肥资源的利用率。据报道,我国年农业浇灌用水利用系数平均约为0.43,而先进国家为0.70~0.80[2]。按照年水利用系数的水平计算,若灌溉用水的利用率提高,那么节省水量可近百亿立方米。

1.2 水肥一体化技术的产生背景

目前,农业领域用水持续增多加剧了整体水资源的匮乏程度;化肥过度使用,导致土壤贫瘠和环境污染日益严重。在这种情况下,国家加大了对农业生产的智慧化和精细化管理,推动并产生了水肥一体化技术。

2 国外水肥一体化技术发展现状

2.1 水肥一体化技术的基本概念

水肥一体化指的是让灌溉和施肥一起作业,同时供给植物水分和营养。从狭义上讲,是将肥料溶解于水中,利用微灌系统进行灌溉,同时达到灌溉与施肥目的,可均匀地满足植物对水分和营养的需求，以此减少重复劳动作业，实现高效率水分和养肥同步化管理的农业技术。

2.2 世界水肥一体化的发展历程分析

1790年前后，欧洲的John Woodward用土壤提取液进行植物种植，此为水肥一体化的初始记录。国际上第一个进行的有关细流灌溉技术的试验可以追寻到1900年之前，但该技术试验真正起源于1950—1960年间[3]。20世纪70年代，随着廉价塑料管大批生产，滴灌技术得以快速发展，滴灌、微喷灌等技术得到加强。

2.2.1 以色列水肥一体化技术的发展历程

在以色列，水肥一体化技术的发展进程非常显著。1950年左右，随着塑料工业的兴起，滴灌也得以蓬勃发展，水肥一体化技术逐渐得到应用。目前，该技术已普遍应用于各个方面，如果园、温室、田地以及植物绿化等，施肥系统也从单一的施肥罐发展到文丘里真空泵和液压驱动的肥料注入器，且伴随着计算机控制系统设备的引入，各种养分的分布均匀度明显提高。在以色列，水肥一体化技术的应用率在50%以上，居世界第一位。

2.2.2 美国水肥一体化技术的发展历程

1913年，第一批滴灌事业在美国起步。因其起步较早，国家扶持力度大、农场认可，现在成为微灌面积最大的国家。灌溉农业中60%的马铃薯、25%的玉米、33%的果树均采用水肥一体化；用于水肥一体化的专用肥料占肥料总量的38%以上。[3]目前，加州的果树种植管理大都采取滴管、渗灌等水肥一体化技术，且已建立了完善的水肥一体化技术应用设备和服务系统，成为世界现代高价值农产品现代化生产的典范。

2.2.3 德国水肥一体化技术的发展历程

1920年，德国农技人员经过多次探索试验，使水通过孔眼流出，大大提高了水资源利用率，实现了水出流的突破。1950年左右，世界塑料工业应运而生，高效灌溉技术也随之迅猛发展，灌水作业与施肥作业被联合应用，成为一种可精准控制土壤水分、养分的新农业技术。

3 我国水肥一体化技术发展现状

我国农业智能化技术的初期发展较为迟缓，基本皆是从美国、墨西哥、以色列等国引进的。由于我国的科学技术水平与发达国家存在一定差距，特别是在工艺、原材料等方面，导致国内在这方面的技术相对落后于较早进入智能化农业发展阶段的国家。

3.1 我国水肥一体化技术的发展历程

1974年，我国从墨西哥引进了滴灌设备，此后，滴灌技术水平得到进一步提高。1988年，我国成功自主研发制造出首代整套滴灌设备。此后，随着国外高新生产技术的引进，我国滴灌作业逐渐形成规模，水肥一体化技术逐渐从理论试验与试点示范领域进入实际应用。到20世纪后期，我国逐步将理论与实践相结合，水肥一体化的技术受到高度关注。

3.2 水肥一体化技术应用存在的不足

首先，想要提高施肥率的农田大部分使用的是水溶性肥料，但如果水溶性肥料不能完全均匀溶解于水，很容易造成施肥不均匀，继而出现烧苗、伤根、幼苗不健康等现象。其次，水溶性肥料的养分含量高，肥效释放所需时间短，在土壤中的留存效果不佳，如果不高精度控制肥料使用量，避免肥料损失，则将降低施肥的经济效益，达不到产量高、质量优、效果好的目的。

4 水肥一体化技术的应用策略

4.1 提高微灌施肥的性能

提高微灌用肥的水溶性，加大对水溶性肥料的研发力度，确保利用科学的配方，同时降低成本，让农民以适宜的价格加以使用；提高微灌施肥设备的性能，在不影响其他功能的前提下，提升其防堵性。从针对性、简洁性和可行性方面对灌溉施肥制度进行完善。

4.2 创新多种水肥一体化模式

开发不同的技术形式，对比研制灌溉设备、检测仪器等，摸索技术具体参数，大力开发多种环境下的水肥一体化技术实施模式，提升水肥一体化技术的使用率，将其优势发挥至最大化。

4.3 大力发展示范基地

建立水肥一体化技术的综合多层次示范网络，形成示范基地。通过互联网连接各地示范基地，在全国范围内进行及时有效的信息传递，实现技术同步。

4.4 结合资本力量促进发展

结合资本力量，形成有效的推广机制。高鹏等[4]认为，应加大水肥一体化技术研发力度，并扩大示范推广范围，提供切实有效的理论支撑和科学化的技术指导；充分发挥农村专业合作社的效用，增强水肥一体化技术应用的产业化和规范化；公司创立以技术服务为主、产品营销为辅的市场销售模式，满足广大农户对系统维护和技术支持等增值服务的需求[4]。

4.5 应用新型现代栽种模式

根据水肥一体化对环境及生产机械各方面的需求，围绕水肥一体化为中心研发农业栽种新模式。确保土壤信息准确收集与实时有效的传递，把握土壤养分信息和植物蓄水状态，通过计算机分析数据，科学合理地制订灌溉方案。

5 结束语

水肥一体化技术的科学运用会扭转我国原有的农业灌溉模式，其作为目前提升水肥利用率的最有效途径，在农业生产中的应用范围越来越广泛，特别是水资源短缺的我国。近年来，我国水肥一体化技术在政府的推进下迅速发展，技术体系框架已经建立，但不够成熟，仍然存在很大的提升空间。结合实际国情，加速开发和研制适用的灌溉施肥设备，完善灌溉施肥制度和栽培措施等，全面提升水肥一体化技术应用水平，是将来重要的任务和努力方向。

参考文献

［1］ 郭嫣. 水资源保护与可持续发展的关系研究［J］. 资源节约与环保，2014

(10): 136.

[2] 穆贤清. 农户参与灌溉管理的制度保障研究 [D]. 杭州：浙江大学, 2005.

[3] 吴勇, 高祥照, 杜森, 等. 大力发展水肥一体化加快建设现代农业 [J]. 中国农业信息, 2011 (12): 19-22.

[4] 高鹏, 简红忠, 魏样, 等. 水肥一体化技术的应用现状与发展前景 [J]. 现代农业科技, 2012 (8): 250, 257.

水肥一体化技术的应用发展建议

随着国家现代化进程的不断推进，我国农业发展水平不断提高，对农业方面的要求也越来越高。水肥一体化是发展现代农业的关键技术中重要一项。正在生长中的农作物只有提高其所需土壤中养分的固定水平，才能够实现集中施肥和平衡施肥。本文介绍了水肥一体化技术在国内外的应用和发展状况，从不同的角度分析了技术的重要点，针对水肥一体化今后的发展前景提出建议。

水肥一体化技术是一种巧妙地融合施肥和灌溉的农业新技术。化肥通过压力系统或自然地形下降上升差，可溶性固体液体肥结合，根据土壤养分含量，作物品种需肥规律和特点，以供肥控水供应管道系统融合成液体肥料和灌溉水配在一起，肥料混合通过管道和滴灌灌水器均匀、定量、定时通过水分渗透到作物根系供给农作物生长和发育，从而使主根部保持疏松土壤和适当的水含量。也可根据不同的植物特性需要施肥，土壤养分状况和环境，蔬菜所需水肥规律进行设计，把水分与养分定时定量，按照一定的比例直接提供给作物。

1 水肥一体化技术

水肥一体化技术从狭义上讲，就是通过在水源或灌溉水中将可溶性肥料均匀混合，借助灌溉管道的输送，运输给农作物。而从广义上讲，就是水和肥供给到作物根区土壤同时进行，满足作物所需所求。水肥一体化技术是水与肥料融合同步的一项农业新技术新方法，它将灌溉水与可溶性或液体肥料结合，通过借助外界压力并同步运输到作物有效根系邻近的土壤，通过监测土壤养分和土壤的干湿程度情况，及根据作物需水、肥规律的特点，经由低压管道滴灌体系稳定、均匀、定量，按适宜比例提供作物适宜水量和肥料。缓慢的滴灌过程中作物吸收水分也吸收养分，有人形象称这种技术为"匙喂"，又称之为"水肥耦合""随水施肥""滴灌施肥"。水肥一体化具有提高水肥利用率、增加产收，减少水肥施用量，保护环境的优势[1]。

1.1 国外水肥一体化技术

在以色列化肥一体化进程尤为经典。20世纪中期，随着工业塑料产业的发展，开始开发利用水肥滴灌集成技术。在今天的以色列，该技术被广泛应用于各个方面：果园、大棚、现场、园林等，灌溉区域面积占一半以上的比例，居世界第一位。水肥一体化技术，被广泛应用在那些缺水干旱和经济发达的国家。

在美国，1913年建立了第一个滴灌项目，目前为止美国是世界上最大的微灌面积国家，60%的马铃薯，25%的玉米，33%的水果使用水肥一体化技术。新型水溶性肥料的研制和应用，农药喷射控制装置，用于水肥一体化的专用肥料占肥料总量的38%。

加州目前已建立了完善的水肥一体化服务体系和设施，水果生产均采用滴管，渗灌水和化肥集成技术。

在德国，20世纪50年代以后，塑料行业的兴起，高效灌溉技术得到了迅速发展，灌溉和施肥的组合很快就发展成一种高精度控制土壤养分和水分的农业新技术。

最近几年在澳洲，水肥一体化技术也发展迅速，2006—2007年设立总额100亿澳元的国家水安全计划，用于发展该技术，并建立土壤墒情监测系统指导施肥[2]。

1.2 国内水肥一体化技术

1998年，我国自主研制的第一代滴灌设备。此后，中国引进了大型灌区先进的生产技术，规模化的灌溉生产在我国逐步实现。施肥集成技术从田间试验和示范应用面积逐步扩大。到了20世纪后期，越来越多于水肥一体化的技术关得到高度重视，技术人员的专业培训，组织开展技术研讨，资金方面也得到了国家拨款支持。2000年水肥一体化的技术指导与培训得到进一步的发展，我国连续5年开展了技术培训课程。将理论技术和实际操作经验结合在一起，加大了微灌施肥的面积。目前，水肥一体化技术发展已经由过去当地试验示范到现在的大规模应用，从中国北方辐射传播到西北干旱地区，中国南方的东北寒温带和亚热带地区。涵盖了水果种植、无土栽培、种植以及蔬菜、花卉、苗木，等大田作物种植和各种农作物，特别是在世界领先水平的西北膜下滴灌施肥技术。继南宁、北京、昆明、临沂、以色列，2014年7月6日中国—以色列施肥一体化首脑会议在西安隆重开幕。

"中国·以色列水肥一体化技术应用国际峰会"是2014年水肥一体化技术推广的盛大会议，已经在我国与以色列举办了六场，会议以政策解读、技术交流、经验分享和实地考察等形式，让全国各地的近4 000多名农业技术推广人员、种植大户对国内外水肥一体化技术有了更加深刻的认识，并受到了深刻的启发[3]。

2 水肥一体化技术的优点

"三节"在水肥一体化技术中主要表现为节水、节肥、节药，"三省"包括省工、省力、省心，"三增长"，包括增产、增收、增效，是现代农业的发展，加快农业转型发展"头号技术"。

水肥一体化技术是现代设施农业技术，其核心是通过滴灌设施、水、肥料和土壤处理直接有效的投放块根作物，以达到省水、保肥、省工、增效的目的[4]。

2.1 节约资源

在节约资源方面，首先为节水，节约用水是通过滴灌设施滴灌的一个基本概念，提高用水频率，减少每次用水数量，根据不同作物和不同生长阶段，水分3~10每平方米，只有沟灌或大水漫灌1/50~1/10，总耗水量只有沟灌或大水漫灌1/5~1/4。其次是节约化肥，全埋不仅能滴灌，而且可使肥料均匀直达作物根部，高效集中施肥，减少了水分流失与蒸发和被土壤固定等损失量，不仅可以节省化肥施用，也可以提高工作效率，在一块土地的设施，每滴水孔均匀地实现全埋滴灌。甚至可以根据用水，化肥和能源作物需要进行施肥，施肥的时间由你轻松控制。还可以减少病害，减少用药区域：很多病害都是由于过度的田间湿度，而水和肥料集成技术不但可以有效控制田间湿度从而减少病害的发生，也对土传病害进行有效的控制。最后，还可以减少杂草，因为整个滴灌埋在土中，表土干燥，不容易滋生杂草。还可以防止土壤板结，传统的灌溉用水，由于重力作用影响，经常野外作业，以及较少的水或大量的水易造成致病微生物，尤其是有氧微生物等方面的原因，容易使土壤板结，影响农作物生长，水肥一体化技术为这些问题解决提供了途径和方案。

2.2 省工省心省力

在省工方面水肥一体化技术不需再单独花时间灌水、施肥，减少农药、除草、中耕，大大节约了工时。在节约成本上来说如水和肥料、农药和人工成本方面降低了生产成本，提高了生产的效率与收益。一至三季生产可收回，设施则可多年使用。

2.3 增产增收

增产增收增效、促进生长、高效生产。除水稻及一些水生作物，大部分作物会因为土壤中的水分多了或少了而影响生长，滴灌使作物根部水分保持在最佳状态，使作物在整个生长周期中保持持续、强劲的生长发育过程，奠定了优质和丰产的基础[5]。

3 存在问题与对策

发展现代集约型农业水化肥一体化设备是先决条件，在中国起步较早，但由于技术、经济等方面的问题，它的发展是近年来较为缓慢，高效节水农业的发展和水资源管理最严格的制度实施，为水肥一体化设备的发展起助力作用，但是还是存在下面的问题[6]。具体体现为：①农户认识不到位，大多数只追求眼前效益。②水权市场现行制度普及率低，再加上农业用水政策，劳动力和灌溉水其他投入成本的大量补贴。成本低，农户节水积极性不高，对水肥一体化推广产生阻力。③现有水肥一体化控制设备存在集成度不高，田间现场数据采集传输有困难，易受环境因子耦合的影响等问题。④目前市场上灌溉施肥和水肥一体化设备种类繁多，微灌灌水器等制造精度较低，配套性差。

针对上述问题，应使水权管理制度健全，阶梯水价的实施促进水权交易，使农民节水积极性得到提高，促进政府补贴和激励措施的实施，积极引导农民。有计划地进行种植结构调整，鼓励农民种植低耗水高产作物，降低种植高消费类型作物。继续加强科研水肥一体化设备，对其结构进一步优化，最大限度地降低成本和人工，开发基于GPRS/SMS智能控制系统，模糊控制和虚拟仪器，改善灌溉和施肥准确性；使用网络应用系统和营销，制造微型化，智能化，低成本的水肥一体化灌溉和施肥设备，积极推广高效节水农业和高产、优质、高效农业[7]。

4 发展前景

水肥一体化技术是保持作物得到充足营养肥料的同时并节约水肥的重要创新。当今世界上淡水资源严重短缺，中国作为人均淡水量只有世界平均水平四分之一的国家，走高效利用水资源的农业发展道路迫在眉睫。中国也是世界上化肥消费量的大国，单位面积施肥量在世界排名居于首位，全国化肥生产需要消耗大量的能源，节约肥料也是节约能源的间接有效实施措施。

水肥一体化技术在较大范围内得到有效推广和应用，具有不可替代的重要意义，其意义不仅在于节约用水本身，随着水肥一体化技术在更大区域的推广和跟进，也是中国从传统农业走向现代化农业的一项革新。

参考文献

[1] 高鹏，简红忠，魏样，等．水肥一体化技术的应用现状与发展前景［J］．现代农业科技，2013（8）：250-250，257．

[2] 吴娜，刘吉利．宁夏水肥一体化存在问题及对策［J］．宁夏农林科技，2013，53（10）：124-126．

[3] 万军，典瑞丽，赵献章，等．烟草水肥耦合技术的研究现状与展望［J］．贵州农业科学，2011，37（11）：68-70．

[4] 夏立忠，韩庆忠，向琳，等．三峡库区柑桔园水肥一体化管理的对策［J］．农业环境与发展，2012，29（6）：12-15．

[5] 赵吉红．水肥一体化技术应用中存在的问题及解决对策［D］．杨凌：西北农林科技大学，2015．

[6] 李茂权，朱帮忠，赵飞，等．"水肥一体化"技术试验示范与应用展望［J］．安徽农学通报（上半月刊），2011（7）：100-101．

[7] 高祥照，杜森，钟永红，等．水肥一体化发展现状与展望［J］．中国农业信息，2015（4）：14-19．

水肥一体化技术在设施温室中的应用分析

我国是一个水资源十分匮乏的国家,传统的灌溉和施肥方式造成了水资源的严重浪费,化肥的不合理使用对环境造成了极大破坏。水肥一体化技术是一种将节水灌溉技术与高效施肥技术相结合的新型农业灌溉技术。水肥一体化技术可以有效节约水资源和化肥,可以提高生产效率和效益,减轻对环境的破坏。

水肥一体化技术是一种现代农业技术,做到了节水灌溉技术与高效施肥技术的结合。从狭义上讲,水肥一体化技术是将肥料按比例溶解于水中,通过田间供水系统将肥料和水均匀地输送到作物的根部区域[1]。从广义上讲,就是将水和肥料同时供应给农作物。

1 水肥一体化技术的主要特点

1.1 节约水资源

在温室大棚中使用水肥一体化技术可以有效减少水分的流失和蒸发,从而提高水资源利用率。采用传统的灌溉方式,水资源利用率为45%左右,浪费了大部分水资源。在水肥一体化技术在的滴灌方式,水资源利用率为95%左右[2],节水效果非常明显,有利于缓解我国水资源稀缺现状。

1.2 提高肥料的利用率

在温室大棚中使用水肥一体化技术便于控制各种肥料的比例、肥液的浓度和灌溉时间。使用滴灌的方式进行灌溉施肥,肥料可直接作用于作物的根区,加速作物对营养物质的吸收。传统的肥料利用率约为35%,而滴灌的肥料利用率在75%以上[3],节肥效果明显,提高了肥料利用率。采用水肥一体化技术可以有效避免人工施肥对肥料的滥用,从而保护环境。

1.3 节约时间和生产成本

水肥一体化设备操作简便,利用温室大棚中铺设的管网供水,可以做到自动化灌溉施肥,免去了人工挖沟、人工撒肥等过程,节约了人工施肥时间,同时节省了生产成本。

2 水肥一体化系统简介

温室大棚中的水肥一体化系统主要由灌溉用水源、首部枢纽工程、输水管网及灌水器4部分构成[4]。

2.1 灌溉用水源

温室大棚水肥一体化系统中常用的灌溉水源有池塘水、井水、水库和河水等水源,也可以根据实际需要在温室大棚周围建设人工蓄水工程，灌溉用水源在蓄水量和水质方面要满足灌溉要求。水源要充足，水质要洁净，没有受到污染[5]。

2.2 首部枢纽工程

温室大棚中的首部枢纽工程（图1）是灌溉系统的重要一环，主要包括为整个系统提供动力的水泵、过滤设备、肥料罐、吸肥装置、控制设备及量测装置等。首部枢纽的作用是对输送过来的水进行过滤、加压。在控制设备的控制下按农户提前预设的比例和浓度进行混肥，最后把混合好的肥液输送到输水管网中，量测装置可以监测设备的运行情况并记录相关流量。

2.3 输水管网

输水管网在温室大棚中的作用是将首部枢纽输送过来的水按照作物的需要输送到每个灌水器中。温室大棚的输水管网由干管、支管和毛管3级管道构成。具体的铺设需要根据温室大棚中作物的种植情况而定，毛管作为整个输水管网的末级管道直接连接灌水器。

2.4 灌水器

灌水器的作用是对输水管网里的水减压，最后以水滴的形式施入作物根部区域。温室大棚的灌水器主要是滴头、滴管带等。

3 水肥一体化技术在农业大棚中的应用

3.1 应用的必要性

虽然中国水资源总量居世界第六位，但水资源分布不均，中国许多地区水资源稀缺，人均拥有量低。传统的灌溉和施肥技术水资源浪费严重，化肥的不合理使用污染了环境。因此，非常有必要推广节水节肥技术，水肥一体化技术的应用实现了节水和节肥的目的，非常值得推广。

3.2 应用存在的问题

水肥一体化技术在温室大棚中的推广和应用效果良好，但也存在一些不可忽视的问题。首先，我国水肥一体化技术的推广和应用还处于起步阶段，相关的配套设施不完善，如水源和供电等基础设施不健全。水肥一体化设备需要一定的空间，传统的温室大棚在建设之初没考虑这一问题，所以可能需要改造或重建。其次，与水肥一体化设备配套的肥料较为缺乏，我国液体肥的研发还处于发展阶段，液体肥在国内的市场份额较小，市面上仍以传统的肥料为主。最后，农业收入相对较低，加之人们只关注短期效益，而水肥综合设备的一次性投资相对较高。因此，农民对实施新技术缺乏积极性。

3.3 应用前景

在温室大棚中应用水肥一体化技术可以实现节水和节肥，有利于我国农业的发展。随着国家政策的不断出台，水肥一体化设备会逐渐普及，相关的配套设施也会更加完善。随着国家有关标准的制定，水肥一体化也会越来越规范。

物联网技术和传感器技术的不断成熟，融合了物联网技术的水肥一体化系统会越来越智能。水肥一体化系统可以根据传感器收集到的数据，为农户提供更加科学的施肥建议。农户可以利用手机等终端得到水肥一体化设备的运行状态以及温室大棚中实时的环境数据，并远程操作水肥一体化设备，融合了信息化的水肥一体化系统会越来越受欢迎。水肥一体化技术的不断发展，可加速我国从传统农业向现代农业的转变。

参考文献

[1] 曹洪祥．葡萄种植中水肥一体化技术推广应用分析［J］．农民致富之友，2018（24）：60．

[2] 张敏．水肥一体化滴灌技术在设施蔬菜中的应用［J］．农业科技通讯，2017（12）：337-338．

[3] 刘磊，吕令华，刘秋兰．温室大棚番茄水肥一体化技术应用效果研究［J］．农业科技通讯，2015（5）：181-183．

[4] 徐卫红．水肥一体化实用新技术［M］．北京：化学工业出版社，2018．

[5] 王丽萍．水肥一体化实用技术［J］．农民致富之友，2019（3）：65．

水肥智能调控技术在设施温室中的应用

在温室蔬菜大棚中采用水肥一体化灌溉设备,通过智能水肥一体化技术根据蔬菜生长所需的营养元素进行有效的需求设计,把水与肥料按照比例进行混合,利用滴灌带将肥水以较小的流量均匀传送至蔬菜根部。所以在温室大棚中应用智能水肥一体化技术可以有效地解决水资源浪费严重以及肥液配比不准确的问题,从而实现蔬菜品质提升、产量提高的目标。

水肥一体化技术从狭义上说,将可溶性肥料同水混合灌溉到作物处,一般而言,就是将水和肥料通过混合后供应至所需作物,保证作物的生长需要[1]。智能水肥一体化技术是指将传统的水肥一体化技术加上现代农业科技借助于新型产品设备对可溶性颗粒或者液体肥料,按照作物生长需要规律以及土壤营养元素含量勾对成相应的营养液,而且可以定量定时的、科学的、准确的提供给作物。

1 当前我国蔬菜温室大棚发展分析

1.1 农业用水资源现状

"我国农田灌溉水有效利用系数远低于 0.7~0.8 的世界先进水平;单位用水的粮食产量不足 $1.2kg/m^3$,而世界先进水平为 $2kg/m^3$ 左右[2]。"现在我国大部分种植区基本上还是采用传统农业灌溉,这样不仅浪费了大量的水资源而且作物的生长也起不到最大的促进作用。而且在我国淡水资源紧缺严重,在农业上还存在灌溉设备老化、节水意识不强等问题,所以运用现代科学技术发展智慧农业灌溉设备势在必行。

1.2 蔬菜温室大棚水肥基本情况

在肥料的使用量上我国也远超其他发达国家,根据中商情报网数据显示:我国单位耕地化肥消费量远超世界平均水平(图1),这说明我国在化肥使用上存在严重的浪费现象。"蔬菜一枝花,全靠肥当家",在我国蔬菜大棚中普遍存在着肥料使用过多,而且科学施肥知识不足,大多数农户普遍认为"肥大水勤,不用问人"。在现实生产中,一些蔬菜温室大棚种植农户没有充分了解科学施肥的作用,不注意科学施肥,出现水资源和肥料浪费问题,没有达到提高产量的效果。所以,实现农业上新旧动能转换运用智能水肥一体化技术就显得尤其重要。

2 智能水肥一体化技术在蔬菜温室大棚中应用的优势

2.1 智能水肥一体化技术

智能水肥一体化技术主要就是通过灌溉设备和施肥装置为基础,使用控制设备按照配方对水和肥料进行比例混合,最终运输至作物。智能水肥一体化技术的系统组成如图

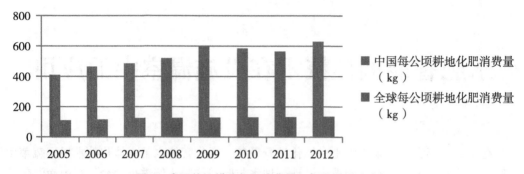

图1 我国单位耕地化肥消费量远超世界平均水平

2 所示。主要包括水源、加压水泵、叠片过滤器、沙石过滤器、流量计、压力表、水肥机智能控制平台、混肥桶以及田间滴灌管等。在棚内进行一些作物常用的传感器安装，采集温度、湿度、CO_2 浓度等与水肥机进行有效的结合使用。由于在水肥机控制设备中使用了文丘里管施肥器、EC 和 pH 值检测计，同时在混肥不充分时，加压泵可以增强其混肥效果。4 个混肥桶可以添加不同微量元素的肥料，通过水肥机的控制能够有效对作物进行灌溉施肥作业。

图2 智能水肥一体化技术装备

2.2 提高水资源与肥料利用率

与传统的大田漫灌方式相比智能水肥一体化技术采用滴灌的技术可以将肥液均匀准备的运送至作物。根据相关实验数据显示：使用水肥一体化技术施肥方法，水的利用率提高了 40%~60%，而且相对于传统的漫灌既省时又省力同时节约了人工成本，非常有利于大规模种植。

智慧的水肥一体化技术在施肥上采用水肥机根据植物所需自动配比的方式，并且使用滴灌带进行灌溉。在这个过程中减少了肥料和药物的挥发流失以及由于养分过剩造成的土壤板结所造成的损失，而且还可以结合当地土专家或者种植户的经验去控制水肥机进行精确的水肥配比。这样施肥非常简便节省了人工并且用肥准确，减少了人为的浪费，相比较于传统技术利用率提高了40%~50%。

2.3 提高作物品质与产量

大多数的农产品蔬菜都会出现一些病虫害的状况，而这些病虫害往往都是由于温室大棚温度过湿、阳光吸收不充足所造成的。利用水肥一体化技术能够有效地控制棚内湿度，可以降低病虫害对种植户造成的损失，另外，这些可以大大提高农产品的质量，起到了增收的作用，减少农药用量与人工打药的成本。

3 智能水肥一体化应用技术要点

3.1 远程控制

智能一体化水肥机可以根据用户调整自主添加灌溉计划，设定时间周期计划，能够实现智能化的控制。在整个水肥机设备中还安装有各种压力、流量传感器，能够有效监测灌溉运行情况，实现一种自动化的灌溉施肥，节约了大量的人力资源。

3.2 水与肥料的混合比例浓度准确

结合温室大棚内安装的各种传感器，通过水肥一体机进行有目的有比例的混肥。在混肥桶内放入农作物所需的微量元素，水肥机就能根据需要自动混肥并且通过滴灌管道定时定量均匀的运送到农作物根部生长区域进行供肥，使得土壤也处于比较松散的状态不至于出现板结的状况，改善农作物的生长状况。

3.3 提高农作物品质与产能

大多数蔬菜会因为土壤中的湿度以及营养物质影响生长，智能水肥一体化技术则可以根据棚内传感器所反馈的湿度等因素去配比一个合适的肥液以及合理的灌溉，使土壤环境保持一个蔬菜所需要的最佳环境，进而使整个作物的生长周期持续保持、旺盛的生长。

4 结束语

蔬菜温室大棚通过智能水肥一体化技术，能够有效的促进蔬菜生产达到提量、增强、提高经济效益，实现整个蔬菜业长远发展。同时也推动了我国农业新旧动能转换，更好更快的加快乡村振兴战略的实施，具有重大战略意义。

参考文献

[1] 张承林．国内外水肥一体化技术概况［C］．2010年中国水溶性肥料高峰论坛论文集．2010．

[2] 陈雷．水利部部长陈雷向全国人大常委会报告农田水利建设工作情况报告［R］．北京：国务院，2012．

网站创建与管理课程建设的探索与实践

课程建设是专业、学科建设的基础，是保证与提高教学质量的决定性因素。网站创建与管理课程组在分析当前课程建设现状的基础上，就课程建设的整体目标与思路、教学内容改革、教学条件建设、教学方法与手段等方面进行积极的探索与实践，促进了教学质量的提高，提升了创新人才的培养质量。

课程建设要以先进的现代高等教育理念为指导、以提高人才培养质量为目标、以教学改革为主线、以课程群建设为主体、以提高教育教学水平为重点，着重加强师资队伍、教学课件、网络教学视频、网络教学资源和教材建设，改革教学内容、教学方法和学生学业考核方法。

课程建设是专业、学科建设的基础，是保证与提高教学质量的决定性因素，对于构建大学生合理的知识结构、能力结构和培养创新精神具有十分重要的意义。网站创建与管理课程组在分析当前课程建设现状的基础上，就课程建设的整体目标与思路、教学内容改革目标与措施、教学条件建设的目标与措施、教学方法与手段、改革目标与措施等方面进行积极的探索与实践，促进了教学质量的提高，提升了创新人才培养质量。

1 课程建设现状

1.1 教学队伍

本课程组目前由5人组成，全部为中青年教师，从年龄结构看，35岁以下2人，35~50岁3人。从职称分布来看，3人具有副高级职称，2人具有中级职称。学历结构上看，2人已取得博士学位，1人为在读博士，2人为硕士。

自2004年以来，课程组先后为本科生开设《网站创建与管理》《计算机网络技术》《网页设计》《网络广告设计与制作》《网络数据库技术》《交互媒体设计》《网络多媒体技术》《Internet网站建设》《网络数据库应用开发》等课程，教师开课能力有了较大的提高。经过多年的教学实践，个人讲授的课程均在3门以上，开课能力指数达85%以上，年平均工作量达到350个以上。

在近几年的教学实践中，课题组针对当前的行业需求和学科发展，考虑社会就业难的实际情况，积极组织教师参加了2011年全国新闻传播学本科人才培养和教学会议，到中国农业大学、中国农科院、清华大学、中国传媒大学、北京邮电大学、青岛市软件园进行调研，到复旦大学新闻学院、南京师范大学新闻传播学院、浙江传媒大学等兄弟院校进行调研学习，了解专业当前的总体形势和发展需要，为开展教学工作和课程建设提供了宝贵经验。课题组重视后续教学，先后有2人取得博士学位，3人取得硕士学位，1人正在攻读博士学位。

教学研究方面，课题组围绕该课程完成了多项教学研究课题，包括校级优秀课程、校级教改项目、多媒体课件立项、实验技术课题等，并取得了较好的实际效果。发表教学研究论文4篇，其中一篇被CSSCI收录，获得山东省优秀教学成果一等奖一项，校级教学成果一等奖一项。

在教风和教学效果方面，课题组的所有教师都在教学一线兢兢业业，一丝不苟，多年来无教学事故发生。学生对课程的评价较高，对课程具有浓厚的兴趣。从教学评价反应的数据来看，自开课以来课程一直保持较高的评价，连续3年来均在90分以上，学生的及格率也保持在95%以上。

1.2 教学内容建设

网站创建与管理作为网络传播学专业学生的一门课程，让学生从Internet和计算机网络的用途入手，通过了解Internet的产生、发展过程，引导学生理解网站在Internet中的重要地位，熟悉网站的各种功能，并结合实验教学，熟悉网站工程中常用的软硬件设备，掌握基本的网站创建与管理技术和方法，具备较强的动手实践能力和逻辑思维能力。主要内容包括：网站的工作原理、网站构成、网站建设方法、网站平台建设、网站内容设计与创意设计、网站的发布、网站管理、信息查询与搜索引擎、商务网站的经营管理等。

在具体授课过程中，既重视理论学习的重要性，又兼顾实践教学的现实性。考虑到授课对象是非计算机专业学生，对部分概念掌握有难度，所以理论的讲授不宜过深，以够用为宜，侧重应用和实践技能的教学，但对部分学习成绩优秀的学生，又要兼顾今后的发展，对重要的理论知识进行特别的说明。针对该课程所涉知识更新快，主流技术不断变化等特点，也适当地增加部分新的教学内容和行业理念，如CSS+DIV，JQuery，信息架构，UI（用户体验）设计等。

在教学内容的组织与安排上，考虑到课程的特点，采取灵活的授课方式，课内和课外结合，理论和实践结合，加强课程的实践教学，实验总学时数达20课时。此外，还鼓励学生参加一些力所能及的项目，在任务驱动中学习。经过近5年的摸索，讨论并制定了相对合理的教学大纲、教学计划、讲义、教案等基本教学文件，经过几轮的应用和反复修改，符合专业培养方案的要求和学校培养应用型人才的总体目标。

1.3 教学方法手段和环节建设

本课程重视理论和实践的结合，根据学生的特点，将课程内容模块化设计，采用任务驱动、探究学习等多种教学方法，重视"以学生为中心"的教学理念。在教学手段上，采用多媒体技术和互联网等现代教育技术手段，在保证教学效果的前提下大大提高了课堂教学的效率，扩展了学生的知识面。在课堂上，较多地采用讨论式的教学方法，启发学生积极发言，调动学生的积极性和主动参与学习的兴趣。

在课程的考核上，采用多种考核方式相结合，既重视期末考试的成绩，又重视平时作业、实验、课堂表现以及考勤的成绩。课程在期末考试中所考察的知识点侧重基本概念的理解、掌握以及用基本的原理、方法去解决实际问题，考察难度适中，覆盖面广，学生成绩连续几年均保持较好的正态分布。同时也在积极探索改革课程的考核方法，综合考虑课程的性质，尝试期末考核以作品为主的考试方法，摆脱笔试环节的繁琐。

在教学环节的质量监控方面，严格按照学校教学质量监控的管理方法，实现学校、学院、学生评价多层次监控体系。根据监控结果和反馈情况及时的调整教学方法。

2 课程建设规划

2.1 课程建设的整体目标与思路整体目标

建成一门具有较高水平师资队伍，教学内容科学合理，教学条件较为完善，教学方法灵活多变的精品课程。学生通过该课程的学习，能够掌握独立策划、设计、建设以及管理一个中小规模农业网站的能力，为将来的专业发展打下良好的基础。

为实现上述目标，课程规划遵循如下思路：①首先提高师资队伍的整体水平。通过参加各类学术会议、课程建设研讨、参观交流访问、短期培训等方式，把握学科发展的重点领域和学术前沿，不断提高自身的学术水平。②优化课程内容和知识体系。通过整理完善教学中的经验、教训，发现教学过程中的成功之处和不足，进一步完善课程内容和知识结构。③加强教学条件建设。完善大纲、授课计划、教案、课件等资源，建立课程网站和教学录像库。④加强教学环节建设。课堂上注重讲课过程的效果，运用现代教学理念和教学方法，调动学习的积极性，培养学生的思维能力和综合创新能力。加强课外实践的教学，提高学生的综合动手能力。

2.2 教学内容改革目标与措施

进一步优化教学内容，使其符合专业教学的需要，在原来课程的基础上进一步优化理论教学内容，使知识更加系统，重点突出在侧重应用方面的一些要点，为进行实验和实践等提供理论指导。具体来讲，就是研究传播学的理论和信息组织理论，紧密结合农业网站的需求，分析农业网站的特征和区别其他类型网站的关键要素，在基础理论的内容方面有所创新。在实验和实践方面，一方面实验侧重基本原理的验证工作，让学生对理论授课中的基本概念掌握，加深理解；另一方面要紧密结合当前的行业发展，根据当前的行业要求和标准进行网站的规划、开发，特别是对农业网站的开发培养一定技术专长。

2.3 教学条件建设的目标与措施

实验教学方面，根据总体目标完善实验指导书的修订，并打印装订成册，供学生在实验室使用。

教材方面，引进国内外一部分优秀教材，作为教学过程中教师备课的参考书。

网络教学环境方面，积极向学校申请更新网络实验室的软硬件设备，重新规划机房的建设工作。

教学资料方面，充分利用网络教学平台，完善基本教学资料，包括大纲、讲义、资料、课件、案例等内容，利用教学平台进行作业布置，答疑、在线提交作业等。

教学基地方面，继续与原来的友好单位保持联系，同时扩大联系范围，寻找建立更多的实习基地。

2.4 教学方法与手段改革目标与措施

在现代教学理念的指导下，积极探索各种有效的教学方法，采用探究式、讨论式、任务驱动式等教学方法。课后布置适当的任务，增加学生动脑思考和动手实践的机会，结合实际情况建立学生的学习兴趣小组,共同完成一个农业网站的建设。来调动学生的学习热情,增强对课程的学习兴趣。

在考核方式上,采用课堂、平时成绩、期末作业、中期作品等多种考核形式,打破传统的仅仅以试卷成绩考核的机制。

3 结束语

课程建设要以先进的现代高等教育理念为指导、以提高人才培养质量为目标、以教学改革为主线、以课程群建设为主体、以提高教育教学水平为重点,着重加强师资队伍、教学课件、网络教学视频、网络教学资源和教材建设,改革教学内容、教学方法和学生学业考核方法。课程组对网站创建与管理课程按照精品课程的标准与要求进行积极的探索与实践,促进了教学质量的提高,提升了创新人才培养质量。

参考文献

[1] 李晓辉,朱勇,周东升,等.Visual FoxPro 数据库与程序设计精品课建设探索 [J]. 农业高等教育,2014 (4):72-74.

[2] 赵正道. 关于精品资源共享课程建设的思考 [J]. 农业网络信息,2014 (5):149-150.

[3] 王丽红,坎杂,江英兰,等.《高等农业机械学》课程建设与改革初探 [J]. 农业网络信息,2014 (11):153-154.

我国农业信息管理系统发展现状及趋势分析

20世纪80年代中期,我国开始研究农业管理信息系统,1990年便开发出棉花生产管理模拟系统,有效地将播种期、种植密度、施肥量和化学调控等结合起来,为棉花高产和优质栽培提供了更加便利的方案。随着社会经济的高速发展,2000年以后我国农业信息化事业取得了从无到有的成绩,农业管理信息系统也在诸多领域得到了应用。基于此,本文简要介绍我国农业信息管理系统的发展现状及趋势。

1 农业管理信息系统发展现状

1.1 种质资源管理

种质资源管理系统在农业管理信息系统中十分常见。以南京农业大学的基于网络化的杂草标本管理信息系统为例,林春华等[1]开发了蔬菜种质资源图文系统,多达81种蔬菜信息被存放在此系统中,有1500多份种质信息被保存于此,该系统实现了多项性状的检索查询和分类统计。

1.2 土地资源管理

土壤信息系统在农业管理信息系统中也很常见,其是在信息化的支持下,将土壤及相关数据按照统一的地理坐标,按照相对应的编码和格式,采集、存储、操纵、修改、分析和综合应用的技术系统。土壤信息系统的建立意义深远,其能够精确快速地对土壤调查和研究的数据进行管理,按照不同的专业需要进行解释和评价,促使土壤信息得到最大化的利用。

1.3 农业机械化和农资企业管理

农业机械化管理司早期建立了农业机械化的管理信息系统,提供了农业农村部农机化管理司与地方农机管理部门及协同单位之间的信息传递的途径,能够合理共享有效资源、数据,及时地提供有价值的信息,提高政府主管部门宏观调控农业农村的科学性。在农资企业中,服务种子企业的管理信息系统针对种子行业的经营规模和发展目标,以及目前的经营状况、仓储、销售实时状况等,提供进货、出货的信息,通过对销售和结算的有效管理,能清楚地分析利润,梳理一系列报表、催款单、出入货单和销售咨询等情况。

2 当前农业管理信息系统的弊端

2.1 农业管理信息系统利用率不高

对于管理、数据、编码等来说,行之有效的标准是亟需统一的。当前的农业信息系统因为标准不统一导致共享困难、缺乏交流,系统的兼容性也比较低。在农业管理信息系统的发展过程中，甚至还会出现重复开发的问题，农业管理信息系统一直处于完善过程中。

2.2 期望值过高，片面追求大而全

当前，许多农业管理信息系统追求大而全，在实际开发过程或应用中，严重缺乏数据的支撑，开发出来的信息管理系统没有充实的内容，缺乏实用性和针对性。在实际生活中，人们往往对农业管理信息系统的效果期望过高，不能很客观地评价该系统的作用。实际上，当前农业管理信息系统缺乏信息资源管理本身的信息化建设，缺少计算机辅助下大数据的应用。

2.3 软件不配套，轻视维护和过程

在农业管理信息系统研发过程中，人们时常陷入一种误区，会把硬件设备的好与坏当作是否具有可行性的标准，对软件开发十分不重视，忽略了软件开发在系统开发中的作用。尤其是在后期的维护中，对系统的维护不够积极，做不到及时创新与发展。在系统开发过程中缺少一个标准，开发团队、开发文档、开发过程都存在质量不一的现象，严重影响了后期的维修、维护。

2.4 农业信息化队伍建设滞后

当下，懂得计算机技术的人很多，但是能够兼备农业专业知识与才能的人很少，农业信息化队伍的建设因为人才资源的缺乏，相对比较薄弱。因此，在研发的过程中，研发团队和专业技术人员脱节比较严重。

3 农业管理信息系统的前景

3.1 系统架构网络化

随着计算机网络技术的发展，网络化成为农业管理信息系统发展的必然趋势，农业管理信息系统的规模和覆盖范围将会不断扩大，传统的桌面型农业管理信息系统已不适用于当今的现代农业发展，网络化的农业管理信息系统在未来更具有发展潜力。

3.2 智能化决策支持

管理信息系统将与模型技术和人工智能技术进一步结合，由传统的提供一般业务管理和信息服务向智能化决策支持方向发展。

3.3 开发方法多元化、更兼容

面向对象的分析、设计和开发是国内外新兴的开发和研究方法，在某种意义上讲，更贴合人类的自然思维，擅长分析一些复杂多样的系统，并且能够巧妙运用软件复用的优势，使开发和管理的方法更兼容、更多元化。

3.4 海量数据处理异军突起

随着计算机软硬件技术的成熟，当今社会人们不断熟悉计算机的应用，管理农业信息系统是一件复杂而长期的工作。目前，农业信息系统所需要分析和管理的数据量庞大，传统的数据库技术已不适用于大数据的海量管理。因此，数据仓库技术出现，除了管理与分析海量的数据，甚至还可以为数据挖掘提供智力支持。

3.5 新兴技术汇聚

当前，许多新技术在农业管理信息系统中越来越受重视，其中一些已成为某些领域

管理信息系统的必要组成部分。全球定位系统、地理信息系统、遥感技术已与农业管理信息系统息息相关。农业生产离不开科研与自然资源，农业生产与地理、人文也有着千丝万缕的联系，其中地理信息系统已成为土地信息系统中至关重要的组成部分。新媒体技术或者当前备受关注的融媒体技术，是将更加丰富的媒体手段集成来展现信息的一种技术手段，这些新技术都将为农业管理信息系统注入新鲜血液。

4 农业管理信息系统的发展对策

4.1 力推农业信息化建设，培养新时代农业信息化人才

很多农业院校、农业科研所都开设了农业信息化、农业工程与信息技术等相关的专业，农业管理信息系统的建设作为其中一门课程也深深地影响着农业院校的发展走向。我国教育部致力于培养既精通农业专业知识又掌握农业管理信息技术的人才，因此，应加强学科建设，大大增加该学科的硕士点和博士点的数量，同时提高教学质量，不能使农业信息学科的学、硕、博之间脱节。同时，也要在以往的传统农科专业中逐步添加一些农业信息技术方面的知识，使更多农业院校的大学生能够掌握农业信息技术的基本手段和方法。

4.2 增强基础资源建设，积极开展农业服务

首先，要加强农业信息基础硬件、软件设施的建设，使农业管理信息系统能尽快以信息网络为载体，快速开展农业信息服务，软硬件齐抓，夯实农业管理信息系统的发展根基；其次，加强农业管理信息基础数据库的建设，虽然我国在已有的农业数据库方面取得了一定的成绩，但是很多成果处于试验阶段，还没有得到完善和巩固；最后，要在已有成果的工作上持续发展、壮大，促进农业信息管理系统的未来应用不是一朝一夕就可以完成的，有时会影响到农业决策支持系统和农业专家系统的发展。只有加强农业信息化的基础建设，积极开展农业信息化服务，才能取得良好的发展效果。

参考文献

[1] 林春华，张文海，谭兆平，等. 南方蔬菜种质资源图文信息系统的研究[J]. 广东农业科学，2000（6）：20-23.

无线传感器网络在环境监测应用领域的研究进展

环境监测是无线传感器网络的重要应用领域，分析了无线传感器网络的节点结构与网络体系结构的特点，概述了无线传感器网络在环境监测应用领域的研究进展，指出了无线传感器网络在环境监测应用中的一些发展方向。

继 Internet（互联网）之后，各种现代技术得到了迅猛发展，无线传感器网络集成了网络通信、传感器与微机电这三大技术，对人们的生活与生产方式产生了重大影响。Internet 彻底改变了人们之间的沟通交流的方式，对人类的生产与生活影响深远，而无线传感器网络将信息与物理世界有机地融为一体，从而改变了自然与人的交互模式。

近几年以来，微机电技术、集成电路与无线通信网络技术发展迅速，这些技术的发展与进步使得成本越来越低、功耗越来越小的多功能微型无线传感器大批量生产，从而推动了在实际生产生活中的应用。无线传感器网络在环境监测、国防军事、目标追踪、卫生医疗等领域应用广泛，在学术界与工业界引起了越来越多的关注。无线传感器网络是信息采集与感知领域的一场革命，将给人们的生活与生产带来巨大的变革[1-3]。

笔者通过分析结构特点，概述了无线传感器网络在环境监测领域的应用研究进展，就典型应用进行剖析，为无线传感器网络在环境监测应用中的发展提供一些建议与新思路。

1 无线传感器网络结构

1.1 无线传感器节点结构

无线传感器节点的结构如图 1 所示，主要包括传感模块（A/D 转换部件、传感部件）、数据存储控制处理模块（存储部件、微处理器部件）、无线通信模块（无线收发部件、网络协议栈）和能量供应模块（锂电池或干电池）这四部分构成，该结构是无线传感器网络的基本功能单元。传感器模块负责采集感知监测范围内的信息，并根据需要对采集的信息进行转化；数据存储控制模块负责协调与控制无线传感器节点的行为，存储采集的数据信息；无线通信模块主要进行无线传感器节点间信息的交互，采集数据的发送以及各种控制与管理命令的接收；能量供应模块一般为电池供电，提供节点运行所需要的能量。

1.2 无线传感器网络体系结构

无线传感器网络体系结构如图 2 所示。从图中可以看出无线传感器网络由汇聚点（sink node，通常一个，也可根据实际应用采用多个）、无线传感器节点（sensor node，包含各类传感器节点，可同构亦可异构）以及面向最终用户的各管理客户端（针对不同的需求采用不同的管理客户端）等构成。在监测区域内一般会部署大量的无线传感器节点，这些节点通常情况下是功能、能量均不同的异构节点，通过自组织的方式形成

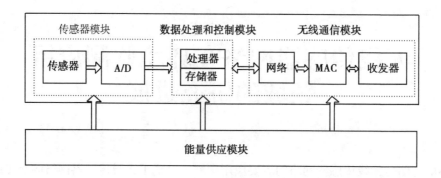

图1 无线传感器网络节点结构

无线通信网，感知采集的数据采取多跳或单跳路由至汇聚节点，汇聚节点通常通过网关（WiFi 网关、以太网网关、卫星网关、3G 网关等）接入 Internet，管理客户端通过 Internet 来访问无线传感器网络采集的感知数据信息，同时面向用户的各管理客户端可对无线传感器节点进行各种配置与管理工作。

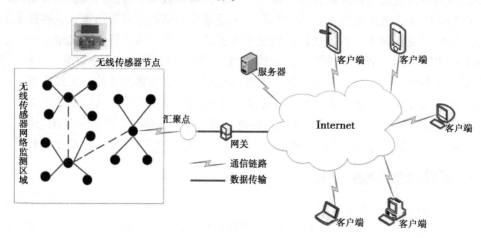

图2 无线传感器网络体系结构

2 应用动态

环境检测领域的应用包括海洋动植物生存环境监测、植物种植环境监测、动物生长环境监测、火山喷发监测、城市道路环境监测等。如在牲畜养殖场中，可用来采集养殖场环境中的各种环境参数，对养殖环境进行反馈控制。在大型农田灌溉区或育苗场，可监测土壤湿度与空气的温湿度以控制灌溉，采集育苗场的各种环境参数并对育苗环境进行控制。

美国加州大学伯克利分校对一颗红杉树的生存环境进行监测，通过将无线传感器网络部署在该树上，采集光合作用、温湿度等信息；该校也开展了代号为"Firebug"用于野外火险监测的研究，用来监测大鸭岛的环境参数；在多山多雨地区部署无线传感器网络，可用于预测、监测泥石流的发生，降低危害。北卡罗莱那大学与哈佛大学合作来对火山爆发情况进行检测，主要是通过无线传感器网络收集次声波与振动信息。新南威尔士大学研究了一种名为"Cane toad"的蛤蟆在澳大利亚的分布情况，主要是通过无线传感器网络采集跟踪完成的。中国的"863"课题"精细养殖中的数据采集与无线传输技术的研究"，重点探索了奶牛个体信息采集，该研究综合运用了无线传感器网络、射频识别、无线局域网以及多种信息技术。

在农业环境监测应用领域，可通过无线传感器网络对各种作物种植环境的 pH 值与各种肥料的含量、空气的温湿度、光照强度、土壤湿度等进行监测，通过长时间的观察与统计，确定农作物的最佳生长环境，同时可对病虫害的预警与防治开展相关工作。在农业应用领域的传感器一般要求成本低、部署简单、维护方便等，这样可通过大量密集部署的方式来监测农作物生长环境，为农作物的最佳生长环境与环境的监测以及保护等提供了新的方式与途径。

无线传感器在动物养殖环境监测有着重要的应用，通过采集动物养殖场的温湿度、氮气浓度、沼气浓度、CO_2 浓度、光照等参数，对养殖环境进行反馈控制，对紧急情况进行预警，降低动物养殖人员的人力成本，为动物提供舒适的生活环境，可帮助养殖人员改进养殖方法，提高养殖动物的品质与质量。

3　结束语

基于无线传感器网络的结构与特点，随着传感器及网络部件的成本进一步降低，无线传感器网络在环境监测应用领域大有可为，必将对海洋动植物生存环境、植物种植环境、动物生长环境、火山喷发监测和城市道路道监测等起到保护和改善作用，对整个生态环境的保护具有重大意义。

参考文献

[1]　YICK J, MUKHERJEE B, GHOSAL D. Wireless sensor network survey [J]. Computer Networks, 2008, 52 (12)：2292-2330.

[2]　李艳琼，曾文，谢国亚. 一维纳米 SnO_2 传感器对变压器油中气体的检测 [J]. 电子元件与材料, 2013, 32 (3)：50-53.

[3]　钱志鸿，王义君. 面向物联网的无线传感器网络综述 [J]. 电子与信息学报, 2013, 35 (1)：215-227.

智慧农业现状与发展建议

智慧农业是智慧经济的重要组成部分，而农业信息化发展是大势所趋。在国家现代农业政策的引导和技术支撑下，我国现代农业实现了突飞猛进的发展。本文主要研究信息化背景下我国智慧农业发展现状及存在的问题，并针对相关问题提出适合国情的建议，为今后智慧农业的发展提供参考。

信息化是 21 世纪飞速发展的标志。在信息化背景下，农业也展现出蓬勃发展的态势。智慧农业是智慧经济的重要组成部分，是依托物联网、云计算以及 3S 技术等现代信息技术与农业生产相融合的产物，可以通过对农业生产环境的智能感知和数据分析，实现农业生产的精准化管理和可视化诊断。随着设施农业、休闲农业以及温室农业等现代农业的兴起，农业生产不再局限于季节变化和时间制约，大大提高了土地和生产资料的利用率。但是，由于劳动力有限，很多资源没有得到得到有效利用。智慧农业的到来有效解决了劳动力不足带来的问题，科技革命的蓬勃发展为智慧农业注入了源源不断的动力。

当前，我国智慧农业发展中存在一系列问题，主要包括高素质农民匮乏、职业农民教育体系还未建立、科研体系不健全、农业科技推广不力、基础设施落后、农机设备现代化程度较低等，亟待我国政府、企业和农业相关人员运用农业知识与技能加大对智慧农业的投入，助力我国农业快速实现现代化。

1 智慧农业的内涵及价值

智慧农业是充分应用现代信息技术的成果，结合农业自身的生产管理方式，集成应用计算机与网络技术、物联网技术、音视频技术、3S 技术、无线通信技术及专家智慧与知识，实现农业可视化远程诊断、远程控制以及灾变预警等智能处理，从而改变传统农业的弊端，促进农业走向信息化和现代化。智慧农业单从字面意思讲是使农业"变得聪明起来"，本质是现代农业的更高阶段。智慧农业通过生产领域的智能化、经营领域的差异性以及服务领域的全方位信息服务，推动农业产业链改造升级，实现农业精细化、高效化与绿色化，保障农产品安全、农业竞争力提升和农业可持续发展[1]。因此，智慧农业是我国农业现代化发展的必然趋势。促进智慧农业的发展，对于建设高水平的现代农业具有很高的价值。

2 智慧农业发展现状

2.1 智慧农业研究进展

2.1.1 国内智慧农业研究

在中国知网（CNKI）以"智慧农业"为关键词检索，从 2012 年开始，截至 2019

年 5 月，搜索到相关发表文献数量共 1 279 篇，如图 1 所示。

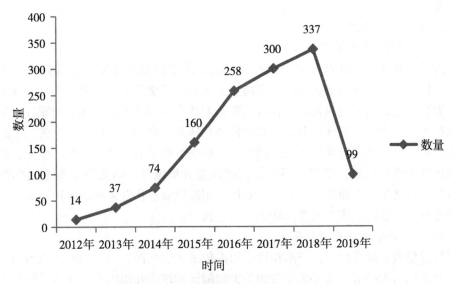

图 1　国内智慧农业相关论文发表数量

从图 1 可以看出，有关智慧农业方面的研究呈现出逐年递增的趋势，且这种趋势变化很大，与国内促进农业现代化发展的政策有关。国家对建设农业信息化的呼声越来越高，因此有关智慧农业的研究也越来越多。

2.1.2　国外智慧农业研究

在 Web of Science 上以"Smart agriculture"为关键词检索，2012—2019 年共有相关文献 529 篇，如图 2 所示。

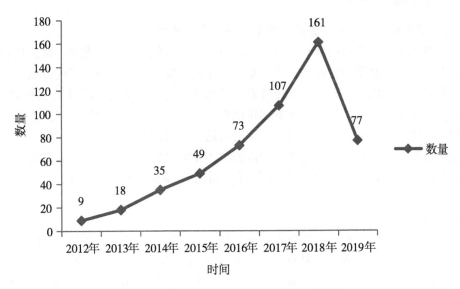

图 2　国外国内智慧农业相关论文发表数量

从图 2 可以看出，国外对于智慧农业的研究数量也呈现上升趋势，说明推进智慧农业是大势所趋。

2.2 智慧农业发展现状

2.2.1 国内智慧农业发展现状

我国近几年大力发展农业信息化和现代化，用于智慧农业发展的基础设施基本完善，一些偏远地区的网络覆盖率有所提高，大大提高了物联网、互联网等在农业上的应用率，智慧农业的发展效果初显。在政府和企业的全力配合下，全国各地积极开展智慧农业示范工程，鼓励农户体验和学习农业网络化体系，农业与金融、经济的联系越来越大，农产品品牌逐渐走向全国各地，网络电商更是方便了农产品的销售。但是，我国的物联网还处于成长阶段，技术并不完善，在数据分析统计、信息传递等方面都存在问题。另外，网络技术方面的人才缺乏、农民应用新型农业作业系统的呼声不高以及应用物联网智慧农业系统的成本高等问题也不容忽视[2]。

2.2.2 国外智慧农业发展现状

国外智慧农业起步较早，技术相比于国内较成熟。当前，智慧农业已成为当今世界现代农业发展的大趋势，世界多个发达国家和地区的政府和组织相继推出了智慧农业发展计划[3]。在国外的农田作业中一般有作物分布稳定、作物规模大、分工明确且细化、产业链完整等特点。例如，以色列严重缺水，通过兴修水利使用先进技术提高生产机械化程度，农业获得迅速发展，智慧农业发展处于世界领先地位，农产品的进出口在世界上也位居前列或中上等，在世界上具有相当的竞争能力。再如，荷兰的现代温室，和植物生产相关的一切因素如温度、湿度、光照、水分以及空气等，都可以由电脑控制，配合传感器检测植物的生理状况，几乎达到了最理想的精细化农业生产。

3 发展智慧农业的优势

3.1 提高农业生产率和资源利用率，创造更大的效益

信息技术本身具有高精度、高效率的特点，将这些优势应用于农业，将提高农业的生产效率。例如，水肥一体化系统等一系列物联网系统，大大地减少了资源浪费。物联网技术在农业中的应用显著提高了传统农业的管理水平，使农业逐渐走向精细化方向。再如，在农业生产环节，利用农业智能传感器实现农业环境信息的实时采集、监测，以及利用智能物联网对采集数据进行远程实时报送，能及时获知作物生产状况，为农作物大田生产和温室大棚精准调控提供科学依据。优化农作物生长环境不仅可获得作物生长的最佳条件[4]，提高农作物的产量和品质，而且可提高水资源、化肥等消耗品的利用率。

3.2 符合创新、绿色发展的理念

好的农业创新理念可以使农业得到突飞猛进的发展，因此创新体制机制推进农业绿色发展是中国农业发展方式的战略选择。国家农业政策明确提出了农业绿色发展的任务。尤其是在资源保护与节约利用上，建立耕地轮作休耕、节约高效农业用水等制度，健全农业生物资源保护与利用体系。智慧农业可以建立农业监测系统、智能节水灌溉系统等，将这些系统运用于农业不仅可减少资源浪费，更是顺应了绿色发展的要求。

3.3　将农户融入智慧农业，实现企业带领农户致富

智慧农业可以改变农民传统的耕作方式以及管理土地及作物的方式。一些企业向农户推广智慧农业工具，不仅能够提升农产品品质，还可以减少劳动力，增产增收，提高作物的生产效率，增加农民收入。

3.4　保障人民生活

将智慧农业应用于农产品溯源问题，可以保障农产品和食品安全。目前，食品安全问题备受关注。在农产品和食品流通领域，集成应用电子标签、条码、传感器网络、移动通信网络和计算机网络等农产品和食品追溯系统，可实现农产品和食品质量跟踪、溯源和可视数字化管理。人们对农产品的生产可以实现透明化，可大大提高农产品和食品的质量安全水平。

4　智慧农业发展建议

4.1　提升农业生产者和操作者的素养

一是知识和文化素养。一般农业操作者都非常有经验，但是这种经验化可能导致他们失去科学判断能力。因此，要增强农民的技能水平，掌握高端技术需要农民有相关的知识支撑和对应的职业技能[5]。农民是土地的主人，只有提高农民的素质，智慧农业才能快速发展。因此，培养人才是智慧农业发展的需要和重要因素。二是思想道德素质。农民的思想要紧跟社会发展潮流，对解读国家农业政策至关重要，在一定程度上影响着农业的发展。

4.2　扩大智慧农业生产规模

智能农业对大规模的农业生产具有很大优势，但由于技术和方案的不足，如今智慧农业只是在部分地区和产田实施。只有使土地形成规模，才能够使智慧农业惠及万家。

4.3　创新智慧农业发展模式

创新是产业发展的灵魂，创新力度是决定智慧农业能否顺利发展的关键因素。因此，要大力创新智慧农业发展模式，从而促进农业高效、可持续发展。

5　展　望

同其他领域的发展一样，智慧农业发展必然经过一个从初级到发展再到成熟的过程。因此，要科学谋划，全局发展，制定出符合我国国情的智慧农业发展方向和方法，统筹各地区的农业发展差异，因地制宜，为智慧农业发展描绘总体发展的蓝图，破解我国农业方面存在的问题。同时，将企业、政府、农业生产单位以及信息化人才联通起来，打造信息化背景下各行业互联互通的局面，共创智慧农业的未来。

参考文献

[1]　刘来毅，张晓娟．关于智慧农业发展现状的探析［J］．辽宁农业职业技术学院学报，2019（2）：58-60．

[2]　苗德伟．智慧农业发展与应用探讨［J］．新农业，2019（2）：33．

[3] 赵春江. 智慧农业发展现状及战略目标研究 [J]. 中国农业文摘—农业工程, 2019 (3): 15-17.

[4] 原涛. 农业科技创新助力智慧农业发展的思考 [J]. 农民致富之友, 2019 (1): 232.

[5] 王帅, 刘莹璐. 郑州市智慧农业发展趋势及建议 [J]. 合作经济与科技, 2019 (4): 16-18.

作物腾发量计算方法比较与评价

首先介绍了四种参考作物腾发量（ET_0）计算方法的发展和应用状况，对各公式的理论依据、优缺点及适宜性做了系统阐述。通过对不同公式 ET_0 计算结果的大量对比和深入分析，提出了精确获得辐射项（ET_{rad}）的建议，同时展开对作物腾发物理和生理机制的深入研究，不断提高计算精度，从而使结果更接近真值。

作物需水量是确定灌溉用水定额的基础，其关键参数是作物的蒸腾蒸发量（腾发量）。作物蒸腾蒸发理论及其计算方法的研究历来受到国内外学者的高度重视[1-2]，如何准确计算作物腾发量已成为作物需水规律研究的热点。

作物腾发量的计算，概括起来主要有两类[3]：一类是直接计算法，如 Jensen-Haise 法（1974）、A 级蒸发皿法、Ivanov 法、Behnke-Makey 法、Stephens-Stewart 法、Blaney-Criddle（1950）、Hargreaves（1974）、VanBavel-Bhsinger，这些方法均为经验公式，即采用主要气象因子与作物腾发量的经验关系进行结果的估算，由于经验公式有较强的区域局限性，其使用范围受到很大限制。另一类是通过参考作物腾发量与作物系数间接确定作物腾发量的计算方法，即参考作物腾发量（ET_0）与作物系数（K_c）相乘，可得到实际作物的腾发量：

$$(ET_c)：ET_c = K_c \cdot ET_0$$

这是目前国际上较通用的作物腾发量的计算方法。

1 ET_0 计算公式的研究进展

国外对 ET_0 计算公式的确认方法基本是通过蒸渗仪进行率定的[4-5]，Modified-Penman 公式（M-P 公式）、Penman-Monteith 公式（P-M 公式）和标准 ASCE-PM 公式在国际上代表了 20 世纪 70、90 年代以及 21 世纪初期三个时期 ET_0 计算公式的主要研究成果。

1.1 M-P 公式

Blaney（1952 年）和 Doorenboos（1977 年）等人在试验中的对该计算模式进行多次修正，得到 Modified-Penman 公式（M-P 公式），被联合国粮农组织（Food and Agriculture Organization，FAO）在 1977 年推荐使用，用其来确定参考作物腾发量[6]。

$$ET_0 = \frac{\dfrac{P_0 \Delta R_n}{P\gamma} + 0.26(e_a - e_d)(1 + 0.54 u_2)}{\dfrac{P_0 \Delta}{P\gamma} + 1}$$

式中，ET_0 为参考作物蒸腾蒸发量（mm·d^{-1}）；P_0 为海平面气压（hPa）；P 为本站气压（hPa）；Δ 为饱和水汽压曲线斜率（kPa·℃$^{-1}$）；R_n 为地表净辐射通量（MJ·m^{-2}·d^{-1}）；γ 为干湿表常数 0.064 6（kPa·℃$^{-1}$）；ET_a 为干燥力项（mm·d^{-1}）；u_2 为 2m 高度处风速（m·s^{-1}）；e_a 和 e_d 分别为饱和水汽压和实际水汽压（kPa）。

该公式将 ET_0 值定义为生长茂盛，高度均一的 8~15cm 青草完全覆盖且土壤供水充足的开阔地面的潜在腾发量。但因参考作物高度变化范围为 8~15cm 将造成空气动力学特性和冠层表面阻力的变化，进而影响计算结果。另外，同一种规定的参考作物，在不同地区、不同气候条件，其表面形态特征的差异也会导致计算结果缺乏可比性。

1.2 P-M 公式

1990 年 3 月，在意大利罗马举行的作物需水量计算方法研讨会上，推荐使用 Monteith（1965）在 Penman 等人的工作基础上以能量平衡和水汽扩散理论为基础，推导出的适用于参考作物腾发量计算的阻力公式，即 Penman-Monteith 公式（P-M 公式）。FAO（1994）按照 P-M 公式的要求，对参考作物腾发量 ET0 重新定义，ET_0 为一种假想的作物冠层的蒸腾蒸发速率，假定作物高度为 0.12m，固定的叶面阻力为 70s·m^{-1}，反射率为 0.23，非常类似于表面开阔、高度一致、生长旺盛、完全覆盖地面而不缺水的绿色草地的蒸腾蒸发速率[7]。

1998 年，在 FAO 出版的《作物腾发量-作物需水量计算指南（FAO 灌溉与排水手册-56）》中，推荐将 P-M 公式作为 ET0 的标准计算方法[8]。其计算公式为：

$$ET_0 = \frac{0.408\Delta(R_n - G) + \gamma \frac{900 u_2}{T + 273}(e_a - e_d)}{\Delta + \gamma(1 + 0.34 u_2)}$$

式中，G 为土壤热通量（MJ·m^{-2}·d^{-1}），T 为 2m 高度处的平均温度（℃），其他参数意义同前。

1.3 标准 ASCE-PM 公式

1992 年，Smith[9] 在总结前人试验的基础上提出了简化 ASCE-PM 公式。该公式主要以小时（h）和一天（d）为步长，同时针对高草与矮草设置了对应的参数 C_n 和 C_d，在实际应用中具有比较高的计算精度。2000 年，这个简化公式被美国工程师协会——灌溉蒸腾与水文学委员会推荐使用，用来计算高草（0.5m 左右，小麦、棉花等）的腾发量及矮草（0.12m 左右，牧场、草坪等）的腾发量。

2005 年，美国工程师协会-环境与水资源机构（ASCE-EWRI）推出标准 ASCE-PM 公式作为美国研究计算 ET0 的最新方法[10]。具体公式如下：

$$ET_0 = \frac{0.408\Delta(R_n - G) + \gamma \frac{C_n u_2}{T + 273}(e_a - e_d)}{\Delta + \gamma(1 + C_d u_2)}$$

式中，C_d 和 C_n 分别为随着作物冠层表面和白天（夜晚）的空气动力阻力而变化的系数；其他参数意义同前。

其中，标准 ASCE-PM 公式中指出：以小时为计算尺度，固定的叶面阻力（rc）白天取 50s·m^{-1}，晚上取 200 s·m^{-1}；以天为计算尺度，rc 取 70s·m^{-1}。

1.4 国内 Penman 修正式

随着 ET_0 计算公式的不断涌现，其适用性也受到广泛关注。我国学者根据我国气候条件、地理位置等实际情况，提出了适合我国的彭曼修正式-国内 Penman 修正式[11]，并在一定范围内得到应用，其结果较为满意。计算公式如下：

$$ET_0(NP) = \frac{\Delta R_n + 0.16\gamma(1+0.41V)(e_a - e_d)}{\Delta + \gamma}$$

式中，V 为 10m 高度处风速（$m \cdot s^{-1}$）；其他参数意义同式前。

2 方法评价

以上各参考作物腾发量 ET_0 计算公式，均是建立在半理论半经验的基础上，具有一定的气候性、地区性，各公式的普遍适用性还没有得到普遍共识。近几十年来，研究者们通过大量的试验，对不同 ET_0 公式的计算结果进行了较详细的对比分析和讨论。

刘钰等[12]经过一系列的研究提出 M-P 公式计算值较 P-M 公式计算值有偏小的趋势；杜尧东等[13]采用辽宁 33 个气象站 30 个月气象资料对 P-M 公式和 M-P 公式进行对比，认为 4~6 月 ET_0（P-M）<ET_0（M-P），7~9 月 ET_0（P-M）>ET_0（M-P）。

毛飞等[14]对泰安和西峰地区 1980—1989 年参考作物腾发量进行研究，结果 M-P 公式和国内 Penman 修正式历年的计算值均低于 P-M 公式计算值，其中国内 Penman 修正式的计算结果低于 P-M 公式约 104~147mm，M-P 公式的计算结果低于 P-M 约 107~120mm。

杨聪等[15]的研究结果显示，M-P 公式的计算值比 P-M 公式的计算值平均偏大 16%左右，而且经过统计分析，它们具有很好的线性关系，即在代表流域内使用 M-P 公式计算出参考作物腾发量后再乘以一个折算系数（如 0.84），即可得到与 P-M 公式的计算值较为接近的结果。

强小嫚[16]为寻求半湿润易旱地区最适宜的 ET_0 计算公式，对以上介绍的 ET_c 计算公式进行试验率定，试验结果分析在苜蓿整个生育期内，M-P 公式、国内 Penman 公式、P-M 公式和标准 ASCE-PM 公式计算值和实测值的变化趋势基本一致，而且呈显著性线性相关关系，相关性最好且偏差最小的是标准 ASCE-PM 公式，其次是 P-M 公式、M-P 公式，最差的是国内 Penman 公式。

由试验结果可以看出，由于实验条件、地区气候、作物品种等综合因素影响，各公式在不同的地区对于不同的作物计算结果差异较大，而且由于在具体计算时所需要的数据资料较多，计算公式较为复杂，在气象数据缺失的情况下或者部分数据精度不高的情况下，计算值经常严重偏离真值，影响了其实际应用。

对于参考作物腾发量的计算，使用不同的公式其计算结果存在差异，这主要是由于各个计算公式中辐射项（ET_{rad}）和空气动力学项（ET_{aero}）选用不同的参数造成的。通过分析各气象因素可知，对于参考作物腾发量的计算，ET_{rad} 起主要决定作用。为准确得到太阳辐射数据，需要通过大量的气象台站的长期观测和统计分析，而目前国内测量太阳辐射尚存在时间短、地区少、数据质量不高等问题，部分气象数据还来自国外。因此，进行太阳辐射数据的研究是一项意义重大且具有基础性的研究工作。

参考文献

[1] 袁志发，周静芋．多元统计分析［M］．北京：科学出版社，2002．

[2] 余建英，何旭宏．数据统计分析与SPSS应用［M］．北京：人民邮电出版社，2003．

[3] 李瑞歌，赵仲麟，张淑利，等．SPSS软件在实验设计与数据处理中的应用［J］．农业网络信息，2016（11）：127-129．

[4] 王文娟．基于SPSS的两因素完全随机实验的方差分析［J］．现代经济信息，2015（8）：422-424．

[5] 盖钧镒．试验统计方法［M］．北京：中国农业出版社，2000．

[6] 世界优秀统计软件SPSS10.0 for Windows实用基础教程［M］．北京：北京希望电子出版社，2001．

[7] 胡竹菁，戴海琦．方差分析的统计检验力和效果大小的常用方法比较［J］．心理学探新，2011，31（3）：254-259．

[8] 何晓群．多元统计分析［M］．北京：中国人民大学出版社，2012．

[9] 张文彤．SPSS统计分析高级教程［M］．北京：高等教育出版社，2004．

[10] 沈其君．SAS统计分析［M］．北京：高等教育出版社，2005．

[11] 陈长生，徐勇勇．重复观测数据单变量方差分析的前提条件的检验［J］．中国卫生统计，2000，17（2）：74-76．

[12] Stephen Olejnik, JamesAlgina. Generalized Eta and Omega SquaredStatistics：Measures of Effect Size for Some Common Research Designs［J］. Psychological Methods, 2003, 8（4）：434-447.

[13] 陈长生，徐勇勇．重复测量的研究设计与统计分析［J］．中国卫生统计，2002，19（2）：124-126．

A clustering protocol based on Virtual Area Partition using Double Cluster Heads scheme

A number of multifunction sensors deployed randomly with limited energy is one character of wireless sensor networks, so energy efficient is a critical factor in sensor network design. In this literature, we present a clustering protocol based on Virtual Area Partition using Double Cluster Heads scheme (VAP-DCH) for wireless sensor networks, which partitions the network and selects a Main Cluster Head (MCH) and a Vice Cluster Head (VCH) in each cluster adaptively. The simulation results show that the protocol we presented improves the system performance, prolongs the network lifetime and transfers more data to the base station.

1 INTRODUCTION

WIRELESS sensor networks (WSNs) have gained world-wide attention in recent years, and have a great number of applications, such as disaster rescuing, environment monitoring and military tracking, etc. In design of WSNs, we must consider the factors such as the object, hardware, cost and other system constraints[1]. The sensors often use tiny batteries with limited energy supply, and deploy in dangerous region, so the recharge or replacement of batteries is impossible. Thus reducing and balancing the energy dissipation to prolong the network lifetime are essential objects for wireless sensor networks[2], while clustering is one of the energy-saving techniques that extending the sensor network's lifespan[3].

A number of low-powered sensor nodes networked together in a WSN, Low-Energy Adaptive Clustering Hierarchy (LEACH) is proposed in[4] that randomly rotating the role of cluster head among all the nodes in the network. Another clustering-based protocol PEGASIS[5] which prolong the network lifetime by using local collaboration among sensor nodes. DHSC is presented in[6], nodes density area is the central, and two cluster heads, a main cluster head which is selected in thick and an assistant cluster head which is selected in thin area. VAP-E[7] based on node partition to balance the loads. The author proposed protocol[8] to select a minimum number of sensors to achieve full k-coverage of a field while guaranteeing connectivity between them and get better performance.

In this paper, we present a clustering protocol based on Virtual Area Partition using Double Cluster Head scheme (VAP-DCH), which is an energy-efficient clustering protocol. The protocol obeys the idea of energy efficient to obtain good performance in terms of network lifespan and data delivery. We use virtual area partition to guarantee the mean value of the number of cluster heads is optimal in each round and select a Main Cluster Head (MCH) and a Vice Cluster Head (VCH) in each cluster to balance the energy consumption. It also improves the running rate of clustering for it has a characteristic of high speed parallel. The simulation results and analysis show that VAP-DCH can achieve better performance in network lifespan and data delivery by virtual area partition and double cluster head scheme that reducing and balancing the energy consumption evenly in the networks.

2 NETWORK MODEL

We suppose there are a number of nodes distributed randomly in a squared region with the base station (BS) located far away from the region. We use heterogeneous network model in this literature to strengthen facticity of wireless sensor networks, the energy of each node is randomly distributed among $[E_0, \alpha E_0]$, where α is a constant and $\alpha > 1$, and E_0 is the minimal energy of sensor nodes.

We assume the sensor network model that has properties as follows:

(1) The base station is located away from the sensing field.

(2) The sensor nodes are energy constrained with initial energy allocated, and all sensor nodes have no mobility.

(3) The nodes have power control units that can vary their transmitted power to save energy.

(4) Each node has location information. There have many localization methods in WSNs[9],[10]. For example, we use the solution presented in[9] to achieve the localizat-ion.

We adopt clustering techniques in wireless sensor networks, which divide all sensor nodes into three kinds respectively, member sensor nodes, main cluster heads (MCHs) and vice cluster heads (VCHs). The MCHs consume more energy, and charge to administrate their member sensor nodes, aggregate the data and send to the VCHs in its cluster. We can use the method similar to[5] to form a greedy chain among VCHs and random select a VCH as chief VCH, the chief VCH collects data that VCHs have received and aggregated, that further minimize the energy dissipation, the chief VCH send the aggregated data to the BS directly. While member nodes only send sensing data to their cluster heads. We suppose that each member node sends 1-bit message to its cluster head in each round. The partial network architectures are as shown in Fig. 1

A typical WSN node is constituted of four major components: a sensor unit, a power supply unit, a data processor unit and a wireless communication unit that consists of transceiver/receiver circuit, antenna and amplifier[11],[13],[14]. Although in a sensor node, the energy is dissipated in all of the components except power supply unit, we mainly consider the energy consumed related to the wireless communication unit since the goal of this literature is to

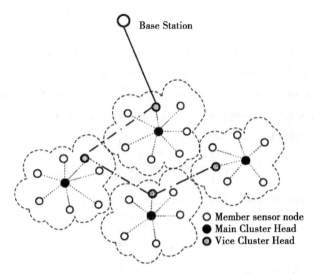

Fig. 1 Partial network architecture using VAP-DCH

present an energy efficient protocol to prolong the network lifetime, and the energy consumed by the cluster heads during the data aggregation is also take into account.

We use a first radio model as used in[4]. In this model, a receiver consumes energy to run radio units, a transceiver consumes energy to run radio units and power amplifier. If we transmit an l-bit message through the distance r, the energy consumption that a transceiver need is:

$$E_{Tx}(l, r) = E_{elec}(r) + E_{amp}(l, r)$$
$$= \begin{cases} l \times E_{Tx} + l \times \varepsilon_{fs} \times r^2, & r \leq r_o \\ l \times E_{Tx} + l \times \varepsilon_{mp} \times r^4, & r > r_o \end{cases} \quad (1)$$

Where r_o is the threshold distance[12] and is given by $r_o = \sqrt{\varepsilon_{fs}/\varepsilon_{mp}}$. The energy consumption that a receiver need to receive an l-bit message is: $E_{Rx}(l) = l \times E_{elec}$.

Where E_{elec} is the energy dissipated per bit to run the transceiver or the receiver circuit, ε_{fs} and ε_{mp} depend on the transceiver amplifier model, the energy for data aggregation is set as $E_{DA} = 5nJ/bit/message$[15]. The transceivers have power control ability that can consume minimum energy by properly setting the power amplifier according the distance and can be turned off to avoid receiving unexpected data.

3 VAP-DCH PROTOCOL

A. Optimum Number Of Clusters

We present an efficient clustering protocol to promote the existing clustering protocols' performance. Use the energy consumption model as described above, the amount energy consumed by all nodes during a round is:

$$E_r = l\left(NE_{DA} + 2NE_{elec} + k\varepsilon_{mp}d_{toBS}^4 + N\varepsilon_{fs}\frac{1}{2\pi}\frac{M^2}{k^2}\right) \quad (2)$$

Where and k is the number of clusters, d_{toBS} is the distance between the BS and the cluster head, and l is the number of bits in each message.

In order to minimize the energy consumption, we can calculate the optimal value of k from E_r to minimize the energy consumption. We set the derivative of the continuous function when $k>0$ about E_r with respect to k to zero

$$E'_r = l\left(\varepsilon_{mp}d_{toBS}^4 - \frac{N}{2\pi}\varepsilon_{fs}\frac{M^2}{k^2}\right) = 0 \quad (3)$$

So we get the optimum number of clusters,

$$k_{opt} = \frac{\sqrt{N}}{\sqrt{2\pi}}\sqrt{\frac{\varepsilon_{fs}}{\varepsilon_{mp}}}\frac{M}{d_{toBS}^2} \quad (4)$$

B. Area Partition Clustering

In a given sensor network, when k_{opt} is determined, we perform virtual area partition on the networks. We part the network into k_{opt} partitions by $2\pi/k_{opt}$ in origin of network region center. It's not that the sensor nodes are divided, but the network is parted and the sensor nodes belong to its partition.

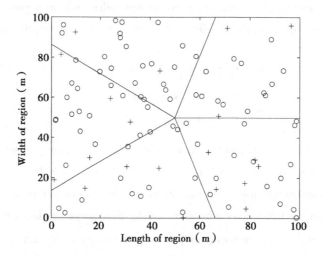

Fig. 2 Virtual area partition clustering with $k_{opt}=5$

As shown in Fig. 2, a 100m×100m region with 100 nodes distributed randomly, we use virtual area partition that divided the network by $k_{opt}=5$.

Most energy efficient clustering protocols[4],[16]-[22] consider the residual energy as an important factor that decide which node to be cluster head in a cluster, the more energy a node has, the more opportunity it is to become a cluster head. That can prolong the network lifetime by balancing the energy consumption contrast to randomly select node as cluster head. However, from (1) we can see that the distance is also a factor that affects the energy dissipation. So based on the consideration above, we define the following probability $p(i)$ for node i to decide which node to become cluster head.

$$p(i) = p_{opt}\left[\lambda\frac{E_i(r)}{\overline{E}(r)} + (1-\lambda)\frac{D(i)}{d_{max} - d_{toBS}}\right] \quad (5)$$

Where p_{opt} is the ratio of optimum number of clusters, and $p_{opt} = k_{opt}/N$, $E_i(r)$ is node i's residual energy during round r, $\overline{E}(r)$ is the average energy of the alive nodes during round r, d_{max} is the maximum distance between node i and the BS, d_{toBS} is the average distance between nodes and the BS, and $D(i) = d_{max} - d_{toBS}(i)$. λ is the adaptive parameter and $\lambda = \frac{1}{1+\alpha}$, $\alpha = \frac{E_i(r)}{E_i}$ α varies from 1 to 0, so λ varies from 0.5 to 1, that shows the residual energy are become more important during the cluster head election.

Obviously, we select the node that has greater $p(i)$ as cluster head, that can prolong the network lifetime due to the node has more energy and the distance between the node and the BS is minimal. We calculate the average energy of the nodes through a piggyback manner, which need not extra cost and make the energy consumption evenly among all nodes.

C. Double Cluster Heads Scheme

After the clusters have formatted and the cluster heads have been selected, we treat the cluster heads as MCH, and select a VCH in each cluster. We use the probability (5) to decide which node to become a VCH. The MCH sets up a TDMA schedule to avoid collision for its member nodes, allows the radio electronics to be turn off at all times except during their transmission period for each member nodes, that has further reduce the energy consumption and enhance the network performance. Once the MCH has finished receiving data from all its member nodes, the MCH performs data aggregation and sends the aggregated data to the VCH within its cluster to balance energy consumption. We use the method similar to[5] to form a greedy chain among VCHs, then random select a VCH as chief VCH. The chief VCH collects data that VCHs have received and aggregated, that further minimize the energy dissipation. The chief VCH send the aggregated data to the BS directly, the current round is finished and the next round will begin.

Table 1 Simulation Parameters

Parameter	Value
network coverage	(0, 0) ~ (100, 100) m
sink location	(50, 175) m
E_0	0.5J
E_{elec}	50nJ/bit

A clustering protocol based on Virtual Area Partition using Double Cluster Heads scheme

(continuation)

Parameter	Value
ε_{fs}	$10 pJ/bit/m^2$
ε_{mp}	$0.0013 pJ/bit/m^4$
E_{DA}	$5 nJ/bit/message$
α	3
Data message size	125 bytes
Data message header	25 bytes

We run the simulations to evaluated the performance of the proposed protocol, there are 100 nodes randomly distributed in a 100m×100m network area with the BS located at (50, 175), the energy of each node is randomly distributed in $[E_0, \alpha E_0]$, where $E_0 = 0.5J$ and $\alpha = 2$. For simplicity, we assume 40% nodes have αE_0 energy, the other nodes have E_0 energy. The data message size was 125 bytes with the packet header was 25 bytes long. The simulation parameters are given in Tabe 1, in which the parameters are as described in section III. The performances of our proposed protocol are compared with clustering protocol LEACH[4] and VAP-E[9].

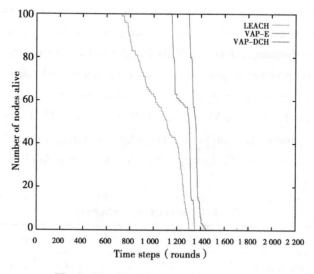

Fig. 3 Number of nodes alive over time

The number of nodes alive over time is illustrated in Fig. 3, we define it system lifetime. Obviously our proposed protocol VAP-DCH can prolong the network lifetime significantly compared to LEACH and VAP-E. This is because our protocol produces better network partitioning, selects double cluster head in a cluster to balance the energy consumption and forms a greedy chain among the SCHs to reduce the energy consumption. On the contrary, in LEACH and VAP-E some nodes have to transmit long distances in order to reach a CH due to weak network clustering or the CH consume more energy for data gathering, aggregation and transmitting, so some nodes dissipate a large amount of energy while transmitting their data to the CH or the base station.

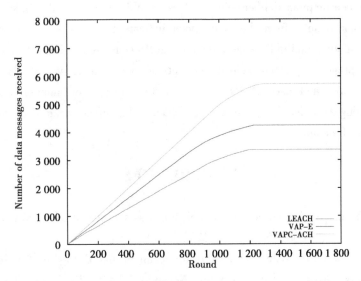

Fig. 4 Total amount of messages received at BS over time

The total data messages received at the BS is shown in Fig. 4. It indicates the effectiveness of the proposed protocol in delivering more data messages than LEACH and VAP-E. Our protocol offers improvement in data delivery by factors of 45% over LEACH and 21% over VAP-E. The reason for this is because the proposed protocol can take advantage of higher energy nodes as MCH by considering the remaining energy of the CH candidates and the distance between the nodes and the BS, use SCH to balance the energy consumption, and form greedy chain among SCHs to further reduce energy dissipation. While LEACH does not take into account the energy of a node when selecting the CH, and may select the CH with insufficient energy to remain alive during the data transfer phase, and VAP-E does not balance the energy dissipation. Our protocol considers the network states information while select CHs and uses SCHs to balance the energy consumption.

As can be seen from these figures, the proposed protocol can balance the energy consumption and cluster heads are located at the place that have more energy, and select SCHs to balance the energy dissipation. LEACH and VAP-E on the other hand, either produce an uneven distribution of cluster heads throughout the sensing field or do not balance the energy consumption and consume more energy by cluster heads. Our proposed protocol produce even partitions and selects SCHs to balance the energy consumption and form greedy chain among SCHs to further reduce the energy dissipation.

4 CONCLUSION

In this paper we have presented a clustering protocol based on Virtual Area Partition using Double Cluster Heads Scheme (VAP-DCH) for wireless sensor networks. We select cluster head considering the network states information such as residual energy and the distance that affect the energy consumption deliberately, and use double cluster head scheme to balance the energy dissipation. Results from the simulations indicate that the proposed protocol gives a higher network lifetime and deliver more data to the BS compared to LEACH and VAP-E. Furthermore, the proposed protocol selects the optimal nodes as cluster heads throughout the sensor network area and uses DCHs to balance the energy consumption. Our future work includes exploring more detailed parameters and factors that affect system performance to further improve energy efficiency.

REFERENCES

[1] J. Yick, B. Mukherjee, D. Ghosal. Wireless sensor network survey [J]. Computer Networks, 2008, 52 (12): 2292-2330.

[2] I. F. Akyildiz, W. Su Y. Sankarasubramaniam and E. Cayirci. Wireless sensor networks: A survey [J]. Computer Networks, 2002, 38 (4): 393-422.

[3] A. A. Abbasi, M. Younis. A survey on clustering algorithms for wireless sensor networks [J]. Computer Communications, 2007, 30 (14-15): 2826-2841.

[4] W. B. Heinzelman, A. P. Chandrakasan, H. Balakrishnan. An Application-Specific Protocol Architecture for Wireless Microsensor Networks [J]. IEEE Trans. Wireless Communications, 2002, 1 (4): 660-770.

[5] X. Qiao, Y. Chen. A Control Algorithm Based on Double Cluster-head for Heterogeneous Wireless Sensor Network [R]. in Proc. International Conference on Industrial and Information Systems, 2010.

[6] X. Yi, L. Deng. A Double Heads Static Cluster Algorithm for Wireless Sensor Networks [R]. in Proc. Conference on Environmental Science and Information Application Technology, 2010.

[7] R. Wang, G. Liu, C. Zheng. A Clustering Algorithm based on Virtual Area Partition for Heterogeneous Wireless Sensor Networks [R]. In Proc. IEEE Mechatronics & Automation, 2007.

[8] H. M. Ammari, S. K. Das. Centralized and Clustered k-Coverage Protocols for Wireless Sensor Networks [J]. IEEE Trans. Computers, 2012, 61 (1): 118-133.

[9] P. Kumar, A. Chaturvedi and M. Kulkarni. Geographical location based hierarchical routing strategy for wireless sensor networks [J]. in Proc. ICDCS, 2012, 9-14.

[10] T. He, C. Huang, B. M. Blum, et al. Stankovic and T. Abdelzaher. Range free localization and its impact on large scale sensor networks [J]. ACM Trans. Embedded Computing Systems, 2005, 4 (4): 877-906.

[11] V. Raghunathan, C. Schurgers, P. Sung, et al. Energy-Aware Wireless Microsensor Networks [J]. IEEE Signal Processing Magazine, 2002, 2: 40-50.

[12] S. Reddy, C. R. Murthy. Dual-Stage Power Management Algorithms for Energy Harvesting Sensors [J]. IEEE Trans. wireless communications, 2012, 11 (4): 1434-1445.

[13] A. B. da Cunha, D. C. da Silva. Behavioral Model of Alkaline Batteries for Wireless Sensor Networks [J]. IEEE Latin America Transactions, 2012, 10 (1): 1295-1304.

[14] A. Wang, W. Heinzelman, A. Chandrakasan. Energy-scalable protocols for battery-operated microsensor networks [R]. *In Proc. 1999 IEEE Workshop Signal Processing Systems (SiPS'99)*, 1999.

[15] J. Lee, W. L. Cheng. Fuzzy-Logic-Based Clustering Approach for Wireless Sensor Networks Using Energy Predication [J]. IEEE Sensors Journal, 2012, 12 (9): 2891-2897.

[16] G. Smaragdakis, I. Matta, A. Bestavros. SEP: A stable election protocol for clustered heterogeneous wireless sensor networks [R]. In 2th International Workshop on Sensor and Actor Network Protocols and Applications, 2004.

[17] S. D. Muruganathan, D. C. F. Ma, R. I. Bhasin, et al. A Centralized Energy-Efficient Routing Protocol for Wireless Sensor Networks [J]. IEEE Radio Communications, 2005, 43 (3): 8-13.

[18] S. Lindsey, C. Raghavendra, K. M. Sivalingam. Data Gathering Algorithms in Sensor Networks using Energy Metrics [J]. IEEE Trans. Parallel and Distributed Systems, 2002, 13 (9): 924-935.

[19] O. Younis, S. Fahmy. HEED: A Hybrid, Energy-Efficient, Distributed Clustering Approach for Ad Hoc Sensor Networks [J]. IEEE Trans. Mobile Computing, 2004, 3 (4): 366-379.

[20] Z. Huang, H. Okada, K. Kobayashi, et al. A study on cluster lifetime in multihop wireless sensor networks with cooperative MISO scheme [J]. Journal of Communications and Networks, 2012, 14 (4): 443-450.

[21] X. Gao, Y. Vanq, D. Zhou. Coverage of communication-based sensor nodes deployed location and energy efficient clustering algorithm in WSN [J]. Journal of Systems Engineering and Electronics, 2012, 21 (4): 698-704.

A clustering protocol with Adaptive Assistant-Aided Cluster Head using Particle Swarm Optimization

Clustering is one of the energy-saving and efficient techniques that extends the sensor network's lifetime. In this literature, we propose and analyze a clustering protocol with Adaptive Assistant-Aided Cluster Head using Particle Swarm Optimization (AAACH-PSO), a protocol architecture for wireless sensor networks using Particle Swarm Optimization (PSO) to select cluster head (CH) and assistant cluster head (AACH) on a needed basis, thus saving more energy and balancing the energy consumption. Our simulation results show that AAACH-PSO can improve system lifespan and data delivery by distributing energy dissipation evenly in the networks.

1 INTRODUCTION

A wireless sensor network (WSN) consists of few tens to thousands of small battery powered multifunctioning devices with limit energy supply[1]. Once deployed, the sensor nodes are often unreachable to users, and the replacement or recharge of batteries are impossible in hostile environment, so energy efficiency is the vital factor for WSNs.

Clustering is one of the energy-saving techniques that extends the sensor network's lifespan[2], and is often formulated as optimization problems[3]. Conventional analytical optimization methods require tremendous computational efforts, which grow up exponentially with the problem scale increases. But for implementation on an individual wireless sensor node, an optimization method that requires moderate even minimal computing and memory resources and yet produces better results is needed. Bio-inspired optimization methods are computationally efficient compared to traditional analytical methods[4].

Particle swarm optimization (PSO) is one of the bio-inspired optimization methods, which is a popular, simple, effective and multidimensional optimization algorithms[5], it also has many good qualities, such as computational efficiency, ease of implementation, high quality of solutions and fast speed of convergence, etc.

An adaptive clustering scheme called Low-Energy Adaptive Clustering Hierarchy (LEACH) is proposed in[6] that randomly rotates the role of cluster head among all the nodes in the network. VAPE[7] based on virtual area partition for heterogeneous wireless sensor networks. Based on GAF, a clustering algorithm[8] is proposed to find the optimum position of the cluster head with a grid for energy saving, and divide the virtual grid dynamically and periodically. DCCG[9] algorithm consists of two levels of clustering: local clustering and global clustering.

The particle swarm optimization has been applied to address WSN issues such as clustering, deployment, localization, and data aggregation[3][10][11]. A protocol using PSO[12] has been proposed in minimizing the intra-cluster distance and optimizing the energy consumption of the network. A cluster-based algorithm using PSO is proposed[13], the election of cluster-heads needs to consider the information of location and energy reserved about nodes and their neighbors.

In this literature, we design a clustering protocol with Adaptive Assistant-Aided Cluster Head using Particle Swarm Optimization (AAACH-PSO), which takes advantage of PSO that can get multimodal optimal values. The selection of cluster heads (CHs) considers state information of the sensor networks, such as reserved energy, location of nodes, their neighbors and the base station. The protocol first generates CHs by selecting the optimal fitness, then based on the cluster's information such as CH's residual energy, the nodes situation and the number of nodes among the cluster to determine whether we need to select an assistant-aided cluster head (AACH) or not, and select the suboptimal fitness as AACH if in need. The objective of the proposed solution is to obtain good performance in terms of system lifespan, energy consumption and application-perceived quality. The analysis and simulation results show that AAACH-PSO can achieve better system lifespan and data delivery by optimizing the CH selection and balancing the energy consumption evenly in the networks.

2 NETWORK MODEL

We consider a total of N sensor nodes are distributed randomly within a $M \times M$ squared field, and assume a sensor network model has properties as follows:

- A fixed base station (BS) is located at the center of the sensor field.
- The sensor nodes are energy constrained with initial energy allocated and distributed randomly.
- The nodes are equipped with power control device to vary their transmitted power and can directly communicate with the BS.
- Each node is aware of its own location information.
- All sensor nodes have no mobility (or move slowly with respect to the BS).

This model uses an AACH scheme adaptively. Consider the partial network structure as show in Fig. 1. Each cluster has a CH which collects data from its members, aggregates and sends it to the AACH if in need. The main features of such architecture are:

- We use PSO to select the best node as CH, so the energy dissipation is evenly and reasonable.

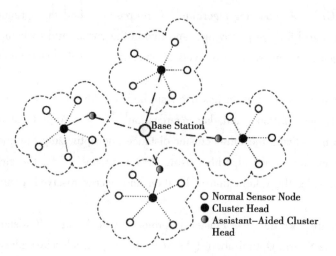

Fig. 1 AAACH-PSO clustering architecture

• All nodes only need to transmit to their nearest immediate CH, so the energy dissipated is minimum.

• Only CH needs to perform additional computation on the data, So energy is again conserved.

• The AACH balances the energy consumption, that further makes energy consumption minimal and evenly.

Sensed data is gathered in a periodic way, and each period is named a round, the period we have defined round consists of three phase. A set-up phase which used AAACH-PSO algorithm to select CH and AACH adaptively, followed by a steady state phase when data collected by the CH, finally a finish phase when data transfered to the BS occurred. In order to minimize the overhead, the steady state should be longer than other phases.

3 AAACH-PSO ALGORITHM

We proposed an adaptive energy-efficient clustering algorithm, aiming to enhance the existing clustering algorithms' performance. In this section, we design AAACH-PSO, present threshold function to decide if we need an AACH and use particle swarm optimization algorithm to select CH and AACH.

3.1 Particle swarm optimization

Through competition and cooperation among the population, population - based optimization approaches usually can find very good solutions efficiently and effectively. There are many search approaches that motivated by evolution we knew in nature, such as Genetic Algorithms, Ant Colony Algorithm, etc. Particle Swarm Optimization is a population based, self-adaptive search optimization technique, and is motivated from the simulation of social behaviors of insects and animals such as fishing schooling and bird flocking. Since its inception by

James Kennedy and Russell Eberhart[14], PSO has gained increasing attention among scientists and researchers as an efficient and robust technique in solving complex and difficult optimization problems[15].

In a standard PSO, a swarm consists of m particles flying at a certain speed in a D-dimension search space, each particle searches around (or towards) a region defined by its personal best position (i.e., theposition giving the best fitness value) and the neighborhood best position (or the entire population). Let us use $v_i = (v_{i1}, v_{i2}, \cdots, v_{id})$, $1 \leq i \leq m$, $1 \leq d \leq D$ to denote the velocity of the ith particle in the swarm, $x_i = (x_{i1}, x_{i2}, \cdots, x_{id})$ as its position, $p_i = (p_{i1}, p_{i2}, \cdots, p_{id})$ the best position it has found so far, and $p_g = (p_{g1}, p_{g2}, \cdots, p_{gd})$ the best position found from its neighborhood (so-called global best). v_i and x_i of the ith particle in the swarm are updated according to the following two equations[16]:

$$v_{id}^{k+1} = \omega v_{id}^k + c_1 \xi (p_{id}^k - x_{id}^k) + c_2 \eta (p_{gd}^k - x_{id}^k) \tag{1}$$

$$x_{id}^{k+1} = x_{id}^k + v_{id}^{k+1} \tag{2}$$

Where ω is the inertia weight, it decides how the particle inherit present velocity, a large inertia weight facilitates the particle's exploration ability (global search ability) while a small inertia weight facilitates the particle's exploitation ability (local search ability). c_1 and c_2 are learning factor (or called acceleration coefficient) which are positive constants normally, the learning factor enables the particle flying towards its personal best position and the global best position. ξ and η are random numbers between 0 and 1. If we consider all the particles in the swarm as neighbor, we get the global version of PSO; if we consider partial particles as neighbors, we get the local version of PSO; and there has different methods to format the neighbors[17].

3.2 The fitness function of AAACH-PSO

The determination of fitness function is closely related with the properties of problem domain, and it determines the algorithm's performance of the optimal solution directly. So it is important to design the fitness function, based on[12][13], we consider the condition that influence the system performance, and do many experiments to explore the better fitness function and better parameters, finally we define the fitness function deliberately. We define the fitness function $f(i)$ for node i as follows.

$$f(i) = \alpha f_1(i) + \beta f_2(i) + \gamma f_3(i) \tag{3}$$

$$f_1(i) = E(i) \Big/ \frac{1}{m} \sum_{i=1}^{m} E(i)$$

$$f_2(i) = \frac{1}{N} \sum_{k=1}^{N} d(i, k) \Big/ \frac{1}{m} \sum_{k=1}^{m} d(i, k)$$

$$f_3(i) = \overline{d_{toBS}(i)} / d_{toBS}(i)$$

$$\alpha + \beta + \gamma = 1, \quad 0 \leq \alpha, \beta, \gamma \leq 1$$

Function $f_1(i)$ is the ratio of node i's energy to the average energy within the cluster,

$E(i)$ is the energy reside in node i, m is the number of member nodes within the cluster. Function $f_2(i)$ is the inverse of average Euclidean distance of nodes within the cluster to node i, N is the number of alive nodes in the network. $d(i,k)$ is the Euclidean distance between node i and node k. $f_3(i)$ is the inverse of the Euclidean distance of node i to the BS. α, β, γ are the weight of each sub function that we get from experiments. The fitness function that defined above considers the energy efficient of the network as quantified by $f_1(i)$; simultaneously minimizing the intra-cluster distance between the CH and its member nodes as quantified by $f_2(i)$; and the distance between node and the BS as quantified by $f_3(i)$. So according the fitness function defined above, we choose the node that has the maximum value as CH, and it is the optimum selection during that round.

3.3 Assistant-aided cluster head setup

After the CH has been selected, it's time to decide whether to select an AACH in the cluster or not. We consider this question as follows, the residual energy of the CH, the distance between the CH and the BS, and the number of nodes among the cluster. Generally if the CH has not more residual energy, the distance is a little far from the BS and the amount of nodes among the cluster is large, on this occasion, we need an AACH to balance the energy dissipation and enhance the system robustness. So we define the threshold function as:

$$T_{AACH} = c \cdot \frac{E_{max} - E_{re}}{E_{max}} \cdot \frac{d_{(CH, BS)}}{d_{avg}} \cdot \frac{N_{Ci}}{N_{avg}} \quad (4)$$

Where c is a control parameter; E_{re} is the residual energy of the CH; E_{max} is the maximum energy of the CH; d_{avg} is the average distance between the BS and nodes in the network; $d_{(CH,BS)}$ is the distance between the CH and the BS; N_{avg} is the average number of nodes in a cluster, and N_{Ci} is the number of nodes in the cluster i.

The decision is made by the CH choosing a random number, if the number is less than the threshold T_{AACH}, the cluster need selects a node as an AACH, the node that has the suboptimal fitness as AACH, and it is the optimum selection during that round.

We first initial clustering use LEACH or VAP-E algorithms, all nodes send information about its current energy and locations to its neighbors in each cluster. The AAACH-PSO algorithm executes in period, each cluster optimally chooses the CH and AACH using PSO. The CH acts as local control center to coordinate the data transmission within the cluster. The CH sets up a TDMA schedule to avoid collision with its member nodes, allowing the radio electronics to be turn off at all times except during their transmission slot for each member nodes. That has further reduce the energy dissipation and enhance the network performance. Once the CH has finished receiving data from all its member nodes, the CH performs data aggregation and sends the aggregated data to the AACH within its cluster if needed or sends to the BS directly. As AAACH is selected through threshold function (4), it's either has more residual energy or close to the BS, that further minimizing and balancing energy consumption.

3.4 AAACH-PSO setup

As clustering is often formulated as optimization problems[3], and PSO is one of bio-inspired optimization methods that is computationally efficient compared to traditional analytical methods[5]. So we take advantage of PSO as used in[12][13] to optimize the clustering.

We use the fitness function (3) as described above to select the optimal solution as CH, and the suboptimal solution as AACH if in need according the cluster state information when the iterations of PSO finished in each round. According to the fitness function, the CH has the optimal fitness value that it is energy sufficient and the energy consumption is minimal, the CH gathers and aggregates the data. The AACH also has more energy and is close to the CH, it gets data from the CH and transmits it to the BS respectively using a fixed spreading code and CSMA (Carrier Sense Multiple Access), a similar approach that used in[18], that balances and minimizes the energy consumption. Our simulation shows that we can achieve good performance and prolong the network lifetime. The detailed steps are asfollows:

We first select the node that closed to the partition center as candidate cluster head, during the subsequence rounds, we use the candidate cluster head that get its member nodes information about current energy and location through a piggybacking manner. So it does not need extra cost and make the energy dissipation evenly distributed among all nodes.

The candidate cluster head runs AAACH-PSO algorithm to select the CH and AACH if in need, Fig. 2 shows the flowchart of AAACH-PSO algorithm.

(1) Initialize position and velocity of each particle randomly.

(2) Calculate the fitness value of each particle using Formula (3).

(3) Find the personal best and global best. The personal best is the position of the particle itself has experienced that has the maximum fitness; and the global best is the position of the particles in the swarm (or subspaces) that have experienced and have the maximum fitness.

(4) Update each particle's position and velocity using (1) and (2).

(5) Limit the change of the particle's position value and velocity value, then map the new updated position with the closest coordinates.

(6) Repeat step 2) to 5) until the maximum number of iterations is reached. Select the best solution as CH, and the suboptimal solution as AACH if in need, the candidate cluster head broadcast an advertisement message using a non-persistent carrier sense multiple access (CSMA) MAC protocol[18]. This message is a short message containing the node's ID and a header that distinguishes this message as an announcement message. The CH node sets up a TDMA schedule and transmits it to the nodes in the cluster, after all nodes in the cluster know the TDMA schedule, the set-up phase is complete and the steady-state can begin.

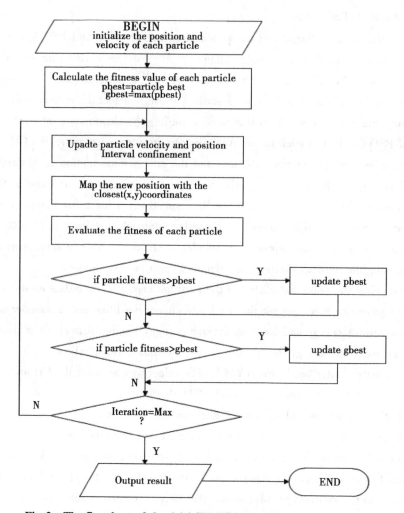

Fig. 2 The flowchart of the AAACH-PSO algorithm for cluster setup

4 SIMUALTION

We evaluate the performance of the proposed protocol, we run the simulations for 100 nodes randomly distributed in a 100m×100m network area with the BS located at ($x=50$, $y=50$). The data message size is 125 bytes with the packet header is 25 bytes long. The simulation parameters of the radio model are as used in[6], where E_{elec} = $50nJ/bit$, ε_{fs} = $10pJ/bit/m^2$, ε_{mp} = $0.0013pJ/bit/m^4$, E_{DA} = $5nJ/bit/message$, as for PSO parameters, we use $m=20$ particles, learning factor $c_1=c_2=2$, $x_{max}=v_{max}=200$, inertia weight from $\omega=0.9$ to 0.4, $\alpha=0.5$, $\beta=0.4$.

The performance of our protocol was compared with the benchmark clustering protocols for WSNs, LEACH, LEACH-C[6] and PSO-C[12]. Fig. 3 illustrates the system lifetime, defined by the number of nodes alive over time for the simulation. Clearly our proposed protocol can

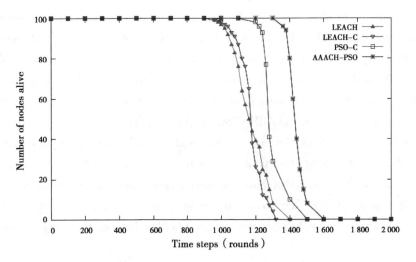

Fig. 3 Number of nodes alive over time

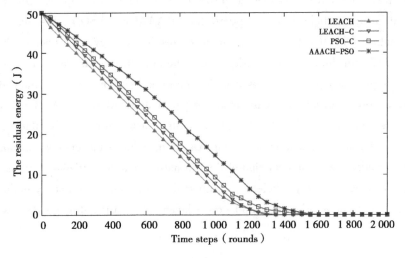

Fig. 4 The residual energy of the network over time

prolong the network lifetime significantly compared to LEACH, LEACH-C and PSO-C. This is because our protocol produces better network partitioning with minimum intra-cluster distance, CHs that are optimally distributed across the network and we define threshold function (4) to determine if we need an AACH to balance the energy dissipation. Thus, the energy consumed by all nodes for communication can be reduced since the distances between non-cluster head nodes and their CHs are shorter and balanced by the AACH evenly. On the contrary, in LEACH, LEACH-C and PSO-C some nodes have to transmit long distances in order to reach a CH due to poor network clustering or the CH consume more energy for data gathering, aggregation and transmitting. As a result, some nodes dissipate a large amount of energy while transmitting their data to the CH or the BS at a time.

Fig. 4 shows the residual energy of the network, the protocol we proposed consumed less energy with compared to LEACH, LEACH-C and PSO-C. As AAACH-PSO consider more conditions that may affect the energy dissipation and select an AACH that balances the energy consumption. While LEACH stochastic selects nodes as CHs, LEACH-C only minimizes the sum of distance and PSO-C emphasizes on minimize intra-cluster distance, that leads to the energy dissipation is more than AAACH-PSO and not evenly.

As can be seen from these figures, the proposed protocol can result in good network partitioning and balance the energy consumption, where CHs are evenly positioned across the network and located at the place that have the best fitness, and select AACH to balance the energy dissipation adaptively. That because our proposed protocol considers better network partition, minimum intra-cluster distance and selects AACH to balance the energy consumption in need.

5 CONCLUSION

In this paper we have presented a clustering protocol with Adaptive Assistant-Aided Cluster Head using Particle Swarm Optimization (AAACH-PSO). We have defined a threshold function (4) that takes into account the residual energy, the distance between the CH and the BS, and the number of nodes among the cluster. Results from the simulations indicate that our proposed protocol gives a higher network lifespan, delivers more data to the BS and the energy consumption is minimal compared to LEACH, LEACH-C and PSO-C. Furthermore, the proposed protocol produces better clustering by evenly allocating the CHs throughout the sensor network area and using AACH to balance the energy consumption. Our future work includes the implementation of PSO in hardware and multi-hop routing among the CHs to further improve energy efficiency. Comparison with other evolutionary optimization algorithm, such as Genetic Algorithm and cross-layer optimization through PSO will also be made.

REFERENCES

[1] J. Yick, B. Mukherjee, D. Ghosal. Wireless sensor network survey [J]. Computer Networks, 2008, 52 (12): 2292-2330.

[2] A. A. Abbasi, M. Younis. A survey on clustering algorithms for wireless sensor networks [J]. Computer Communications, 2007, 30 (14-15): 2826-2841.

[3] R. V. Kulkarni, G. K. Venayagamoorthy. Particle Swarm Optimization in Wireless Sensor Networks: A Brief Survey [J]. IEEE Trans. Systems, Man, and Cybernetics, 2011, 41 (2): 262-267.

[4] R. Xu, J. Xu, D. C. Wunsch. A Comparison Study of Validity Indices on Swarm-

Intelligence-Based Clustering [J]. IEEE Trans. Systems, Man, and Cybernetics, 2012, 42 (4): 1243-1256.

[5] Y. Valle, G. K. Venayagamoorthy, S. Mohagheghi, et al. Particle swarm optimization: Basic concepts, variants and applications in power systems [J]. IEEE Trans. Evol. Comput., 2008, 12 (2): 171-195.

[6] W. B. Heinzelman, A. P. Chandrakasan, H. Balakrishnan. An Application-Specific Protocol Architecture for Wireless Microsensor Networks [J]. IEEE Trans. Wireless Commun., 2002, 1 (4): 660-770.

[7] R. Wang, G. Liu, C. Zheng. A Clustering Algorithm based on Virtual Area Partition for Heterogeneous Wireless Sensor Networks [R]. In Proc. IEEE Mechatronics & Automation'07, 2007.

[8] X. Qi, C. Qiu. An Improvement of GAF for Lifetime Elongation in Wireless Sensor Networks [J]. Journal of JCIT, AICIT, 2010, 5 (7): 112-119.

[9] J. Huang, J. Zhang. Distributed Dual Cluster Algorithm Based on Grid for Sensor Streams [J]. Journal of JDCTA, AICIT, 2010, 4 (9): 225-233.

[10] M. Soliman, G. Tan. Conditional Sensor Deployment Using Evolutionary Algorithms [J]. Journal of JCIT, AICIT, 2010, 5 (2): 146-154.

[11] M. Romoozil, H. Ebrahimpour-komleh. A Positioning Method in Wireless Sensor Networks Using Genetic Algorithms [J]. Journal of JDCTA, AICIT, 2010, 4 (9): 174-179.

[12] N. M. Abdul Latiff, C. C. Tsimenidis, B. S. Sharif. Energy-aware clustering for wireless sensor networks using particle swarm optimization [R]. Proc. IEEE PIMRC'07, 2007.

[13] Y. Liang, H. Yu, P. Zeng. Optimization of clusterbased routing protocols in wireless sensor network using PSO [J]. Control and Decision, 2006, 21: 453-456.

[14] J. Kennedy, R. Eberhart. Particle swarm optimization [R]. Proc. IEEE Int. Conf. Neural Netw., 1995.

[15] R. Eberhart, Y. Shi. Particle swarm optimization: development, applications and resources [R]. IEEE Int. Congress on Evolutionary Computation, 2001.

[16] Y. Shi, R. Eberhart. A modified particle swarm optimization [R]. IEEE Int. Congress on Evolutionary Computation, 1998.

[17] T. Blackwell. A Study of Collapse in Bare Bones Particle Swarm Optimization [J]. IEEE Trans. Evol. Comput., 2012, 16 (3): 354-372.

[18] K. Pahlavan, A. Levesque. Wireless Information Networks [M]. New York: Wiley, USA, 1995.

An Adaptive Assistant-Aided Clustering Protocol using Niching Particle Swarm Optimization

Wireless sensor networks request robust energy efficient communications. In thisliterature, we propose and analyse an Adaptive Assistant-Aided Clustering Protocol using Niching Particle Swarm Optimization (AAAC-NPSO), a protocol scheme for wireless sensor networks that obeys the idea of energy efficient with application-specific to obtain good performance. Our simulation results show that AAAC-NPSO can improve system lifespan and data delivery by optimizing energy dissipation in the networks.

1 Introduction

Wireless sensor networks (WSNs) havegained more attention in recent years, especially with the development of Micro-Electro-Mechanical Systems (MEMS) and very-large-scale integration (VLSI) technology which have geared up the progress of wireless sensors[1]. A wireless sensor network consists of hundreds to thousands of small battery powered multifunctioning devices with limit energy supply. Once deployed, the sensor nodes are often unreachable to users, and the recharge or replacement of batteries are impracticable, so energy efficient is the vital factor for WSNs.

Clustering is one of the energy-efficient techniques that prolonging the sensor network's lifetime[2], and is often formulated as optimization problems[3]. Canonical analytical optimization methods require abundant computational resources, and grow up exponentially with the problem scale increases. But for implementation on a tiny wireless sensor node, we need an optimization method that requires moderate even minimal computing and memory resources and yet produce better results. Bio-inspired optimization methods are computationally efficient compared to canonical analytical methods[4-5].

Particleswarm optimization (PSO) is a bio-inspired optimization, and it is a popular, simple, effective and multidimensional optimization algorithm[6], it also has many good qualities, such as computational efficiency, ease of implementation, high quality of solutions and speed of convergence etc. Niching method can be incorporated into PSO to promote and maintain formation of multiple stable subpopulations within a single population, with an aim to locate multiple optimal or suboptimal solutions[7].

A clustering scheme called Low-Energy Adaptive Clustering Hierarchy (LEACH) is proposed in[8] that randomly rotating the role of cluster head among all the nodes in the network. SEP[9] considers node's heterogeneity and based on weighted election probabilities to yields longer stability. An approach that periodically elects cluster heads and assistant nodes together is proposed in[10]. The authors propose multi-assistant cluster heads[11] to cope with the power hungry issues. BEES[12] is a lightweight bio-inspired backbone construction protocol, it can help mitigate many of the typical challenges in sensor networks by allowing the developments of simpler network protocols. The application of PSO algorithm used to solve sensor networks clustering has been proposed in[13], the authors attempted to equalize the number of nodes in each cluster to minimize the energy consumption. A protocol using PSO[14] has been presented in minimizing the intra-cluster distance and optimizing the energy consumption of the network.

In this literature, we design an Adaptive Assistant-Aided Clustering protocol using Niching Particle Swarm Optimization (AAAC-NPSO), which takes advantage of niching PSO that can get multimodal optimal values. The selection of cluster heads considers the sensor network's state information, such as reserved energy, location of nodes and their neighbors, etc. The protocol first generates cluster head (CH) by select the optimal fitness, then based on the cluster's information such as CH's residual energy, the nodes situation and the number of nodes among the cluster to determine if we need to select an assistant-aided cluster head (AACH), and select the suboptimal fitness as assistant-aided cluster head in need. The CH and AACH are in charge of data collection, data aggregation and transmit the data to the base station respectively. The protocol abides by the idea of energy efficient to obtain good performance in terms of system lifespan and data delivery. Our simulation results and analysis show that AAAC-NPSO can achieve better system lifespan and data delivery by optimizing the cluster head selection and balance the energy consumption evenly in the networks.

2 Niching Particle Swarm Optimization

Through competition and cooperation among the population, population-based optimization approaches usually can find very good solutions efficiently and effectively. Particle Swarm Optimization (PSO) is one of the population-based optimization and motivated from the simulation of social behaviors of insects and animals such as fishing schooling and bird flocking. Since its inception by James Kennedy and Russell Eberhart[15], PSO has gained increasing attention among scientists and researchers as an efficient and robust technique in solving complex and difficult optimization problems.

Niching methods are of great value when the object is to locate a single/ multiple global optimum. Since a niching PSO searches for multiple optima in parallel, the probability of getting trapped on a local optimum may be reduced[7].

PSO is population based like an EA, however, PSO differs from EAs in the way it manipulates each particle in the population. Other EAs use evolutionary methods such as mutation and crossover, PSO updates each particle's position in the search space, according to its present velocity, its previous best positions it has found, and the best position found by its neighbors. So each particle searches around (or towards) a region defined by its personal best position and the neighborhood best position. Let us use $v_i = (v_{i1}, v_{i2}, \cdots, v_{id})$, $1 \leq i \leq m$, $1 \leq d \leq D$ to denote the velocity of the ith particle in the swarm, $x_i = (x_{i1}, x_{i2}, \cdots, x_{id})$ its position, $p_i = (p_{i1}, p_{i2}, \cdots, p_{id})$ the best position it has found so far, and $p_n = (p_{n1}, p_{n2}, \cdots, p_{nd})$ the best position found from its neighborhood (so-called global best). v_i and x_i of the ith particle in the swarm are updated according to the following two equations[17]:

$$v_{id}^{k+1} = \omega v_{id}^k + c_1 \xi (p_{id}^k - x_{id}^k) + c_2 \eta (p_{nd}^k - x_{id}^k) \tag{1}$$

$$x_{id}^{k+1} = x_{id}^k + v_{id}^{k+1} \tag{2}$$

Where ω is the inertia weight, it decides how the particle inherit present velocity, c_1 and c_2 are learning factor (or called acceleration coefficient), ξ and η are random numbers between 0 and 1. If we consider all the particles in the swarm as neighbor, we get the global version of PSO; if we consider partial particles as neighbors, we get the local version of PSO, and there has different methods to format the neighbors (we use the similar method as described in[7] to format the niching).

3 AAAC-NPSO Protocol

We proposedan AAAC - PSO protocol, aiming to enhance the existing clustering algorithms' performance. We present threshold function to decide if we need an assistant-aided cluster head and use particle swarm optimization algorithm to select cluster head and assistant cluster head if in need.

Fitness Function of NPSO

Thedetermination of fitness function is closely related with the properties of problem domain, and it determines the algorithm's performance of the optimal solution directly. So it is important to design the fitness function, based on[14-16], we consider the condition that influence the system performance and define the fitness function deliberately. First, the fitness function should consider the residual energy of the node, because if it becomes the cluster head, it consumes more energy than the member nodes. Second, the average distance between the cluster head and its member nodes should be considered due to the cluster head responsible for gathering data from its member nodes directly.

Based on the above considerations, we define the fitness function $f(i)$ for node i:

$$f(i) = \alpha f_1(i) + (1 - \alpha) f_2(i) \tag{3}$$

$$f_1(i) = E(i) / \frac{1}{m} \sum_{i=1}^{m} E(i)$$

$$f_2(i) = \frac{1}{m-1} \sum_{k=1, k \neq i}^{m} \frac{E_i \times d(i, k)}{d(i, k) + 1}$$

$$0 \leq \alpha \leq 1$$

Function $f_1(i)$ is the ratio of node i's energy to the average residual energy within the cluster, $E(i)$ is the energy reside in node i, m is the number of member nodes within the cluster. Function $f_2(i)$ considers the residual energy and the distance between the node and other nodes among the cluster comprehensively. α is the weight of functions that we get from experiments. The fitness function that defined above has considered the energy efficient of the network and minimized the intra-cluster distance between the cluster head and its member nodes. So according the fitness function defined above, we choose the node that has the maximum value as cluster head, and it is the optimum selection during that round.

Assistant-Aided Cluster Heads Setup

After the cluster head has been selected, it's time to decide whether to select an AACH in the cluster or not. We consider from the following several aspects, the residual energy of the CH, the distance between the CH and the BS, and the number of nodes among the cluster. Generally if cluster head has little residual energy, the distance is far from the BS and the amount of nodes among the cluster is large, on this occasion, we need an AACH to balance the energy dissipation and enhance the system robustness. So we define the threshold function as:

$$T_{AACH} = c \cdot \frac{E_{max} - E_{re}}{E_{max}} \cdot \frac{d_{(CH, BS)}}{d_{avg}} \cdot \frac{N_{Ci}}{N_{avg}}$$

Where c is control parameter; E_{re} is the residual energy of the cluster head; E_{max} is the maximum energy of the cluster head; d_{avg} is the average distance between the BS and nodes in the network; $d_{(CH, BS)}$ is the distance between the CH and the BS; N_{avg} is the average number of nodes in a cluster; N_{Ci} is the number of nodes in the cluster i. The decision is made by the CH choosing a random number, if the number is less than the threshold T_{AACH}, the cluster need to select a node as $AACH$, we select the node that has the suboptimal fitness as $AACH$, and it is the optimum selection during that round.

AAAC-NPSO Setup

We select the optimal solution as CH, and the suboptimal solution as AACH if in need according the cluster state information when the iterations of NPSO finished in each round. According to the fitness function, the CH has the optimal fitness value that it is energy sufficient and the energy consumption is minimal, the AACH also has more energy and close to the CH, then CH and AACH gathered and aggregated the data, and transmit it to the base station respectively. Our simulation shows that we can achieve good performance and prolong the network lifetime.

We first use clustering protocol such as LEACH[7] to part the network field and the cluster heads as temporary cluster heads. All nodes send information about its current energy and locations to its temporary cluster head. The AACH-NPSO algorithm executes in period, each cluster optimally chooses the CH and AACH using NPSO. The CH acts as local control center to coordinate the data transmission within the cluster. The CH sets up a TDMA schedule to avoid collision for its member nodes, allows the radio electronics to be turn off at all times except during their transmission slot for each member nodes[16], that has further reduce the energy dissipation and enhance the network performance. Once the CH has finished receiving data from all its member nodes, the CH performs data aggregation and sends the aggregated data to the AACH within its cluster to further minimizing energy consumption if needed or sends to the BS directly. Fig. 1 shows the flowchart of AAAC-NPSO protocol.

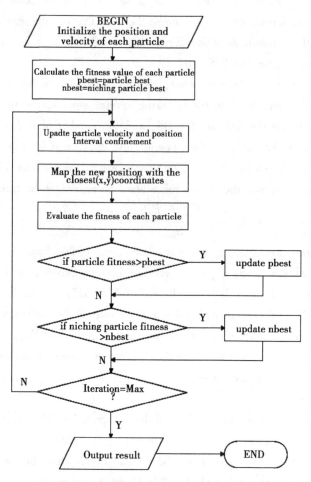

Fig. 1 Flowchart of AACH-NPSO protocol

Thetemporary cluster head runs AAAC-NPSO algorithm to select the cluster head and assistant cluster head if in need. The detailed steps are as follows:

(1) Initialize position and velocity of each particle randomly.

(2) Calculate the fitness value of each particle using Formula (3).

(3) Find the personal best for each particle. The personal best is the position of the particle itself has experience that has the maximum fitness; and the global best is the position of all particles that have experience and have the maximum fitness.

(4) Update each particle's position and velocity using Formula (1) and (2).

(5) Limit the change of the particle's position value and velocity value, then map the new

updated position with the closest coordinates.

Repeat step (2) to (5) until the maximum number of iterations is reached. Select the best solution as CH, and the suboptimal solution as AACH if in need.

4 Simulation

We evaluated the performance of the proposed protocol, we run the simulations for 100 nodes randomly distributed in a $100m \times 100m$ network area with the BS located at $(x = 50, y = 50)$. The data message size was 125 bytes with the packet header was 25 bytes long. The simulation parameters of the energy consumption model are described in[8-9], and for PSO parameters, we used $m = 20$ particles, $c_1 = c_2 = 2$, $x_{max} = v_{max} = 200$, inertia weight from $\omega = 0.9$ to 0.4, the weight of function from $\alpha = 0.2$ to 0.9. The performance of our protocol was compared with the benchmark clustering protocols for wireless sensor networks LEACH[8] and PSO-C[14].

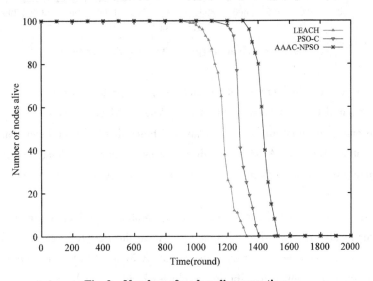

Fig. 2 Number of nodes alive over time

Fig. 2 illustrates the system lifetime defined by the number of nodes alive over time. Obviously our proposed protocol can prolong the network lifetime significantly compared to LEACH and PSO-C. This is because our protocol produces better network partitioning with minimum intra-cluster distance, cluster heads that are optimally distributed across the network and we define threshold function to determine if we need an assistant-aided cluster head to balance the energy dissipation. On the contrary, in LEACH and PSO-C some nodes have to transmit long distances in order to reach a cluster head due to poor network clustering or the cluster head consume more energy for data gathering, aggregation and transmitting. As a result, some nodes dissipate a large amount of energy while transmitting their data to a faraway cluster head or the base station at a time.

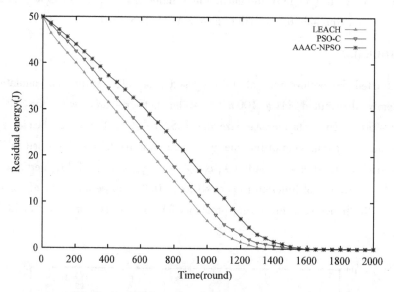

Fig. 3 Residual energy of the network over time

Fig. 3 shows the residual energy of the network, the protocol we proposed consumed less energy with compared to LEACH and PSO-C. As AAAC-NPSO consider more conditions that may affect the energy dissipation and select an assistant-aided cluster head that balanced the energy consumption. While LEACH stochastic select nodes as cluster head and PSO-C emphasize on minimize intra-cluster distance, that lead to the energy dissipation is more than AAAC-NPSO and not evenly.

As can be seen from these figures, the proposed protocol can balance the energy consumption and cluster heads are located at the place that have the best fitness, and select assistant-aided cluster head to balance the energy dissipation if in need. LEACH and PSO-C on the other hand, produce an uneven distribution of cluster heads throughout the sensor field or consume more energy by cluster head. Our proposed protocol minimizes intra-cluster distance and selects assistant cluster heads to balance the energy consumption if in need.

5 Conclusion

In this paper we have presenteda Adaptive Assistant-Aided Clustering protocol for wireless sensor networks using Niching Particle Swarm Optimization (AAAC-NPSO). We have defined a threshold function that takes into account the residual energy, the distance between the cluster head and the BS, and the number of nodes among the cluster to decide whether we need an assistant-aided cluster head in a cluster. Results from the simulations indicate that the proposed protocol using NPSO algorithm gives a higher network lifespan and the energy consumption is minimal compared to LEACH and PSO-C. Furthermore, the proposed protocol selects the optimal nodes as cluster heads throughout the sensor network area and uses assistant-aided cluster head to balance the energy consumption. Our future work includes the implementation of NPSO in hardware and multi-hop routing among the cluster head nodes to further improve energy efficiency. Comparison with other evolutionary optimization algorithm, such as Genetic Algorithm and cross-layer optimization through PSO will also be made.

REFERENCES

[1] J. Yick, B. Mukherjee, D. Ghosal. Wireless sensor network survey [J]. Computer Networks, 2008, 52 (12): 2292-2330.

[2] A. Abbasi, M. Younis. A survey on clustering algorithms for wireless sensor networks [J]. Computer Communications, 2007, 30 (14-15): 2826-2841.

[3] R. Kulkarni, G. Venayagamoorthy. Particle Swarm Optimization in Wireless Sensor Networks: A Brief Survey [J]. IEEE Trans. Systems, Man & Cybernetics, 2011, 41 (2): 262-267.

[4] N. Latiff, C. Tsimenidis, B. Sharif. Performance comparison of optimization algorithms for clustering in wireless sensor networks [R]. In Proc. IEEE Int. Conf. Mobile Ad Hoc Sens. Syst., 2007.

[5] R. Xu, J. Xu. A Comparison Study of Validity Indices on Swarm-Intelligence-Based Clustering [J]. IEEE Trans. Systems, Man & Cybernetics, 2012, 42 (4): 1243-1256.

[6] Y. Valle, G. Venayagamoorthy, et al.. Particle swarm optimization: Basic concepts, variants and applications in power systems [J]. IEEE Trans. Evol. Comput., 2008, 12 (2): 171-195.

[7] X. Li. Niching Without Niching Parameters: Particle Swarm Optimization Using a Ring Topology [J]. IEEE Trans. Evol. Comput., 2010, 14 (1): 150-169.

[8] W. B. Heinzelman, A. P. Chandrakasan, H. Balakrishnan. An Application-Specific Protocol Architecture for Wireless Microsensor Networks [J]. IEEE Trans. Wireless Commun., 2002, 1 (4): 660-770.

[9] G. Smaragdakis, I. Matta, A. Bestavros. SEP: A stable election protocol for clustered heterogeneous wireless sensor networks [R]. in Proc IEEE SANPA'04, 2004.

[10] M. H. Yeo et al.. An Energy Efficient Distributed Clustering Approach with Assistant Nodes in Wireless Sensor Networks [R]. in Proc. IEEE Radio and Wireless Symposium, 2008.

[11] S. Jabbar, et al.. Threshold Based Load Balancing Protocol for Energy Efficient Routing in WSN [R]. in Proc. IEEE ICACT'11, 2011.

[12] H. S. AbdelSalam, S. Olariu. BEES: Bio-inspired backbone Selection in

Wireless Sensor Networks [J]. IEEE Trans. Parallel and Distrib. Sys., 2012, 23 (1): 44-51.

[13] J. Tillet, R. Rao, F. Sahin. Cluster-head identification in ad hoc sensor networks using particle swarm optimization [R]. IEEE International Conference on Personal Wireless Communications, 2002.

[14] N. Latiff, C. Tsimenidis, B. Sharif. Energy-aware clustering for wireless sensor networks using particle swarm optimization [R]. In Proc IEEE PIMRC, 2007.

[15] J. Kennedy, R. Eberhart. Particle swarm optimization [R]. In Proc. IEEE Int. Conf. Neural Netw., 1995.

[16] K. Pahlavan, A. Levesque. Wireless Information Networks [M]. Wiley: USA, 1995.

[17] T. Blackwell. A Study of Collapse in Bare Bones Particle Swarm Optimization [J]. IEEE Trans. Evolutionary Computation, 2012, 16 (3): 354-372.

An Adaptive Clustering Protocol using Niching Particle Swarm Optimization

Clustering is an hierarchical topology control method, and it is also an energy-saving and energy efficient technique that extends the sensor network's lifetime. In this paper, we propose and analyze an adaptive clustering protocol using niching particle swarm optimization (ACP-NPSO), a protocol architecture that uses NPSO to cluster the wireless sensor networks adaptively and efficiently, thus saving energy, balancing energy consumption and enhancing the system's robustness. The simulation results indicate that our proposed protocol ACP-NPSO can enhance system lifespan, accelerate the convergence speed, and deliver more data by distributing energy dissipation evenly in the networks.

1 INTRODUCTION

A number of battery powered sensors are deployed in monitoring areas. Those sensor nodes are powered by small batteries, so the energy supply is limited [1]. As those sensors are often deployed in hostile areas, and are therefore unreachable to users, it's difficult (or impossible) to recharge or replace the batteries, so energy efficiency is a vital factor we must consider during the design of a wireless sensor network (WSN).

Clustering is an energy-saving and efficient technique which can extend the lifetime of WSN[2], and usually clustering is formulated into optimization problems to get the best solutions[3]. Conventional optimization such as analytical methods incur a higher computation overhead, and with the increase in scale of the problem, the computation grows exponentially. But for application on a wireless sensor node, we need an optimization algorithm that demands moderate even minimal memory and computing resource, and the optimization algorithm can generate better results[4,5]. Bio-inspired optimization methods are computationally efficient methods that we want to compare to the conventional optimization algorithms[6,7].

As a bio-inspired optimization algorithm, particle swarm optimization (PSO)[8] has many good qualities, such as rapid convergence, operational ease, and efficient computation. It is also simple, effective and multidimensional, so it has been widely used and has become popular in optimization areas. Niching is also inspired by nature and it is a significant multi-modal optimization method[9]. In natural ecosystems, all kinds of species take on different roles to survive through competition and selection. Niching technique can be integrated into PSO to obtain the steady formation of multiple sub-populations within an individual population, helping to find multiple optimal solutions[10,11].

In this paper, we take advantage of niching PSO and design an adaptive clustering protocol using niching particle swarm optimization (ACP-NPSO). The protocol clusters the network adaptively and selects cluster heads (CHs) considering the state information of the sensor network deliberate, it first considers the network's information to decide the optimal number of clusters dynamically and then uses Niching PSO to select the optimal CH set adaptively. The objective of the proposed solution is to obtain good performance in terms of system lifespan, energy consumption and application-perceived quality. The analysis and simulation results show that ACP-NPSO can achieve better system lifespan, obtain fast convergence speed, and deliver more data by clustering the network adaptively and selecting the optimal cluster heads set, thus minimizing and balancing the energy consumption in the networks.

The outline of the paper is as follows. Section II surveys the related work. Section III describes the system model. Section IV introduces the niching PSO algorithm and elaborates the design of ACP-NPSO in detail. Section V provides the simulation results followed by the conclusion in Section VI.

2 RELATED WORK

A large number of low and small battery-powered sensors have been proposed, and conventional methods, such as direct transmission, are avoided due to their inefficiency. A low-energy adaptive clustering hierarchy (LEACH) was proposed in[12]. LEACH is an adaptive clustering scheme and the role of cluster head is rotated randomly among the network. SEP[13] is based on the node's weighted election probabilities to choose cluster head on the basis of the remaining energy in the sensor nodes, it is suitable for heterogeneity network and obtains longer stability. BEES[14] allows the development of simpler network protocols, so it helps to mitigate many typical challenges in sensor networks, and it is a lightweight bio-inspired protocol.

Particle swarm optimization has been applied to resolve WSN issues such as clustering, data fusion, localization, and deployment[3]. The authors used PSO to solve the clustering problem of a wireless sensor network, and attempted to minimize energy dissipation by equalizing the number of sensors in each cluster[15]. Another clustering protocol using PSO to optimize the energy dissipation and minimize the intra-cluster distance, thus got good performance[16]. The authors proposed a cluster-based protocol based on PSO, the network considered the reminder energy and location about the node and their neighbours during cluster selection[17]. A dynamic clustering optimization problems has been investigated, the protocol used a hierarchical clustering technique to search and determine multiple optimum[18]. Our design takes advantage of PSO, considers more network states information deliberately, and uses Niching technology to avoid getting trapped on a local optimum.

Some techniques, such as the distributed sampling rate control method[19], the load-balanced algorithm[20], and others[21], could be used to balance and minimize the energy consumption further. In practical applications, we consider more factors that may affect the system performance, but in this paper, we emphasize clustering and hierarchical topology control methods.

3 SYSTEM MODEL

3.1 Network model

We consider a total amount of N sensor nodes being distributed randomly within a $M \times M$ squared field, and assume a sensor network model has properties as follows:

- A fixed base station (BS) is situated at the center of the sensing area.
- The sensor nodes are powered by batteries with initial energy assigned.
- The sensors can communicate with the BS directly.
- All sensors has power control abilities to vary their transmitted power according to the distance.
- All sensor nodes are aware of their location. There has been many existing work on WSN's localization[22,23]. For instance, we can use the method proposed in[22] to get the localization.
- Each sensor moves slowly with respect to the BS (or has no mobility to move).

The wireless sensors gather the sensed data in a periodic way, and we define each period as a round, with each round including three phases: a set-up phase, during which the network uses the ACP-NPSO to select CHs and divides the network into clusters; a steady state phase follows when the CHs gather and aggregate the sensed data; and finally a finish phase when the CHs transfer the aggregated data to the BS. The periodic way the network runs is depicted in Fig. 1.

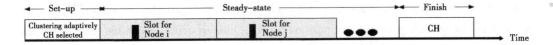

Fig. 1 Time line showing the ACP-NPSO periodic method of operation

3.2 Radio model

The radio model we use is as described in[12]. In this radio model, a transmitter consumes energy to run a power amplifier and radio units, and a receiver consumes energy to run radio units. In order to transfer data reliably, the amplifier needs to consume energy to keep an acceptable signal-to-noise ratio (SNR), we use a multipath fading channel model (r^4 power loss) when $r > r_o$, and a free space channel model (r^2 power loss) when $r \leqslant r_o$, and r_o is the threshold distance[24]. A transmitter consumes energy to transmit an l-bit message along a distance r using this radio model as shown:

$$E_{Tx}(l, r) = E_{Tx}(l) + E_{amp}(l, r)$$

$$= \begin{cases} l \times E_{Tx} + l \times \varepsilon_{fs} \times r^2, & r \leq r_o \\ l \times E_{Tx} + l \times \varepsilon_{mp} \times r^4, & r > r_o \end{cases} \quad (1)$$

The threshold distance r_o is given by $r_o = \sqrt{\varepsilon_{fs}/\varepsilon_{mp}}$, The receiver consumes to receive an l-bit message is: $E_{Rx}(l) = l \times E_{Rx}$.

The receiver can be turned off to avoid receiving unwanted messages, and the transmitter can set the power amplifier according to the communication distanceto minimize the energy dissipation. The parameters of communication energy consumption are set as $E_{Tx} = E_{Rx} = 50nJ/bit$, $\varepsilon_{fs} = 10pJ/bit/m^2$, $\varepsilon_{mp} = 0.0013pJ/bit/m^4$, the data aggregation energy consumption is set as $E_{DA} = 5nJ/bit/message$ [25].

4 ACP-NPSO ALGORITHM

We present an adaptive energy-efficient clustering protocol to improve the existing clustering protocols' performance. In this section, we designACP-NPSO in detail, present the fitness function and adopt niching particle swarm optimization to cluster the network.

4.1 Niching particle swarm optimization

By means of cooperation and competition among the population, population-based optimization methods can usually find solutions effectively and efficiently. There are a lot of approaches that are motivated by evolution in nature, such as genetic algorithms (GAs), and the ant colony algorithm (ACA). Particle swarm optimization (PSO) is derived from the simulation of social behaviors of animals and insects such as bird flocking. It is a population-based optimization method. PSO has been widely investigated among researchers and scientists since it was put forward by Russell Eberhart and James Kennedy[26]. it is a robust and efficient method in solving difficult and complex optimization problems[27].

In a standard PSO, within a D-dimension search space, a swar m contains m particles flying among the space with a given speed. Each particle flies towards (or around) a region that defines the particle personal best position (i.e., the position at which the particle has the best fitness value) and the neighborhood best position (or the entire population best position). We use $v_i = (v_{i1}, v_{i2}, \cdots, v_{id})$, $1 \leq i \leq m$, $1 \leq d \leq D$, to stand for the i th particle's velocity, the particle's position is $x_i = (x_{i1}, x_{i2}, \cdots, x_{id})$, the best position it has found is $p_i = (p_{i1}, p_{i2}, \cdots, p_{iD})$, and the best position that the entire swarm (or its neighborhood) has found is $p_g = (p_{g1}, p_{g2}, \cdots, p_{gD})$; the i th particle's velocity v_i and position x_i are updated according to the equations as follows[28]:

$$v_{id}^{k+1} = \omega v_{id}^k + c_1 \xi (p_{id}^k - x_{id}^k) + c_2 \eta (p_{gd}^k - x_{id}^k) \quad (2)$$

$$x_{id}^{k+1} = x_{id}^k + v_{id}^{k+1} \quad (3)$$

Where ω is the inertia weight, which determines the present velocity the particle has inherited; a small inertia weight promotes the particle's local search ability (exploitation ability), a large inertia weight promotes the particle's global search ability (exploration ability).

c_1 and c_2 are acceleration coefficients (also called learning factors), and they are normally positive constants. The acceleration coefficient prompts the particle flying towards the neighborhood (or global) best position and its personal best position. ξ and η are stochastic numbers between 1 and 0. If we take partial particles as neighbors, we get the local version of PSO; if we take the entire particles as neighbors, we get the global version of PSO, There are many methods to formthe neighborhood[29,32], and niching is a simple and popular method we can use.

Niching is inspired by nature, different species must compete to survive and take on various roles in natural ecosystems. Different species evolve to different "subspaces" (or niches) that can sustain different life types, instead of evolving a single population, so the natural ecosystems evolve different species (or subspaces, sub-populations) to form different niches[9,10].

When we want to locate multiple global optima, the niching method is valuable and has many good qualities. Since a niching PSO searches in parallel for multipleoptima, the chance that the algorithm falls into a local optimum is greatly reduced[11].

4.2 Optimum number of clusters

Assume N nodes are distributed randomly in an $A = M \times M$ square meter area, the radio model is as described in Section Ⅲ, the amount energy consumed by all sensors during a round is[12,13]:

$$E_{round} = l \left\{ NE_{DA} + 2NE_{elec} + \varepsilon_{fs} \left(k d_{toBS}^2 + N \frac{M^2}{2\pi k} \right) \right\} \quad (4)$$

Where k is the number of CHs, d_{toBS} is the Euclidean distance from the BS to the CH, and l is the bits that are contained in a data message.

We can find the optimal value of k to minimize the energy consumed during a round, thus saving more energy in the network. We set the derivative of the continuous function about E_{round} with respect to k to zero (when $k>0$)

$$E'_{round} = l\varepsilon_{fs} \left(d_{toBS}^2 - \frac{N}{2\pi} \frac{M^2}{k^2} \right) = 0$$

So the optimum number of clusters is obtained from the derivation.

$$k_{opt} = \sqrt{\frac{N}{2\pi}} \frac{M}{d_{toBS}} \quad (5)$$

Now we prove the k_{opt} is exactly the optimal number we want. First we define a function $f(k) = (k - k_{opt}) \times E'_{round}$, as we know $(k - k_{opt})^2 \times (k + k_{opt}) \geq 0$. The proof process is shown as follows:

$$(k - k_{opt})^2 \times (k + k_{opt}) \geq 0$$
$$\Rightarrow (k - k_{opt}) \times (k^2 - k_{opt}^2)^2 \geq 0$$
$$\Rightarrow \left(k - \sqrt{\frac{N}{2\pi}} \frac{M}{d_{toBS}^2} \right) \times \left(k^2 - \frac{N}{2\pi} \frac{M^2}{d_{toBS}^2} \right) \geq 0$$

$$\Rightarrow \left(k - \sqrt{\frac{N}{2\pi}} \frac{M}{d_{toBS}^2}\right) \times \left(d_{toBS}^2 - \frac{N}{2\pi} \frac{M^2}{k_{toBS}^2}\right) \geqslant 0$$

$$\Rightarrow \left(k - \sqrt{\frac{N}{2\pi}} \frac{M}{d_{toBS}^2}\right) \times l\varepsilon_{fs} \left(d_{toBS}^2 - \frac{N}{2\pi} \frac{M^2}{k_{toBS}^2}\right) \geqslant 0$$

$$\Rightarrow (k - k_{opt}) \times E'_{round} \geqslant 0$$

$$\Rightarrow f(k) \geqslant 0$$

Let us suppose $k \neq k_{opt}$, so $f(k) > 0$, that is the sufficient condition which satisfied the continuous function existing minimum functional value, so k_{opt} is just the optimal number of clusters we want.

Where d_{toBS} is the average distance between the BS and CHs, it can be given by[13]:

$d_{toBS} = \int_A \sqrt{x^2 + y^2} \frac{1}{A} dA = 0.765 \frac{M}{2}$, so the optimal number of clusters is: $k_{opt} = \frac{1}{0.765} \sqrt{\frac{2N}{\pi}}$.

We can see from the above, the optimal number of clusters is decided by the number of sensor nodes N in the given scenario. If we add nodes to a sensor field, k_{opt} will increase; and if we take away some nodes or some nodes die due to runing out of energy, k_{opt} will decrease correspondingly. So, we are on the basis of the number of nodes in the network to decide k_{opt}, and part of the network into k_{opt} clusters adaptively.

4.3 The fitness function of ACP-NPSO

The fitness function determines the optimal solution directly, thus affecting the protocol's performance. The fitness function is related closely with the problem domain's properties, so the design of the fitness function is very important, based on[15-17], we deliberately consider the conditions that may affect or influence the protocol performance, and explore better fitness function and parameters by doing many experiments, we finally define the fitness function f as follows:

$$f = \alpha f_1 + \beta f_2 + \gamma f_3 \tag{6}$$

$$f_1 = \sum_{k=1}^{k_{opt}} \sum_{\forall n_i \in C_{p,k}} d(n_i, CH_{p,k}) / |C_{p,k}|$$

$$f_2 = \sum_{i=1}^{N} E(n_i) / \sum_{k=1}^{k_{opt}} E(CH_{p,k})$$

$$f_3 = \frac{1}{k_{opt}} \sum_{k=1}^{k_{opt}} d(CH_{p,k}, BS)$$

$$\alpha + \beta + \gamma = 1, \; 0 \leqslant \alpha, \beta, \gamma \leqslant 1$$

Where $|C_{p,k}|$ is the number of sensors that belong to cluster C_k of particle p, so f_1 is the sum of average distance of each nodes to their cluster heads. N stands for the number of living nodes, function f_2 is the proportion of the total reminder energy of all living nodes ni ($i = 1, 2, \cdots, N$) with the total reminder energy of the cluster heads candidates in the running round. f_3 is the average distance of cluster heads candidates to the BS. α, β, γ are the weight of each sub function, the weight values are obtained from experiments.

The fitness function we have defined above takes the energy efficient into account as quantified by f_1; minimize the intra-cluster distance between the member sensors and the CH as quantified by f_2; and consider the distance between the BS and CHs as the CHs consume energy to transmit the aggregation data that the CHs have gathered to the BS at the end of each round. So we choose the particle that has the minimum fitness value as CH according to the fitness function, it is the optimal selection and the energy consumption is minimal during that round.

Since k_{opt} is determined according to the number of sensor nodes, we use Niching PSO to select k_{opt} CHs and part the network adaptively. The ACP-NPSO protocol executes in a periodical way, the CH serves as the control centre to harmonize the member nodes actions within the cluster. The CH sets up a TDMA schedule among the cluster to avoid transmission collision, allows the radio electronics to be turn off except during its own sensing and transmission slot, that enhance the network performance and reduce the energy consumption further.

4.4 ACP-NPSO setup

Clustering is usually formulated into an optimization problem[3], and niching PSO is one of bio-inspired optimization methods that has many good qualities such as convergence rapid and computation efficiency[6,7]. We take advantage of PSO as used in[15-18] to optimize the clustering problems.

We use the fitness function (6) as defined above to select the optimal set of CHs and their associated member nodes. According to the fitness function, the CH hasthe optimal fitness consumes minimal energy and has sufficient energy, the CH collects and aggregates the sensing data, then transmits it to the BS using carrier sense multiple access (CSMA) and a fixed spreading code[30], that optimizes andminimizes the energy dissipation. The simulation results show that the protocol can prolong the network lifetime and obtain good performance.

Fig. 2 shows the flowchart of the ACP-NPSO algorithm. Assume in a WSN there are N sensor nodes, we can calculate the predetermined number of optimalclusters, then we cluster the network as follows:

1. Initialize m particles and selected k_{opt} cluster heads randomly.

2. Using the formula (5) to calculate the fitness value of each particle.

• For each living node n_i ($i=1, 2, \cdots, N$), calculate distance $d(n_i, CH_{p,k})$ between node n_i and all cluster heads $CH_{p,k}$

• Assign node n_i to cluster head $CH_{p,k}$ where

$$d(n_i, CH_{p,k}) = \min_{\forall k=1, 2, \cdots, k_{opt}} \{d(n_i, CH_{p,k})\}$$

• Using (5) to calculate the fitness function.

3. Find niching and personal best for each particle. The personal best is the best position

An Adaptive Clustering Protocol using Niching Particle Swarm Optimization

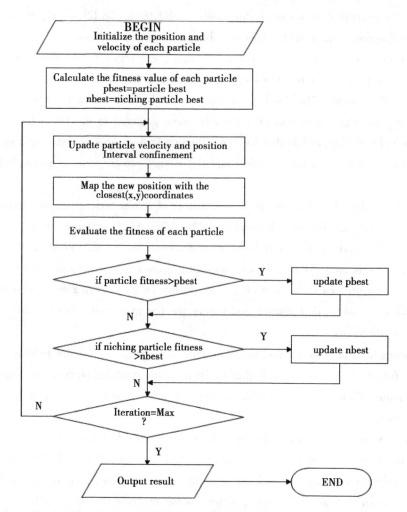

Fig. 2 The flowchart of the ACP-NPSO algorithm for cluster setup

that the particle itself has passed and has the optimal fitness; and the niching best is the best position that the particles in the niche have passed and have the optimal fitness. There is a lot of existing work on niching methods[9-11]. For example, we can use the solution presented in[9] to achieve the niching method:

4. Use (2) and (3) to update each particle's velocity and position.

5. Limit the particle's velocity value and position value (or the change), then map the sensor's position to the closest coordinates.

6. Repeat steps 2 to 5 until the protocol reaches the maximum number of iterations.

The protocol identifies the optimal set of cluster heads and their member nodes, the base station transmits all cluster head ID information to all nodes in the network. The nodes that have been chosen as cluster heads works as control centers to coordinate the actions of the sensors in its cluster. Using a non-persistent CSMA MAC protocol[30], an advertisement message is broadcasted by the cluster head; this message is distinguished from the announcement message because it is a short message that contains a header and the node's ID. The CH sets up a TDMA schedule[31] and transmits it to the member nodes in the cluster using the broadcast method. When all nodes know the TDMA schedule in the cluster, the set-up phase is finished and the network comes into the steady-state.

5 SIMULATION

In this section, we evaluate the performance of the protocol we have proposed. Let's suppose there are 100 nodes distributed in a 100m×100m square meter area randomly. The BS is situated at ($x=50$, $y=50$). The data message length is 125 bytes with 25 bytes packet header. The radio model's parameters are as described in Section Ⅲ, and the niching PSO parameters are as follows: $m=20$ particles, learning factor $c_1=c_2=2$, $x_{max}=v_{max}=200$, inertia weight shifts from $\omega=0.9$ to $\omega=0.4$, $\alpha=0.4$, $\beta=0.5$, and we adopt the similar niching method as in[9]. The performance of our proposed protocol ACP-NPSO is compared with the existing clustering protocols, LEACH, LEACH-C[12], and PSO-C[16].

Fig. 3 depicts the system lifetime, we define the number of nodes alive over time as the system lifetime. We can see that our protocol can obviously prolong the network lifetime compared to other clustering protocols. That is because our proposed protocol ACP-NPSO optimizes select CHs that have minimum intra-cluster distance and the distance between the BS and the CHs. This further minimizes the energy consumption and the dissipates the energy evenly among the network due to the optimal selection of CHs. While in LEACH and LEACH-C, the CHs distribute randomly among the network, and some nodes have a long distance to the CHs, PSO-C does not consider the energy dissipation that is consumed by the CHs transmitting data to the BS, so all that lead to some nodes or CHs consuming more energy during the data gathering and transmitting.

The system lifetime objective is as shown in Table 1; obviously the lifetime objective of ACP-NPSO is longer than LEACH, LEACH-C and PSO-C, that because our protocol uses niching PSO to get optimized network clustering and selects the best nodes as cluster heads. While LEACH and LEACH-C part the network unevenly and select nodes randomly as cluster heads, PSO-C does not use the niching method, and does not consider the distance between the CHs and the BS.

The convergence speed comparisons are as shown in Table 2, we can see that ACP-NPSO has a faster speed of convergence than PSO-C, because ACP-NPSO uses the niching technique to get optimization. We integrate the niching technique into PSO to accelerate the formation of multiple sub-population in a single population and maintain the stable of the sub-population, so the convergence speed is fast. LEACH and LEACH-C don't use optimization method to resolve the clustering problem, so the clusters distribute randomly in the network, and sometimes more or less than the optimal number of clusters. ACP-NPSO uses optimization

An Adaptive Clustering Protocol using Niching Particle Swarm Optimization

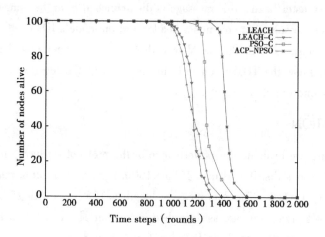

Fig. 3 Number of nodes alive over time

method, and decides the optimal number of clusters according the network states, then divides the network into clusters using niching PSO to find the optimal solutions.

Table 1 System lifetime objective

Protocol	Lifetime objective (Rounds)
LEACH	1 304
LEACH-C	1 342
PSO-C	1 421
ACP-NPSO	1 518

Table 2 Convergence speed comparisons

Particle number	The iteration number that PSO-C convergence	The iteration number that ACP-NPSO convergence
20	589	378
30	518	326
40	407	254
50	326	203

The total data messages received at the BS is as depicted in Fig. 4, clearly our proposed protocol ACP-NPSO can deliver more data to the BS compared to LEACH, LEACH-C, and PSO-C. That is because ACP-NPSO selects the optimal nodes that have more energy as CHs and the CHs distribute evenly among the network, thus the CHs have sufficient energy to act as local control center and the energy consumption is balanced, so the CHs transfer more data to the BS. On the contrary, LEACH and LEACH-C do not consider the reminder energy during the CHs selection, and may select some nodes as CHs that have insufficient energy to remain alive until the transmit phase. PSO-C does not consider the distance between the CHs and the BS, and we use the niching method to avoid getting trapped on a local optimum.

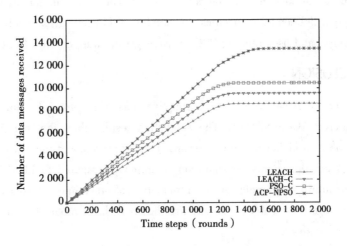

Fig. 4 Total data messages received at the BS over time

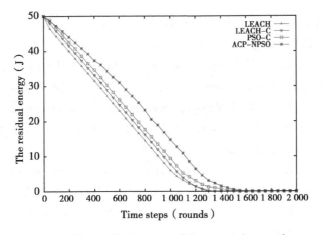

Fig. 5 The residual energy of the network over time

Fig. 5 shows the residual energy of the network, compared to LEACH, LEACH-C, and PSO-C. Our proposed protocol ACP-NPSO consumes less energy. That is because ACP-NPSO considers more network states that may influence the energy consumption and uses niching PSO to choose the optimal cluster head set. While LEACH and LEACH-C select nodes as CHs randomly, PSO-C only minimizes the intra-cluster distance, thus leading to the energy consumption being more than ACP-NPSO, and the energy dissipation not being evenly distributed among the network.

We did comprehensive evaluations on our protocol from the perspectives of system lifetime, lifetime objective, convergence speed, data messages received, and residual energy. As can be seen from these figures and tables, the proposed protocol ACP-NPSO can result in good network clustering, balance the energy consumption and accelerate convergence speed, where CHs are located at the places that have the best fitness and are evenly positioned across the network. This is due to the fact that our proposed protocol considers the conditions that affect the energy consumption deliberately, decides the number of clusters dynamically and clusters the network adaptively, and uses niching PSO to avoid getting trapped on a local optimum.

6　CONCLUSION

In this paper we have presented an adaptive clustering protocol using niching particle swarm optimization (ACP-NPSO). The simulation results show that our proposed protocol ACP-NPSO achieves a higher network lifespan, obtains fast convergence speed, delivers more data to the BS, and the energy consumption is minimal compared to LEACH, LEACH-C, and PSO-C. Our future work includes the introduction of other bio-inspired optimization such as the ant colony algorithm (ACA) to optimize the cross-layer protocol, and the development of niching PSO in hardware.

REFERENCES

[1]　Yick, J., B.Mukherjee, D.Ghosal. Wireless sensor network survey [J]. Comput. Netw, 2008, 52 (12): 2292-2330.

[2]　Abbasi, A.A.and M.Younis. A survey on clustering algorithms for wireless sensor networks [J]. Comp.Comm, 2007, 30 (14-15): 2826-2841.

[3]　Kulkarni, R.V., G.K.Venayagamoorthy. Particle swarm optimization in wireless sensor networks: A brief survey [J]. IEEE Trans.Syst.Man Cybern.-Appl.Rev, 2011, 41 (2): 262-267.

[4]　Li, W., J.Taheri, A.Y.Zomaya, et al. Nature-inspired computing for autonomic wireless sensor networks [R]. In Hamid S-A and A.Y.Zomaya (Eds.) Large scale network-centric distributed systems, JohnWiley & Sons, Inc., Hoboken, New York, USA., 2013.

[5]　Ma, D., J.Ma, P.Xu, et al. An adaptive node partition clustering protocol using particle swarm optimization [R]. Proc. IEEE ICCA' 13, Hangzhou, China, 2013.

[6]　Latiff, N., C.Tsimenidis, B.Sharif. Performance comparison of optimization algorithms for clustering in wireless sensor networks [R]. Proc.IEEE Int.Conf.Mobile Ad Hoc Sens.Syst., Pisa, Italy, 2007.

[7]　Xu, R., J.Xu, D.C.Wunsch. A comparison study of validity indices on swarm-intelligence-based clustering [J]. IEEE Trans.Syst.Man Cybern.Part B-Cybern., 2012, 42 (4): 1243-1256.

[8] Valle, Y., G.K.Venayagamoorthy, S.Mohagheghi, et al. Particle swarm optimization: Basic concepts, variants and applications in power systems [J]. IEEE Trans.Evol.Comput., 2008, 12 (2): 171-195.

[9] Li, X.Niching without niching parameters: Particle swarmoptimization using a ring topology [J]. IEEE Trans.Evol.Comput, 2010, 14 (1): 150-169.

[10] Brits, A.E.R., F.van den Bergh. A niching particle swarmoptimizer [R]. Proc. Asia-Pacif.Conf.Simul.Evol.Learn, Singapore, Singapore, 2002.

[11] Engelbrecht, A., B.Masiye, G.Pampara. Niching ability of basic particle swarm optimization algorithms [R]. Proc.IEEE Swarm Intell.Symp., Pasadena, California, USA, 2005.

[12] Heinzelman, W.B., A.P.Chandrakasan, H.Balakrishnan. An application-specific protocol architecture for wireless microsensor networks [J]. IEEE Trans.Wirel. Commun., 2002, 1 (4): 660-770.

[13] Smaragdakis, G., I. Matta, A. Bestavros. SEP: A stable election protocol for clustered heterogeneous wireless sensor networks [R]. Proc IEEE SANPA'04, Boston, USA, 2004.

[14] AbdelSalam, H.S., S.Olariu. BEES: Bio-inspired backbone selection in wireless sensor networks [J]. IEEE Trans.Parallel Distrib.Syst., 2012, 23 (1): 44-51.

[15] Tillet, J., R.Rao, F.Sahin. Cluster-head identification in ad hoc sensor networks using particle swarm optimization [R]. Proc. IEEE PWC'02, New Delhi, India, 2002.

[16] Abdul Latiff, N.M., C.C.Tsimenidis, B.S.Sharif. Energy–aware clustering for wireless sensor networks using particle swarm optimization [R]. Proc.IEEE PIMRC'07, Athens, Greece, 2007.

[17] Liang, Y., H.Yu, P.Zeng. Optimization of cluster–based routing protocols in wireless sensor network using PSO [R]. Control Decis., 2006.

[18] Yang, S.X., C.H.Li. A clustering particle swarm optimizer for locating and tracking multiple optima in dynamic environments [J]. IEEE Trans.Evol.Comput., 2010, 14 (6): 959-974.

[19] Zhang, Y., S.He, J.Chen, et al. Distributed sampling rate control for rechargeable sensor nodes with limited battery capacity [J]. IEEE Trans. Wirel. Commun., 2013, 12 (6): 3096-3106.

[20] Feng, Y., W.Zhang, X.Tan, et al. Research on the algorithm of load-balanced hierarchical topology control for WSN [J]. Appl.Math.Inf.Sci., 2013, 7 (1): 347-352.

[21] Shi, L., Y.Yuan, J. Chen. Finite horizon LQR control with limited controller-system communication [J]. IEEE Trans. Autom. Control, 2013, 58 (7):

1835-1841.

[22] He, T., C.Huang, B.M.Blum, et al. Range-free localization schemes for large scale sensor networks [R]. Proc.MobiCom'03, San Diego, USA, 2003.

[23] Kumar, P., A.Chaturvedi, M.Kulkarni. Geographical location based hierarchical routing strategy for wireless sensor networks [R]. Proc.ICDCS'12, Coimbatore, India, 2012.

[24] Rappaport, T.. Wireless Communications: Principles & Practice [M]. Prentice-Hall, Upper Saddle River, New Jersey, USA, 1996.

[25] Wang A., W.Heinzelman, A.Chandrakasan. Energy-scalable protocols for battery-operated microsensor networks [R]. Proc.IEEE SiPS'99, Taipei, Taiwan, 1999.

[26] Kennedy, J., R.Eberhart. Particle swarm optimization [R]. Proc. IEEE Int. Conf.Neural Netw., Perth, Australia, 1995.

[27] Eberhart, R., Y.Shi.Particle swarm optimization: development, applications, resources [R]. IEEE Int.Cong.Evol.Computat., Seoul, Korea, 2001.

[28] Shi, Y., R.Eberhart. A modified particle swarm optimization [R]. IEEE Int. Cong.Evol.Computat., Anchorage, AK, USA, 1998.

[29] Li C.H., S.X.Yang, T.T.Nguyen. A self-learning particle swarm optimizer for global optimization problems [J]. IEEE Trans.Syst.Man Cybern.Part B-Cybern., 2012, 42 (3): 627-646.

[30] Pahlavan, K., A.Levesque. Wireless Information Networks [M]. Wiley: New York, 1995.

[31] Cheng, P., F.Zhang, J.Chen, et al. A distributed TDMA scheduling algorithm for target tracking in ultrasonic sensor networks [J]. IEEE Trans.Ind.Electron., 2013, 60 (9): 3836-3845.

[32] Blackwell T.. A study of collapse in bare bones particle swarm optimization [J]. IEEE Trans.Evol.Comput., 2012, 16 (3): 354-372.

An Adaptive Node Partition Clustering protocol using Particle Swarm Optimization

Wireless Sensor Networks (WSNs) are networks of autonomous nodes used for monitoring an environment, and clustering is one of the most popular and effective approach for WSNs that require scalability and robustness. In this paper, we propose and analyze an Adaptive Node Partition Clustering protocol using Particle Swarm Optimization (ANPC-PSO), a protocol that partitions the network field adaptively and selects cluster heads (CHs) consider the networks states information. The results of performance evaluation show that ANPC-PSO can improve system lifetime and data delivery by distributing energy dissipation evenly in the networks.

1 INTRODUCTION

Wireless Sensor Networks (WSNs) consist of few tens to thousands of small battery powered multifunctioning devices with limit energy supply, and have a great number of applications in hazardous circumstance exploration, seismic activity monitoring, natural disaster rescue[1]. Under this circumstances, the sensor nodes are often unreachable, so once deployed, the replacement or recharge of batteries are impossible in hostile environment, so energy efficiency is the vital factor for WSNs.

Clustering is one of the energy-saving techniques that extends the sensor network's lifetime[2], and is often formulated as optimization problems[3]. Traditional analytical optimization methods require tremendous computational efforts, which grow up exponentially with the problem scale increases. But for implementation on an individual wireless sensor node, an optimization method that requires moderate even minimal computing and memory resources and yet produces better results is needed. Bio-inspired optimization methods are computationally efficient compared to traditional analytical methods[4][5].

Particle swarm optimization (PSO) is one of the bio-inspired optimization methods, which is a popular, simple, effective and multidimensional optimization algorithms, it also has many good qualities, such as computational efficiency, ease of implementation, high quality of solutions and fast speed of convergence [6].

In this paper, we design an Adaptive Node Partition Clustering protocol using Particle Swarm Optimization (ANPC-PSO), which takes advantage of PSO that can get multimodal optimal values. The protocol partitions the network adaptively and generates cluster heads

(CHs) considers the sensor network's state information, such as reserved energy, location of nodes, their neighbors and the base station. The objective of the proposed solution is to obtain good performance in terms of system lifetime, data delivery and application-perceived quality.

The outline of the paper is as follows. Section II surveys the related work. Section III elaborates the design of ANPC-PSO in detail. Section IV provides the simulation results followed by the conclusion in Section V.

2 RELATED WORK

Typically, a WSN has little or no infrastructure, and often consists of a number of sensor nodes working together to monitor a region to obtain data about the environment, conventional techniques such as direct transmissions and Minimum Transmission Energy (MTE) routing protocol are inefficient. An adaptive clustering scheme called Low-Energy Adaptive Clustering Hierarchy (LEACH) is proposed in[7] that randomly rotating the role of a cluster head among all the nodes in the network. VAP-E[8] is based on virtual area partition for heterogeneous wireless sensor networks. DNR[9] introduces a distributed system-level diagnosis algorithm that allows every node of a partitionable arbitrary topology network to determine which portions of the network are reachable and unreachable.

The particle swarm optimization has been applied to address WSN issues such as clustering, localization, deployment and data aggregation[3]. A protocol using PSO[10] has been proposed, it has the objective of minimizing the intra-cluster distance and optimizing the energy consumption of the network. It[11] investigates a clustering particle swarm optimizer for dynamic optimization problems, this algorithm employs a hierarchical clustering method to locate and track multiple peaks. Our protocol takes advantage of PSO, considers more network states information deliberately for clustering, and partitions the network adaptively to further improve energy efficiency and balance the energy dissipation.

3 ANPC-PSO ALGORITHM

A. Particle Swarm Optimization

Through competitions and cooperation among the population, population-based optimization approaches usually can find very good solutions efficiently and effectively. Particle Swarm Optimization is a population based, self-adaptive search optimization technique, which motivated from the simulation of social behaviors of insects and animals such as fishing schooling and bird flocking. Since its inception by James Kennedy and Russell Eberhart in 1995[12], PSO has gained increasing attention among researchers as an efficient and robust technique in solving optimization problems.

PSO is population based like an evolutionary algorithm (EA), however, PSO differs from EAs in the way it manipulates each particle in the population. Other EAs use evolutionary methods such as mutation and crossover, PSO updates each particle's position in the search space, according to its present velocity, its previous best positions it has found, and the best position found by its neighbor[13].

In a standard PSO, a swarm consists ofm particles flying at a certain speed in a D-dimension search space, each particle searches around (or towards) a region defined by its personal best position (i.e., the position giving the best fitness value) and the neighborhood best position (or the entire population). Let us use $v_i = (v_{i1}, v_{i2}, \cdots, v_{id})$, $1 \leq i \leq m$, $1 \leq d \leq D$ to denote the velocity of the i particle in the swarm, $x_i = (x_{i1}, x_{i2}, \cdots, x_{id})$ as position, $p_i = (p_{i1}, p_{i2}, \cdots, p_{id})$ the best position it has found so far, and $p_g = (p_{g1}, p_{g2}, \cdots, p_{gd})$ the best position found from its neighborhood (so-called global best). v_i and x_i of the i th particle in the swarm are updated according to the following two equations[14]:

$$v_{id}^{k+1} = \omega v_{id}^k + c_1\xi(p_{id}^k - x_{id}^k) + c_2\eta(p_{gd}^k - x_{id}^k) \quad (1)$$

$$x_{id}^{k+1} = x_{id}^k + v_{id}^{k+1} \quad (2)$$

Where ω is the inertia weight, it decides how the particle inherit present velocity, a large inertia weight facilitates the particle's exploration ability (global search ability) while a small inertia weight facilitates the particle's exploitation ability (local search ability). c_1 and c_2 are learning factor (or called acceleration coefficient) which are positive constants normally, the learning factor enables the particle flying towards its personal best position and the global best position. ξ and η are random numbers between 0 and 1. If we consider all the particles in the swarm as neighbor, we get the global version of PSO; if we consider partial particles as neighbors, we get the local version of PSO; and there has different methods to format the neighbors[16].

B. Optimum Number of Clusters

Using the radio model as described in section III, as to minimize energy consumption, we can find the optimal value of k to minimize energy dissipation during each round. We get the optimum number of clusters[7]:

$$k_{opt} = \frac{\sqrt{N}}{\sqrt{2\pi}}\sqrt{\frac{\varepsilon_{fs}}{\varepsilon_{mp}}}\frac{M}{d_{toBS}^2} \quad (3)$$

Where d_{toBS} is the distance from the CH to the BS, and k_{opt} is the optimum number of clusters.

C. The fitness function of ANPC-PSO

The determination of fitness function is closely related with the properties of the problem, and it determines the algorithm's performance of the optimal solution directly. So it is important to design the fitness function, based on[3][10], we consider the conditions that influence the system performance, and do many experiments to explore the better fitness function and better parameters, finally we define the fitness function deliberately. First, the fitness function should consider the residual energy of the node, because if it becomes the CH, it consumes more energy than the member nodes. Second, we consider the distance among the cluster and the node's residual energy comprehensively, because in a cluster, we hope the node that has more residual energy and the distance between the node and other nodes are minimal. Then the distance between the CH and the BS should be considered due to the CH is responsible for transmitting the gathered datato the BS.

We define the fitness function $f(i)$ for node i based on the above considerations.

$$f(i) = \alpha f_1(i) + \beta f_2(i) + \gamma f_3(i) \tag{4}$$

$$f_1(i) = E(i) / \frac{1}{m} \sum_{i=1}^{m} E(i)$$

$$f_2(i) = \frac{1}{m-1} \sum_{k=1, k \neq i}^{m} \frac{E_i \times d(i, k)}{d(i, k) + 1}$$

$$f_3(i) = \frac{D(i)}{d_{max} - d_{toBS}}$$

$$0 \leq \alpha, \beta, \gamma \leq 1, \alpha + \beta + \gamma = 1$$

Function $f_1(i)$ is the ratio of node i's energy to the average energy within the cluster, $E(i)$ is the residual energy of node i, m is the number of member nodes within the cluster. $f_2(i)$ considers the residual energy and the distance between the node and other nodes among the cluster comprehensively. d_{max} is the maximum Euclidean distance between node i and the BS, d_{toBS} is the average Euclidean distance between the nodes in the cluster and the BS, while $d(i) = d_{max} - d_{toBS}(i)$. So if a node near the BS, $f_3(i)$ has a higher value. α, β, γ are the weights of each sub functions that we get from experiments. The fitness function that defined above has simultaneously maximum the residual energy, minimizing the intra-cluster distance and the distance between the CH and the BS. So according the fitness function defined above, we choose the node that has the maximum value as CH, and it is the optimum selection during that round.

As clustering is often formulated as optimization problems[3], and PSO is one of bio-inspired optimization methods that is computationally efficient compared to traditional analytical methods[4][5], so we take advantage of PSO as used in[10][11] to optimize the clustering.

We use fitness function (4) as described above to select the optimal solution as CH when the iterations of PSO finished in each round. According to the fitness function, the CH has the optimal fitness value is energy sufficient and the energy consumption is minimal.

We first partition the networks into k_{opt} part adaptively, all nodes send information about its current energies and locations to its neighbors in each partition. The ANPC-PSO algorithm executes in period, each partition optimally chooses the CH using PSO. The CH acts as local control center to coordinate the data transmission within the cluster. We select the node that closed to the partition center as candidate cluster head, during the subsequence rounds, we use the candidate cluster head that gets its member nodes information about current energy and location through a piggybacking manner. So it does not need extra cost and make the energy loss evenly distributed among all nodes. The candidate cluster head runs ANPC-PSO algorithm to select the CH, Fig. 1 shows the flowchart of ANPC-PSO algorithm.

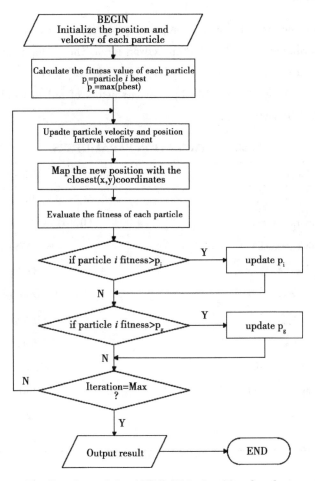

Fig. 1 The flowchart of the ANPC-PSO algorithm for cluster setup

(1) Initialize position and velocity of each particle randomly.

(2) Calculate the fitness value of each particle using Formula (5).

(3) Find the personal and global best position. The personal best position is the particle has experienced that has the maximum fitness, and the global best is the position among all the particles that has the maximum fitness.

(4) Update each particle's position and velocity using (2) and (3).

(5) Limit the change of the particle's position value and velocity value, then map the new updated position with the closest coordinates.

(6) Repeat step (2) to (5) until the maximum number of iterations is reached. Select the best solution as CH, the candidate cluster head broadcast an advertisement message using a non-persistent carrier sense multiple access (CSMA) MAC protocol[15].

4 SIMULATION

We evaluate the performance of the proposed protocol, we run the simulations for 100 nodes randomly distributed in a 100m×100m network area. The performance of our protocol is compared with the clustering protocols for WSNs, LEACH, LEACH-C[7] and VAP-E[8]. The simulation parameters are given in Table 1, the parameters of radio model are as in[7], and PSO are as described in section Ⅲ.

Table 1 SIMULATION PARAMETERS

Parameter	Value
N	100
Network coverage	(0, 0) ~ (100, 100) m
Sink location	(50, 175) m
E_{Tx}, E_{Rx}	$50nJ/bit$
ε_{fs}	$10pJ/bit/m^2$
ε_{mp}	$0.0013pJ/bit/m^4$
E_{DA}	$5nJ/bit/message$
Data message size	125 bytes
Data message header	25 bytes
c_1, c_2	2
ω	0.9~0.4

Fig. 2 illustrates the number of nodes alive over time, we can define it the system lifetime. Obviously our proposed protocol can prolong the network lifetime significantly compared to LEACH, LEACH-C and VAP-E. This is because our protocol produces better network partitioning and the CHs that are optimally distributed across the network, thus the energy consumed by all nodes for communication can be reduced. While LEACH, LEACH-C and VAP-E some nodes have to transmit long distances in order to reach a CH due to poor network clustering or the CH consume more energy for data gathering, aggregation and transmitting.

The total data messages received at the BS is shown in Fig. 3. The plot clearly indicates the effectiveness of the proposed protocol in delivering more data messages than LEACH, LEACH-C and VAP-E. Our protocol offers improvement in data delivery over LEACH, LEACH-C and VAP-E. The reason for this is because the proposed protocol can take advantage of higher energy nodes as CHs by considering the remaining energy of the CH, intra-cluster distance and the distance to the BS. In contrast LEACH and LEACH-C do not take into account the energy of a node when selecting the CH, and may select the CH with insufficient energy to remain alive

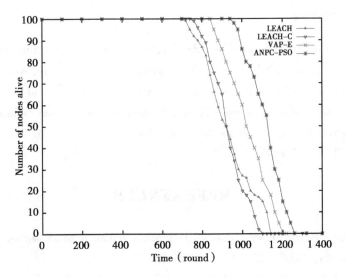

Fig. 2　Number of nodes alive over time

during the data transfer phase, and though VAP-E consider the energy of a node and the distance to the BS, it's still randomly select the CH.

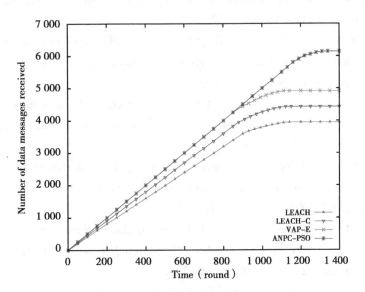

Fig. 3　Total data messages received at the BS over time

As can be seen from these figures, the proposed protocol can result in good network partitioning and balance the energy consumption, where CHs are evenly positioned across the network and located at the place that have the best fitness. That's because our proposed protocol considers better network partition, choose the best solution as CH according the network state information to minimize and balance the energy consumption.

5 CONCLUSION

In this paper we have presented an Adaptive Node Partition Clustering protocol using Particle Swarm Optimization (ANPC-PSO). We partition the network effectively and adaptively, consider the nodes residual energy, the distance among intra-cluster and to the BS and using PSO to select the best solution as the CH. Results from the simulations indicate that the proposed protocol gives a higher network lifespan, delivers more data to the BS compared to LEACH, LEACH-C and VAP-E.

REFERENCES

[1] J.Yick, B.Mukherjee, D.Ghosal. Wireless sensor network survey [J]. Computer Networks, 2008, 52 (12): 2292-2330.

[2] A.A.Abbasi, M.Younis. A survey on clustering algorithms for wireless sensor networks [J]. Computer Communications, 2007, 30 (14-15): 2826-2841.

[3] R.V.Kulkarni, G.K.Venayagamoorthy. Particle Swarm Optimization in Wireless Sensor Networks: A Brief Survey [J]. IEEE Trans.Systems, Man, and Cybernetics, 2011, 41 (2): 262-267.

[4] N.M.A.Latiff, C.C.Tsimenidis, B.S.Sharif. Performance comparison of optimization algorithms for clustering in wireless sensor networks [R]. in Proc.IEEE Int.Conf. Mobile Ad Hoc Sens.Syst., 2007.

[5] R.Xu, J.Xu, D.C.Wunsch. A Comparison Study of Validity Indices on Swarm-Intelligence-Based Clustering [J]. IEEE Trans.Systems, Man, and Cybernetics, 2012, 42 (4): 1243-1256.

[6] Y.Valle, G.K.Venayagamoorthy, S.Mohagheghi, et al. Particle swarm optimization: Basic concepts, variants and applications in power systems [J]. IEEE Trans.Evol. Comput., 2008, 12 (2): 171-195.

[7] W.B.Heinzelman, A.P.Chandrakasan, H.Balakrishnan. An Application-Specific Protocol Architecture for Wireless Microsensor Networks [J]. IEEE Trans.Wireless Commun., 2002, 1 (4): 660-770.

[8] R.Wang, G.Liu, C.Zheng. A Clustering Algorithm based on Virtual Area Partition for Heterogeneous Wireless Sensor Networks [R]. in Proc.IEEE Mechatronics & Automation, 2007.

[9] E.P.DuarteJr, A.Weber, K.V.O.Fonseca. Distributed Diagnosis of Dynamic Events in Partitionable Arbitrary Topology Networks [J]. IEEE Trans.Parallel and Distrib. Sys., 2012, 23 (8): 1415-1426.

[10] N.M.AbdulLatiff, C.C.Tsimenidis, B.S.Sharif. Energy aware clustering for wireless sensor networks using particle swarm optimization [R]. in Proc.IEEE PIMRC'

07, 2007.

[11] S. X. Yang, C. H. Li. A Clustering Particle Swarm Optimizer for Locating and Tracking Multiple Optima in Dynamic Environments [J]. IEEE Trans.Evol.Comput., 2010, 14 (6): 959-974.

[12] J. Kennedy, R. Eberhart. Particle swarm optimization [R]. in Proc. IEEE Int. Conf.Neural Netw., 1995.

[13] C.H.Li, S.X.Yang, T.T.Nguyen. A Self-Learning Particle Swarm Optimizer for Global Optimization Problems [J]. IEEE Trans. Systems, Man, and Cybernetics, 2012, 42 (3): 627-646.

[14] Y.Shi, R.Eberhart. A modified particle swarm optimization [R]. IEEE Int.Congress on Evolutionary Computation, 1998.

[15] K.Pahlavan, A.Levesque. Wireless Information Networks [M]. New York: Wiley, USA, 1995.

[16] T.Blackwell. A Study of Collapse in Bare Bones Particle Swarm Optimization [J]. IEEE Trans.Evol.Comput., 2012, 16 (3): 354-372.

An Adaptive Virtual Area Partition Clustering Protocol using Particle Swarm Optimization

Wireless Sensor Networks (WSNs) are networks of autonomous nodes used for monitoring an environment, and clustering is one of the most popular and effective approach for WSNs that requires scalability and robustness. In this paper, we propose and analyze an Adaptive Virtual Area Partition Clustering Protocol using Particle Swarm Optimization (AVAP-PSO), a protocol that partitions the network field adaptively and selects the cluster heads considering the networks states information. It obeys the idea of energy efficient to obtain good performance in terms of system lifetime, energy consumption and data delivery. The results of performance evaluation show that AVAP-PSO can improve system lifetime by distributing energy dissipation evenly in the networks.

1 INTRODUCTION

Wireless sensor networks (WSNs) are application-specified networks, and have a great number of applications in hazardous circumstance exploration, seismic activity monitoring, natural disaster rescue, military target tracking and surveillance. This has been enabled by the availability of sensors that are smaller, cheaper, and intelligent. These sensors are equipped with wireless interfaces with which they can communicate with one another to form a network. So in the design of a WSN, we must consider the factors such as the object, application environment, hardware, cost and other system constraints [1].

A wireless sensor network consists of few tens to thousands of small battery powered multi-functioning devices with limit energy supply. Once deployed, the sensor nodes are often unreachable to users, and the replacement or recharge of batteries are impossible, so energy efficient is the vital factor for WSNs. Clustering is one of the energy-saving techniques that extending the sensor network's lifetime[2], and is often formulated as optimization problems[3]. Traditional analytical optimization methods require tremendous computational efforts, which grow up exponentially with the problem scale increases. But for implementation on an individual wireless sensor node, an optimization method that requires moderate even minimal computing and memory resources and yet produces betterresults is needed. Bio-inspired optimization methods are computationally efficient compared to traditional analytical methods[4][5].

Particle swarm optimization (PSO) is one of the bio-inspired optimization methods, which is a popular, simple, effective and multidimensional optimization algorithms, it also has many good qualities, such as computational efficiency, ease of implementation, high quality of solutions and fast speed of convergence[6].

In this paper, we design an Adaptive Virtual Area Partition Clustering Protocol using Particle Swarm Optimization (AVAP-PSO), which takes advantage of Particle Swarm Optimization that can get multimodal optimal values. The protocol partitions the network adaptively and generates cluster heads considers the sensor network's state information, such as reserved energy, location of nodes, their neighbors and the base station. The protocol accords the idea of energy efficient with application specific data fusion to obtain good performance in terms of system lifetime, energy consumption and application-perceived quality. Our simulations show that AVAP-PSO can improve system lifetime and energy dissipating by optimizing the cluster head selection and balance the energy consumption evenlyin the networks.

2 RELATED WORK

Typically, a WSN has little or no infrastructures, and often consists of a number of sensor nodes working together to monitor a region to obtain data about the environment, conventional techniques such as direct transmissions and Minimum Transmission Energy (MTE) routing protocol have to be avoided due to their inefficiency. To alleviate this deficiency, an adaptive clustering scheme called Low-Energy Adaptive Clustering Hierarchy (LEACH) is proposed in[7] that randomly rotating the role of a cluster head among all the nodes in the network. VAP-E[8] based on virtual area partition (VAP-E) for heterogeneous wireless sensor networks. DCCG[9] algorithm consists of two levels of clustering: local clustering and global clustering. EDC[10] aims at avoiding temporary failure of sensor node after cluster head died in WSN, and set a backup path for every ordinary sensor node. BEES[11] is a lightweight bio-inspired backbone construction protocol, it can help mitigate many of the typical challenges in sensor networks by allowing the development of simpler network protocols.

The particle swarm optimization (PSO) has been applied to address WSNs issues such as clustering, localization, deployment and data aggregation[3][12]. The application of PSO algorithm to solve the problem of sensor network clustering has been proposed before in[13]. The authors attempted to equalize the number of nodes and optimize to select the cluster heads in each cluster in order to minimize the energy expended by the nodes. A protocol using PSO[14] has been proposed, it has the objective of minimizing the intra-cluster distance and optimizing the energy consumption of the network. A new cluster-based algorithm using PSO[15] is proposed, the election of cluster-heads needs to consider the information of location and energy reserved about candidates and their neighbors. It investigates a clustering particle swarm optimizer (PSO) for dynamic optimization problems, this algorithm employs a hierarchical clustering method to locate and track multiple peaks[16].

3 SYSTEM MODEL

3.1 The network model

In this article we consider a total of N sensor nodes dispersed in a $M \times M$ square field and assume the following properties about the sensor network:

• A fixed base station is located away from the sensing field and has no energy constraints.

• The sensor nodes are energy constrained with initial energy allocated and distributed randomly.

• The nodes are equipped with power control capabilities to vary their transmitted power.

• Every node has location information.

• All sensor nodes have no mobility.

This model uses adaptive virtual area partition cluster scheme. Consider the partial network structure show in Fig. 1. Each cluster has a Cluster Head which collects data from its members and aggregates, then sends to the base station directly.

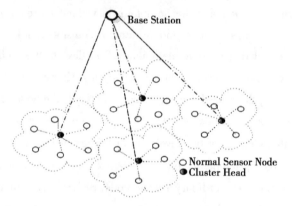

Fig. 1 AVAP-PSO clustering architecture

Sensed data is gathered in a periodic way, and each period is named a round, the period we have defined round consists of three phase, a set-up phase which uses PSO algorithm, when the cluster head is selected, follows by a steady state phase when data gathered, finally a finish phase, when data transfers to the base station occurred. As shown in Fig. 2, in order to minimize energy consumption, the steady state should be longer than other phases.

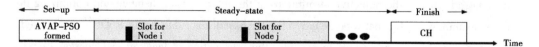

Fig. 2 Time line showing AVAP-PSO operation. Data transmissions are explicitly scheduled to avoid collision and increase the total time of each member node can remain in sleep state

3.2 The radio model

A typical WSN node is constituted of four major components: a data processor unit, a sensor unit, a wireless communication unit that consists of transceiver/receiver circuit, antenna and amplifier, and a power supply unit[17]. Though in a sensor node, the energy is dissipated in all of the first three components, we mainly consider the energy consumed related to the wireless communication unit since the goal of this paper is to propose an energy efficient protocol to prolong the network lifetime, and the energy consumed by the cluster head nodes during the data fusion is also take into account.

We use the first radio model as discussed in[7]. In this model, a receiver dissipates energy to run radio units, a transmitter dissipates energy to run radio units and power amplifier. The amplifier requires energy to maintain an acceptable Signal-to-Noise Ratio (SNR) to transfer data reliably, we use free space channel model (r^2 power loss) when $r \leq r_o$, and multipath fading channel model (r^4 power loss) when $r > r_o$, where r_o is the threshold distance[18]. Using this two channel model, the radio expends to transmit an l-bits message a distance r are shown as follows:

$$E_{Tx}(l, r) = E_{Tx}(l) + E_{amp}(l, r)$$
$$= \begin{cases} l \times E_{Tx} + l \times \varepsilon_{fs} \times r^2, & r \leq r_o \\ l \times E_{Tx} + l \times \varepsilon_{mp} \times r^4, & r > r_o \end{cases} \quad (1)$$

Where the threshold distance r_o is given by $r_o = \sqrt{\varepsilon_{fs}/\varepsilon_{mp}}$

The radio expends to receive an l-bit message is: $E_{Rx}(l) = l \times E_{Rx}$

The radios have power control ability that can expend minimum energy by properly setting the power amplifier according the distance and can be turned off to avoid receiving unintended messages. The communication energy parameters for all experiment are set as: $E_{Tx} = E_{Rx} = 50nJ/bit$, $\varepsilon_{fs} = 10pJ/bit/m^2$, $\varepsilon_{mp} = 0.0013pJ/bit/m^4$, the energy for data aggregation is set as $E_{DA} = 5nJ/bit/message$ [19].

Data receiving is also a high energy consumption operation, with these parameters, when $r^2 = 500$ and $l = 2000$ using the free space channel model, the energy dissipate in the amplifier equals to energy dissipation of the radio electronics units, hence, the cost to transmit a message is twice the cost of receive. So the number of transmission and reception should be minimized to lessen the energy consumption. The energy consumption varies with the distance in transmission, it is also important to limit transmission distance to save energy. We assumed the radio channel is symmetric so that the energy required to transmit a message from node i to node j is the same as the energy required to transmit a message from node j to node i for a given SNR (Signal-to-Noise Ratio).

4 AVAP-PSO ALGORITHM

4.1 Particle Swarm Optimization

Many search approaches that motivated by evolution as seen in nature, through competitions and cooperation among the population, population-based optimization approaches usually can find very good solutions efficiently and effectively. Particle Swarm Optimization (PSO) is a population based, self-adaptive search optimization technique, which motivated from the simulation of social behaviors of insects and animals such as fishing schooling and bird flocking. Since its inception by James Kennedy and Russell Eberhart in 1995[20], PSO has gained increasing attention among scientists and researchers as an efficient and robust technique in solving complex and difficult optimization problems.

PSO is population based like an EA, however, PSO differs from EAs in the way it manipulates each particle in the population. Other EAs use evolutionary methods such as crossover and mutation, PSO updates each particle's position in the search space, according to its present velocity, its previous best positions it has found, and the best position found by its neighbor[21].

In a conventional PSO, a swarm consists of m particles flying at a certain speed in a D-dimension search space, during searching, each particle considers its personal best position (i.e., the position giving the best fitness value), and the position of best particle from its neighborhood (or the entire population). So each particle searches around (or towards) a region defined by its personal best position and the neighborhood best position. Let us use $v_i = (v_{i1}, v_{i2}, \cdots, v_{id})$, $1 \leq i \leq m$, $1 \leq d \leq D$ to denote the velocity of the i th particle in the swarm, $x_i = (x_{i1}, x_{i2}, \cdots, x_{id})$ its position, $p_i = (p_{i1}, p_{i2}, \cdots, p_{id})$ the best position it has found so far, and $p_g = (p_{g1}, p_{g2}, \cdots, p_{gd})$ the best position found from its neighborhood (so-called global best). v_i and x_i of the i th particle in the swarm are updated according to the following two equations[20][21]:

$$v_{id}^{k+1} = \omega v_{id}^k + c_1 \xi (p_{id}^k - x_{id}^k) + c_2 \eta (p_{gd}^k - x_{id}^k) \quad (2)$$

$$x_{id}^{k+1} = x_{id}^k + v_{id}^{k+1} \quad (3)$$

Where ω is the inertia weight, it decides how the particle inherit present velocity, a large inertia weight facilitates the particle's exploration ability (global search ability) while a small inertia weight facilitates the particle's exploitation ability (local search ability). c_1 and c_2 are learning factor (or called acceleration coefficient) which are positive constants normally, the learning factor enables the particle flying towards its personal best position and the global best position. ξ and η are random numbers between 0 and 1. If we consider all the particles in the swarm as neighbors, we get the global version of PSO; if we consider partial particles as neighbors, we get the local version of PSO, and there has different methods to format the neighbors[23].

4.2 Optimum number of cluster heads

We proposed an energy-efficient clustering algorithm, aiming to enhance the existing clustering algorithms' performance. Using the radio model as described in section 3, the total energy dissipated in the network during a round is:

$$E_{round} = l\left(NE_{DA} + 2NE_{elec} + k\varepsilon_{mp}d_{toBS}^4 + N\varepsilon_{fs}\frac{1}{2\pi}\frac{M^2}{k^2}\right) \quad (4)$$

Where d_{toBS} is the distance between the cluster head and the BS, and k is the number of cluster heads, and l is the number of bits in each data message.

As to minimize energy consumption, we can find the optimal value of k from (4) to minimize energy dissipation during each round. We set the derivative of the continuous function when k>0 about E_{round} with respect to k to zero $E'_{round} = l\left(\varepsilon_{mp}d_{toBS}^4 - \frac{N}{2\pi}\varepsilon_{fs}\frac{M^2}{k^2}\right) = 0$, so we get the optimum number of cluster heads, $k_{opt} = \frac{\sqrt{N}}{\sqrt{2\pi}}\sqrt{\frac{\varepsilon_{fs}}{\varepsilon_{mp}}}\frac{M}{d_{toBS}^2}$.

4.3 The fitness function of PSO

The determination of fitness function is closely related with the properties of the problem, and it determines the algorithm's performance of the optimal solution directly. So it is important to design the fitness function, based on[3][14][15], we consider the condition that influence the system performance, and do many experiments to explore the better fitness function and better parameters, finally we define the fitness function deliberately. First, the fitness function should consider the residual energy of the node, because if it becomes the cluster head, it consumes more energy than the member nodes. Then the distance between the cluster head and the BS should be considered due to the cluster head responsible for transmitting data to the BS directly.

We define the fitness function $f(i)$ for node i based on the above considerations.

$$f(i) = \alpha f_1(i) + (1-\alpha)f_2(i) \quad (5)$$

$$f_1(i) = E(i) / \frac{1}{m}\sum_{i=1}^{m}E(i), \quad f_2(i) = \frac{D(i)}{d_{max} - d_{toBS}}, \quad a = \frac{1}{1+\lambda}, \quad \lambda = \frac{E_i(r)}{E_i}$$

Function $f_1(i)$ is the ratio of node i's energy to the average energy within the cluster, $E(i)$ is the energy reside in node i, m is the number of nodes within the cluster. d_{max} is the Euclidean maximum distance between node i and the BS, d_{toBS} is the average Euclidean distance between nodes and the BS, while $D(i) = d_{max} - d_{toBS}(i)$. So if a node near the BS, $f_2(i)$ has a higher value. α is the weight and λ varies from 1 to 0, so α varies from 0.5 to 1, that shows the residual energy are become more important during the cluster head election process. The fitness function that defined above has simultaneously maximum the residual energy and minimizing the distance between the cluster head and the BS. So according the fitness function defined above, we choose the node that has the maximum value as cluster head, and it is the optimum selection during that round.

4.4 AVAP-PSO setup

We select the optimal solution as CH when the iterations of PSO finished in each round. According to the fitness function, the CH has the optimal fitness value that it is energy sufficient and the energy consumption is minimal, then the CHs aggregated the data that have gathered and transmit it to the base station using a fixed spreading code and CSMA (Carrier Sense Multiple Access), a similar approach that used in[22], that balance and minimize the energy consumption. Our simulation shows that we can achieve good performance and prolong the network lifetime.

Since k_{opt} is determined, we part the network field into k_{opt} partitions adaptively, all nodes send information about its current energy and locations to its neighbors in each partition. The AVAP-PSO algorithm executes in period, each partition optimally chooses the CH using AVAP-PSO. The CH acts as local control center to coordinate the data transmission within the cluster. The CH sets up a TDMA schedule to avoid collision within its member nodes, allows the radio electronics to be turn off at all times except during their transmission slot for each member nodes, that has further reduce the energy dissipation and enhance the network performance. Once the CH has finished receiving data from all its member nodes, the CH performs data aggregation and sends the aggregated data to the base station.

We first select the node that closed to the partition center as candidate cluster head, during the subsequence rounds, we use the candidate cluster head that gets its member nodes information about current energy and location through a piggybacking manner, so it does not need extra cost and make the energy loss evenly distributed among all nodes. The candidate cluster head runs AVAP-PSO algorithm to select the cluster head:

(1) Initialize position and velocity of each particle randomly.

(2) Calculate the fitness value of each particle using Formula (5).

(3) Find the personal and global best position. The personal best is the position of the particle itself has experienced that has the maximum fitness, and the global best is the position of the particle that has the maximum fitness.

(4) Update each particle's position and velocity using (2) and (3).

(5) Limit the change of the particle's position value and velocity value, then map the new updated position with the closest coordinates.

(6) Repeat step (2) to (5) until the maximum number of iterations is reached. Select the best solution as CH, the candidate cluster head broadcast an advertisement message using a non-persistent carrier sense multiple access (CSMA) MAC protocol[22]. This message is a short message containing the node's ID and a header that distinguishes this message as an announcement message. The CH node sets up a TDMA schedule and transmits it to the nodes in the cluster, after all nodes in the cluster know the TDMA schedule, the set-up phase is complete and the steady-state can begin.

5　SIMULATION

We evaluated the performance of the proposed protocol, assuming there are 100 nodes randomly distributed in a 200m×200m network area with the BS located at (x=100, y=250).

The performance of our protocol are compared with the clustering protocols for wireless sensor networks, LEACH, LEACH-C[7] and VAP-E[8]. The simulation parameters of the radio model are as described in section Ⅲ, and for PSO parameters, we use $m = 20$ particles, learning factor $c_1 = c_2 = 2$, $x_{max} = v_{max} = 200$, inertia weight ω from 0.9 to 0.4.

Fig. 3 illustrates the number of nodes alive over time, we can define it the system lifetime for the simulation. Obviously our proposed protocol can prolong the network lifetime significantly compared to LEACH, LEACH-C and VAP-E. This is because our protocol produces better network partitioning and the cluster heads that are optimally distributed across the network, thus the energy consumed by all nodes for communication can be reduced. On the contrary, in LEACH, LEACH-C and VAP-E some nodes have to transmit long distances in order to reach a cluster head due to poor network clustering or the cluster head consume more energy for data gathering, aggregation and transmitting. As a result, some nodes dissipate a large amount of energy while transmitting their data to the cluster head or the base station.

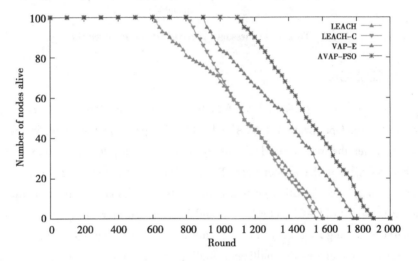

Fig. 3　Number of nodes alive over time

The total data messages received at the base station is shown in Fig. 4. The plot clearly indicates the effectiveness of the proposed protocol in delivering more data messages than LEACH, LEACH-C and VAP-E. Our protocol offers improvement in data delivery by factors of 52% over LEACH, 29% over LEACH-C and 18% over VAP-E. The reason for this is because the proposed protocol can take advantage of higher energy nodes as a cluster head by considering the remaining energy of the cluster head and balance the energy consumption. Hence, more data messages are sent to the base station, and it is unlikely that the cluster head will run out of energy before the steady state phase ends. In contrast LEACH and LEACH-C do not take into account the energy of a node when selecting the cluster head, and may select the cluster head with insufficient energy to remain alive during the data transfer phase, and though VAP-E consider the energy of a node and the distance to the BS, it's still randomly select the cluster head. While our protocol partition the network evenly, consider the nodes residual energy and the distance to the BS, and select the optimal solution as cluster head using PSO.

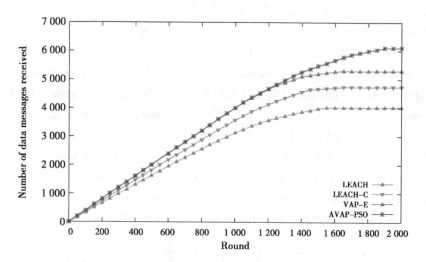

Fig. 4 Total data messages received at the BS over time

6 CONCLUSION

In this paper we have presented an Adaptive Virtual Area Partition Clustering Protocol using Particle Swarm Optimization (AVAP-PSO). We partition the network effectively and adaptively, consider the nodes residual energy and the nodes location, and using PSO to select the best solution as the cluster head. Results from the simulations indicate that the proposed protocol using PSO algorithm gives a higher network lifespan, delivers more data to the base station compared to LEACH, LEACH C and VAP-E. Furthermore, the proposed protocol produces better clustering by evenly allocating the cluster heads throughout the sensor network area. Our future work includes the multi-hop routing among the cluster head nodes and the implementation of PSO in hardware to further improve energy efficiency.

REFERENCES

[1] J. Yick, B. Mukherjee, D. Ghosal. Wireless sensor network survey [J]. Computer Networks, 2008, 52 (12): 2292-2330.

[2] A. Abbasi, M. Younis. A survey on clustering algorithms for wireless sensor networks [J]. Computer Communications, 2007, 30 (14-15): 2826-2841.

[3] R. Kulkarni, G. Venayagamoorthy. Particle Swarm Optimization in Wireless Sensor Networks: A Brief Survey [J]. IEEE Trans. Systems, Man & Cybernetics-PART C: Application And Reviews, 2011, 41 (2): 262-267.

[4] N. Latiff, C. Tsimenidis, B. Sharif. Performance comparison of optimization algorithms

for clustering in wireless sensor networks [R]. In Proc. IEEE Int. Conf. Mobile Ad Hoc Sens. Syst., 2007.

[5] R. Xu, J. Xu. A Comparison Study of Validity Indices on Swarm-Intelligence-Based Clustering [J]. IEEE Trans. Systems, Man & Cybernetics-Part B: Cybernetics, 2012, 42 (4): 1243-1256.

[6] Y. Valle, G. Venayagamoorthy, S. Mohagheghi, et al. Particle swarm optimization: Basic concepts, variants and applications in power systems [J]. IEEE Trans. Evol. Comput., 2008, 12 (2): 171-195.

[7] W. B. Heinzelman, A. P. Chandrakasan, H. Balakrishnan. An Application-Specific Protocol Architecture for Wireless Microsensor Networks [J]. IEEE Trans. Wireless Commun., 2002, 1 (4): 660-770.

[8] R. Wang, G. Liu, C. Zheng. A Clustering Algorithm based on Virtual Area Partition for Heterogeneous Wireless Sensor Networks [R]. In Proc. IEEE Mechatronics & Automation, 2007.

[9] J. Huang, J. Zhang. Distributed Dual Cluster Algorithm Based on Grid for Sensor Streams [J]. Journal of JDCTA, AICIT, 2010, 4 (9): 225-233.

[10] D. Ding, F. Liu, Q. Li, et al. An Improved Clustering Algorithm Based on Backup Path [J]. Journal of AISS, AICIT, 2012, 4 (8): 207-216.

[11] H. AbdelSalam, S. Olariu. BEES: Bio-inspired backbone Selection in Wireless Sensor Networks [J]. IEEE Trans. Parallel and Distrib. Sys., 2012, 23 (1): 44-51.

[12] M. Romoozil, H. Ebrahimpour-komleh. A Positioning Method in Wireless Sensor Networks Using Genetic Algorithms [J]. Journal of JDCTA, AICIT, 2010, 4 (9): 174-179.

[13] J. Tillet, R. Rao, F. Sahin. Cluster-head identification in ad hoc sensor networks using particle swarm optimization [R]. IEEE International Conference on Personal Wireless Communications, 2002.

[14] N. Latiff, C. Tsimenidis, B. Sharif. Energy-aware clustering for wireless sensor networks using particle swarm optimization [R]. In Proc IEEE PIMRC, 2007.

[15] Y. Liang, H. Yu, P. Zeng. Optimization of cluster-based routing protocols in wireless sensor network using PSO [J]. Control and Decision, 2006, 21: 453-456.

[16] S. Yang, C. Li. A Clustering Particle Swarm Optimizer for Locating and Tracking Multiple Optima in Dynamic Environments [J]. IEEE Trans. Evolutionary Computation, 2010, 14 (6): 959-974.

[17] V. Raghunathan, C. Schurgers, S. Park, et al. Energy-Aware Wireless Microsensor Networks [J]. IEEE Sig. Proc. Mag., 2002, 1 (2): 40-50.

[18] T. Rappaport. Wireless Communications: Principles & Practice [R]. Prentice-Hall, USA, 1996.

[19] A. Wang, W. Heinzelman, A. Chandrakasan.Energy-scalable protocols for battery-operated microsensor networks [R]. In Proc. IEEE SiPS, 1999.

[20] J. Kennedy, R.Eberhart.Particle swarm optimization [R]. In Proc.IEEE Int.Conf. Neural Netw., 1995.

[21] C. Li, S. Yang, T. Nguyen. A Self-Learning Particle Swarm Optimizer for Global Optimization Problems [J]. IEEE Trans. Systems, Man & Cybernetics-Part B: Cybernetics, 2012, 42 (3): 627-646.

[22] K.Pahlavan, A.Levesque.Wireless Information Networks [M]. Wiley, USA, 1995.

[23] T. Blackwell. A Study of Collapse in Bare Bones Particle Swarm Optimization [J]. IEEE Trans [J]. Evolutionary Computation, 2012, 16 (3): 354-372.

An Adaptive Virtual Area Partition Clustering Routing protocol using Ant Colony Optimization

Clustering is an energy efficient techniques that extends the network's lifespan, and routing is an efficient techniques that reduces the communication delay and balances the network traffic loads. In this paper, we propose and analyze an Adaptive Virtual Area Partition Clustering Routing protocol using Ant Colony Optimization (AVAPCR-ACO), a protocol that clusters the network field adaptively, then uses ant colony optimization to build a routing path among cluster heads. The simulation results show that AVAPCR-ACO can improve system lifetime obviously.

1 INTRODUCTION

Wireless sensor networks (WSNs) are application-specified networks, and have a great number of applications such as military surveillance, natural disaster rescue etc. This has been enabled by the availability of sensors that are smaller, cheaper, and intelligent. These sensors are equipped with antennas with which they can communicate each others to form a network. So when we design aWSN, we must take the factors such as the object, application environment, hardware, cost and other system constraints into consideration[1].

A wireless sensor network consists of hundreds to thousands of small multifunctioning devices with limit energy supply. Once they are deployed, the sensor nodes are often unreachable to users, the replacement or recharge of batteries are impossible, so energy efficient is the vital factor for WSNs. Clustering is one of the energy efficient techniques that extending the sensor network's lifetime[2], we cluster the network using virtual area partition scheme adaptively; and routing is one of the efficient techniques that reduces the communication delay and balances the network traffic loads, we takes advantage of ant colony optimization to build a routing path among the cluster heads.

Ant Colony Optimization (ACO) has been inspired by the behavior of real ant colonies, in particular, by their foraging behavior[3],[4]. It has the advantage of robust, excellent distributed calculated mechanism, easy to combine with other methods[5], etc.

In this paper, we design an Adaptive Virtual Area Partition Clustering Routing protocol using Ant Colony Optimization (AVAPCR-ACO), which uses virtual area partition scheme to cluster the network, and takes advantage of Ant Colony Optimization to build a routing path among the cluster heads. The protocol clusters the network adaptively, then builds a routing path among the cluster heads to reduce the communication delay, balance the network traffic loads. The protocol accords the idea of energy efficient to obtain good performance in terms of energy consumption and system lifetime. Our simulation results show that AVAPCR-ACO can improve system lifetime and minimize energy dissipating obviously.

2 RELATED WORK

Low-Energy Adaptive Clustering Hierarchy (LEACH) is proposed in[6] that randomly rotating the role of a cluster head in the network among all the nodes, it is an adaptive clustering scheme. VAP-E[7] is based on virtual area partition for heterogeneous wireless sensor networks, but it is not partitions the network adaptively.

Ant Colony Optimization based routing algorithms have been used for improving the system performance. By finding the maximum number of connected covers, it[8] proposes an ACO-based scheme to prolong the lifespan of heterogeneous WSNs. The authors[9] have investigated a routing protocol and used ant colony optimization (ACO) algorithm to this protocol for homogeneous WSNs. A routing protocol[10] considers remainder of node along the path and the path energy cost concurrently, and tries finding the trade off between them.

Our design clusters the network adaptively and takes advantage of ACO, considers the network states information deliberately, and selects the best path to the BS to further improve energy efficiency and balance the energy dissipation.

3 AVAPCR-ACO ALGORITHM

We proposed an adaptive energy-efficient clustering routing algorithm, aiming to enhance the existing clustering algorithms' performance. In this section, we design AVAPRC-ACO, partitions the network into clusters, then use ant colony optimization to build routing path among the cluster heads. The flow chart of AVAPCR-ACO is as shown in Fig. 1.

3.1 Adaptively Virtual Area Partition Clustering

The optimal number of clusters is determined by $K = \frac{\sqrt{N}}{\sqrt{2\pi}} \sqrt{\frac{\varepsilon_{fs}}{\varepsilon_{mp}}} \frac{M}{d_{toBS}^2}$ [6], then we perform virtual area partition clustering, the network is partitioned into clustering adaptively according the network states. As shown in Fig. 2, a 100m×100m region with 100 nodes distributed randomly, we use virtual area partition clustering that divided the network by $K=5$.

3.2 Ant Colony Optimization

Ant Colony Optimization (ACO)[3],[4],[5] is a population-based approach which has been applied to NP-hard combinatorial optimization problems successfully[8],[9]. ACO has been inspired by the behavior of real ant colonies, especially by their foraging behavior. Based on the trails of a chemical substance which called pheromone that real ants use for communication,

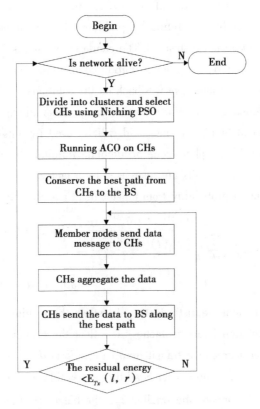

Fig. 1　The flow chart of AVAPRC-ACO

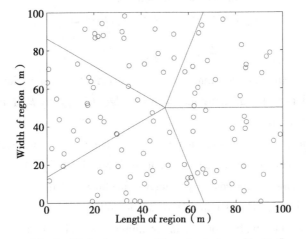

Fig. 2　Virtual area partition clustering with $K=5$

each ant communicates indirectly among a colony of ants. The pheromone trails are a kind of distributed numeric information which is modified by the ants to reflect their experience accumulated while solving a particular problem. The ACO meta-heuristic has been presented to provide a unifying framework for most applications of ant algorithms[4],[5] to combinatorial optimization problems. ACO algorithms make use of ants to construct candidate solutions iteratively of a combinatorial optimization problem. Guiding by problem-dependent heuristic information and pheromone trails, the solution is constructed. The candidate solutions constructed by individual ant starts with an empty solution, then adds solution components iteratively until we get an integrated candidate solution. The ants give feedback on the solutions by depositing pheromone on it after complete the solution construction. Solution components which are used by many ants will receive a higher amount of pheromone, and will be used by ants in the future iterations more likely. We should avoid the pheromone accumulate infinitely before the pheromone trails get enhanced, so all pheromone trails are decreased by a factor.

We define the transition probability from cluster head i to j for the k th ant as:

$$p_{ij}^k(t) = \begin{cases} \dfrac{[\tau_{ij}(t)]^\alpha \cdot [\eta_{ij}]^\beta}{\sum_{k \in allowed_k}[\tau_{ik}(t)]^\alpha \cdot [\eta_{ik}]^\beta}, & j \in allowed_k \\ 0, & otherwise \end{cases} \quad (1)$$

Where α is the pheromone heuristic factor, it reflect the residual pheromones importance of degree, β is the expectation heuristic factor, it reflect the expectation's importance of degree. η_{ij} is the expectation degree that transit from i to j. $allowed_k = \{C - tabu_k\}$ is the cluster head that ant k can select (i.e. the cluster head that ant k didnt visit). $\eta_{ij}(t) = 1/d_{ij} = [(x_i - x_j)^2 + (y_i - y_j)^2]^{1/2}$, that means the smaller d_{ij}, the bigger $\eta_{ij}(t)$ for ant k.

In order to avoiding the residual pheromone overwhelm the heuristic information, we use the following formula to update the residual pheromone.

$$\tau_{ij}(t+1) = (1-\rho) \cdot \tau_{ij}(t) + \Delta\tau_{ij} \quad (2)$$

$$\Delta\tau_{ij}(t) = \sum_{k=1}^m \Delta\tau_{ij}^k(t) \quad (3)$$

Where ρ is the coefficient such that $(1-\rho)$ represents the evaporation of trail. $\Delta\tau_{ij}$ is the quantity per unit of length of trail substance laid on edge (i, j) by k th ant.

3.3 Optimal Routing Path

In order to reduce the communication delay and balance the network traffic loads, we adopt the multi-hop communications between cluster heads. In this paper, we run ACO among the cluster heads to achieve this goal. Fig. 3 shows the flowchart of AVAPCR-ACO algorithm for optimal routing selection. The detailed steps are as follows:

(1) Initialize parameters. Set the maximum number of cycles Nc max, the number of cycles $N_c = 0$, $\tau_{ij}(t) = const$, and $\Delta\tau_{ij}(t) = 0$.

(2) Place m ants on n cluster heads randomly.

(3) $N_c \leftarrow N_c + 1$.

(4) Set the index of Tabu tables $k = 1$.

(5) $k = k+1$.

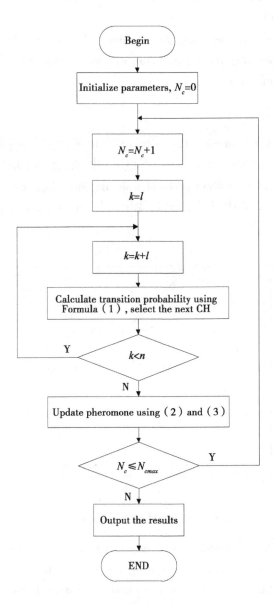

Fig. 3　The flowchart of AVAPCR-ACO for routing selection

(6) Calculate the transition probability using Formula (1) for ants to select cluster head j, $j \in \{C\text{-}tabu_k\}$.

(7) Select the maximum transition probability, move ant to this cluster head, and add this cluster head to the tabu table.

(8) If ant does not visit all the cluster head, that is $k<n$, go to step (5); otherwise go to step (9).

(9) Update the pheromone using Formula (2) and (3).

(10) Output the results if the maximum number of cycles is reached; otherwise clear the tabu and go to step (3).

We select the optimal routing using ACO among CHs, this is effective and efficient for it optimized the inter-clustering routing mechanism, thus reduce the energy consumption and balance the network traffic loads.

4 SIMULATION

We appraised the performance of the proposed protocol, let's suppose there are 100 nodes randomly distributed in a 100m×100m network field with the BS located at ($x=50$, $y=175$). We compares the performance of our protocol with the clustering protocols for wireless sensor networks, LEACH-C[6], VAP-E[7], ACO-MNCC[8]. The simulation parameters of the radio model are as used in[6],[7],[8].

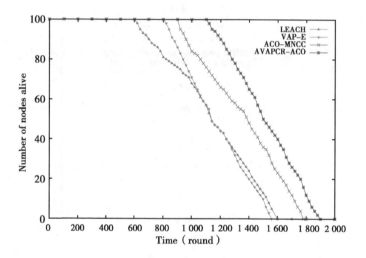

Fig. 4 Number of nodes alive over time

The number of nodes alive over time is shown in Fig. 4, we can define it the system lifetime. Obviously our proposed protocol can prolong the network lifetime compared to LEACH-C, VAP-E and ACO-MNCC. This is because our protocol produces better network partitioning as we partitions the network adaptively, and uses ACO to select the optimal routing path among the CHS. While in LEACH-C, VAP-E and ACO-MNCC some nodes have to transmit long distances in order to reach a cluster head due to poor network clustering or the cluster head consume more energy for data gathering and transmitting, so some nodes dissipate a large amount of energy.

Fig. 5 illustrates the total data messages received at the base station, it clearly indicates the effectiveness of the proposed protocol in delivering more data messages than LEACH-C, VAP-E and ACO-MNCC. The reason is that the proposed protocol can take advantage of virtual area partition to form clusters and use ACO to form an optimal routing path, thus bal-

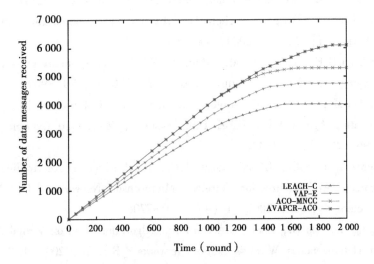

Fig. 5 Total data messages received at the BS over time

ance the energy consumption and send more data messages to the base station. On the contrary, LEACH-C do not take the energy of a node when selecting the cluster head into account, and may select the cluster head with minimal energy that have not enough energy to remain alive during that round, though VAP-E considers the distance to the BS and the energy of a node, its still randomly select the cluster head, ACO-MNCC only emphasis use ACO to find the maximum number of disjoint connected covers. While our protocol partition the network adaptively and build the optimal routing using ACO.

5 CONCLUSION

In this literature we have presented an Adaptive Virtual Area Partition Clustering Routing protocol using Ant Colony Optimization (AVAPCR-ACO). We partition the network effectively and adaptively and using ACO to build an optimal routing path among the cluster heads. Results from the simulations indicate that the proposed protocol using virtual are partition and ACO algorithm gives a higher network lifespan, delivers more data to the base station compared to LEACH-C, VAP-E and ACO-MNCC. Our future work includes the implicit parallelism of the ACO framework can be utilized for for reducing the computational time and tackling large-scale WSNs.

REFERENCES

[1] Yick J., Mukherjee B., Ghosal D.Wireless sensor network survey [J]. Computer Networks, 2008, 52 (12): 2292-2330.

[2] Abbasi A., Younis M.A survey on clustering algorithms for wireless sensor networks

[J]. Computer Communications, 2007, 30 (14-15): 2826-2841.
[3] Bonabeau E., Dorigo M., Theraulaz G.Inspiration for optimization from social insect behavior [J]. Nature, 2000, 406 (6): 39-42.
[4] Jackson D.E., Holcombe M., Ratnieks F.F.W.Trail geometry gives polarity to ant foraging networks [J]. Nature, 2004, 432 (7019): 907-909.
[5] Dorigo M., Maniezzo V., Colorni A.Ant System: Optimization by a Colony of Co-operating Agents [J]. IEEE Trans. Systems, Man, and Cybernetics – Part B, 1996, 26 (1): 29-41.
[6] Heinzelman W.B., Chandrakasan A.P., Balakrishnan H.An Application-Specific Protocol Architecture for Wireless Microsensor Networks [J]. IEEE Trans. Wireless Commun, 2002, 1 (4): 660-770.
[7] Wang R., Liu G., Zheng C.A Clustering Algorithm based on Virtual Area Partition for Heterogeneous Wireless Sensor Networks [R]. In: 2007 IEEE International Conference on Mechatronics & Automation, IEEE Press, Harbin, 2007.
[8] Lin Y., Zhang J., Chung H.S., et al.An Ant Colony Optimization Approach for Maximizing the Lifetime of Heterogeneous Wireless Sensor Networks [J]. IEEE Trans.Systems, Man, and Cybernetics-Part C, 2012, 42 (3): 408-420.
[9] Okdem S., Karaboga D.Routing in wireless sensor networks using ant colony optimization.In: 1st NASA/ESA Conf.Adapt.Hardware Syst, 2006.
[10] Zhong Y.P., Huang P.W., Wang B. Maximum lifetime routing based on ant colony algorithm for wireless sensor networks [R]. In: Proc.IET Conf.Wireless, Mobile, Sensor Networks, 2007.

An Efficient Node Partition Clustering protocol using Niching Particle Swarm Optimization

Wireless Sensor Networks (WSNs) consist of many small battery powered nodes with limit energy supply, so energy efficient is a vital factor for WSNs. Clustering is one of the energy efficient techniques that extend the sensor network's lifetime. In this paper, we present and analyze an Efficient Node Partition Clustering protocol using Niching Particle Swarm Optimization (ENPC-NPSO), a protocol that partitions the network field efficiently and the selection of cluster heads (CHs) considers the networks states information. The results of performance evaluation show that ENPC-NPSO can improve system lifetime and data delivery by distributing energy dissipation evenly in the networks.

1 INTRODUCTION

Wireless sensor networks are application-specified networks, and have a great number of applications in hazardous circumstance exploration, seismic activity monitoring, natural disaster rescue, military target tracking and surveillance[1]. A wireless sensor network consists of few tens to thousands of small battery powered multifunctioning devices with limit energy supply. Once deployed, the sensor nodes often unreachable to users, and the replacement or recharge of batteries are impossible, so energy efficient is the vital factor for WSNs. Clustering is one of the energy-saving techniques that extending the sensor network's lifetime[2], and is often formulated as optimization problems[3]. Traditional analytical optimization methods require tremendous computational efforts, which grow up exponentially with the problem scale increases. But for implementation on an individual wireless sensor node, an optimization method that requires moderate even minimal computing and memory resources and yet produces better results is needed. Bioinspired optimization methods are computationally efficient compared to traditional analytical methods[4].

Particle swarm optimization (PSO) is one of the bioinspired optimization methods, which is a popular, simple, effective and multidimensional optimization algorithms, it also has many good qualities, such as computational efficiency, ease of implementation, high quality of solutions and speed of convergence[5]. Niching is an important multimodal optimization technique and is inspired by nature. In natural ecosystems, individual species must compete to survive by taking on different roles. Niching method can be incorporated into PSO to promote and maintain formation of multiple stable subpopulations within a single population, with an aim to locate multiple optimal or suboptimal solutions[6].

A clustering protocol called Low-Energy Adaptive Clustering Hierarchy (LEACH) is proposed in[7] that randomly rotating the role of cluster head among all the nodes in the network. BCDCP[8] utilizes the base station to set up clusters and routing paths, perform randomized rotation of cluster heads. ACNP-ED[9] based on node partition to balance the loads. The particle swarm optimization (PSO) has been applied to address WSN issues such as node localization and clustering, localization, deployment and data aggregation[3]. The application of PSO to solve the problem of sensor network clustering has been proposed before in[10], the authors attempted to equalize the number of nodes and the cluster heads in each cluster in order to minimize the energy expended by the nodes. A protocol using PSO has been proposed in[11], it has the objective of minimizing the intra-cluster distance and optimizing the energy consumption of the network.

In this paper, we design an Efficient Node Partition Clustering protocol using Niching Particle Swarm Optimization (ENPC-NPSO), which takes advantage of Niching Particle Swarm Optimization that can get multimodal optimal values. The protocol partitions the network efficiently and generates cluster heads (CHs) considers the sensor network's state information, such as reserved energy, location of nodes, etc. The protocol accords the idea of energy efficient with application-specific data fusion to obtain good performance in terms of system lifetime, data delivery and application-perceived quality. Our simulations show that ENPC-NPSO can improve system lifetime and transfer more data to the BS by optimizing the CH selection and balance the energy consumption evenly in the networks.

2 SYSTEM MODEL

A. Network Model

In this article we consider a total of N sensor nodes dispersed in a square field and assume the following properties about the sensor network:

- A fixed base station is located away from the sensor nodes and has no energy constraints.
- The sensor nodes are energy constrained with initial energy allocated and distributed randomly.
- The nodes are equipped with power control capabilities to vary their transmitted power.
- Every node has location information.
- All sensor nodes have no mobility.

This model uses node partition cluster scheme. Each cluster has a CH which collects data from its members and aggregated the data, then send to the base station directly.

B. Radio Model

We use the first radio model as discussed in[7]. In this model, a receiver dissipates energy to run radio units, a transmitter dissipates energy to run radio units and power amplifier. The amplifier requires energy to maintain an acceptable Signal-to-Noise Ratio (SNR) to transfer data reliably, we use free space channel model (r^2 power loss) when $r \leqslant r_o$, and multipath fading channel model (r^4 power loss) when $r > r_o$, where r_o is the threshold distance[12]. Using this two channel model, the radio expends to transmit an l-bit message a distance r are shown as follows:

$$E_{Tx}(l, r) = E_{Tx}(l) + E_{amp}(l, r)$$
$$= \begin{cases} l \times E_{Tx} + l \times \varepsilon_{fs} \times r^2, & r \leqslant r_o \\ l \times E_{Tx} + l \times \varepsilon_{mp} \times r^4, & r > r_o \end{cases} \quad (1)$$

Where the threshold distance r_o. The radio expends to receive an l-bit message is:

$$E_{Rx}(l) = l \times E_{Rx}$$

3 ENPC-NPSO ALGORITHM

A. Niching Particle Swarm Optimization

Population-based optimization approaches usually can find very good solutions efficiently and effectively. Particle Swarm Optimization (PSO) is a population based, self-adaptive search optimization technique, which motivated from the simulation of social behaviors of insects and animals such as fishing schooling and bird flocking. Since its inception by James Kennedy and Russell Eberhart in 1995[13], PSO has gained increasing attention among scientists and researchers as an efficient and robust technique in solving complex and difficult optimization problems.

In a conventional PSO, a swarm consists of m particles flying at a certain speed in a D-dimension search space, during searching, each particle considers its personal best position (i. e., the position giving the best fitness value), and the position of best particle from its neighborhood (or the entire population). So each particle searches around (or towards) a region defined by its personal best position and the neighborhood best position. Let us use $v_i = (v_{i1}, v_{i2}, \cdots, v_{id})$, $1 \leqslant i \leqslant m$, $1 \leqslant d \leqslant D$ to denote the velocity of the i th particle in the swarm, $x_i = (x_{i1}, x_{i2}, \cdots, x_{id})$ its position, $p_i = (p_{i1}, p_{i2}, \cdots, p_{id})$ the best position it has found so far, and $p_n = (p_{n1}, p_{n2}, \cdots, p_{nd})$ the best position found from the niching it belongs (we use the similar niching method as described in[6]. v_i and x_i of the i th particle in the swarm are updated according to the following two equations:

$$v_{id}^{k+1} = \omega v_{id}^k + c_1 \xi (p_{id}^k - x_{id}^k) + c_2 \eta (p_{nd}^k - x_{id}^k) \quad (2)$$

$$x_{id}^{k+1} = x_{id}^k + v_{id}^{k+1} \quad (3)$$

Where ω is the inertia weight, it decides how the particle inherit present velocity, a large inertia weight facilitates the particle's exploration ability (global search ability) while a small inertia weight facilitates the particle's exploitation ability (local search ability). c_1 and c_2 are learning factor (or called acceleration coefficient) which are positive constants normally, the learning factor enables the particle flying towards its personal best position and the global best position. ξ and η are random numbers between 0 and 1[15].

Niching methods are of great value when the object is to locate a single/ multiple global optima. Since a niching PSO searches for multiple optima in parallel, the probability of getting trapped on a local optimum may be reduced[6].

B. Fitness Function of NPSO

The determination of fitness function is closely related with the properties of the problem, and it determines the algorithm's performance of the optimal solution directly. So it is important to design the fitness function, based on[3][10][11], we consider the condition that influence the system performance and define the fitness function deliberately. First, the fitness function should consider the residual energy of the node, because if it becomes the CH, it consumes more energy than the member nodes. Then the average distance between the CH and its member nodes should be considered due to the CH responsible for gathering data from its member nodes directly.

We define the fitness function $f(i)$ for node i based on the above considerations.

$$f(i) = \alpha f_1(i) + (1 - \alpha) f_2(i) \quad (4)$$

$$f_1(i) = \frac{E(i)}{\frac{1}{m}\sum_{i=1}^{m} E(i)}$$

$$f_2(i) = \frac{1}{N}\sum_{k=1}^{N} d(i, k) \Big/ \frac{1}{m}\sum_{k=1}^{m} d(i, k)$$

$$a = \frac{1}{1 + \lambda}, \quad \lambda = \frac{E_i(r)}{E_i}$$

Function $f_1(i)$ is the ratio of node i's energy to the average energy within the cluster, $E(i)$ is the energy reside in node i, m is the number of member nodes within the cluster. Function $f_2(i)$ is the ratio of average Euclidean distance among the network to the average Euclidean distance within the cluster to node i, N is the number of alive nodes in the network, $d(i, k)$ is the Euclidean distance between node i and node k. α is the weight and λ varies from 1 to 0, so α varies from 0.5 to 1, that shows the residual energy are become more important during the CH election process. The fitness function that defined above has simultaneously maximum the residual energy and minimizing the distance among the cluster. So according the fitness function defined above, we choose the node that has the maximum value as CH, and it is the optimum selection during that round.

C. ENPC-NPSO Setup

We select the optimal solution as CH when the iterations of NPSO finished in each round. According to the fitness function, the CH has the optimal fitness value that it is energy sufficient and the energy consumption is minimal, then the CHs aggregated the data that have gathered and transmit it to the base station using a fixed spreading code and CSMA (Carrier Sense Multiple Access), a similar approach that used in[14], that balance and minimize the energy consumption. Our simulation shows that we can achieve good performance and prolong the network lifespan.

We first use clustering protocol such as LEACH[7], BCDCP[8] to part the network field, and the cluster heads are as candidate cluster heads. All nodes send information about its current energy and locations to its candidate cluster head in each partition. The ENPC-NPSO algorithm executes in period, each partition optimally chooses the CH using ENPC-NPSO. The CH acts as local control center to coordinate the data transmission within the cluster. The CH sets up a TDMA schedule to avoid collision with its member nodes, allows the radio electronics to be turn off at all times except during their transmission slot for each member nodes, that has further reduce the energy dissipation and enhance the network performance. Once the CH has finished receiving data from all its member nodes, the CH performs data aggregation and sends the aggregated data to the base station.

The candidate cluster head runs ENPC-NPSO algorithm to select the CH:

(1) Initialize position and velocity of each particle randomly.

(2) Calculate the fitness value of each particle using Formula (4).

(3) Find the personal and niching best position. The personal best is the position of the particle itself has experience that has the maximum fitness, and the niching best is the position of the particle among the niching that has the maximum fitness. We use the similar method as described in[6] to form niching.

(4) Update each particle's position and velocity using Formula (2) and (3).

(5) Limit the change of the particle's position value and velocity value, then map the new updated position with the closest coordinates.

(6) Repeat step (2) to (5) until the maximum number of iterations is reached. Select the best solution as CH, the candidate cluster head broadcast an advertisement message using a non-persistent carrier sense multiple access (CSMA) MAC protocol[14]. This message is a short message containing the node's ID and a header that distinguishes this message as an announcement message. The CH node sets up a TDMA schedule and transmits it to the nodes in the cluster, after all nodes in the cluster know the TDMA schedule, the set-up phase is complete and the steady-state can begin.

4 SIMULATION

We evaluated the performance of the proposed protocol, assuming there are 100 nodes randomly distributed in a 100m×100m network area with the BS located at ($x=50$, $y=175$). The performance of our protocol was compared with the clustering protocols for wireless sensor networks, LEACH[7] and ACNP-ED[9]. The simulation parameters are given in Table 1, in which the parameters are as described above.

Table 1 SIMULATION PARAMETERS

Parameter	Value
Network coverage	(0, 0) ~ (100, 100) m
Sink location	(50, 175) m
E_{Tx}, E_{Rx}	$50 nJ/bit$
ε_{fs}	$10 pJ/bit/m^2$
ε_{mp}	$0.0013 pJ/bit/m^4$
E_{DA}	$5 nJ/bit/message$
Data message size	500 bytes
Data message header	25 bytes
m	100
c_1, c_2	2
ω	0.9~0.4

Fig. 1 illustrates the number of nodes alive over time, we can define it the system lifetime for the simulation. Obviously our proposed protocol can prolong the network lifetime significantly compared to LEACH and ACNP-ED. This is because our protocol produces better network partitioning and the CHs that are optimally distributed across the network, thus the energy consumed by all nodes for communication can be reduced. On the contrary, in LEACH and ACNP-ED some nodes have to transmit long distances in order to reach a CH due to poor network clustering or the CH consume more energy for data gathering, aggregation and transmitting. As a result, some nodes dissipate a large amount of energy while transmitting their data to the CH or the base station.

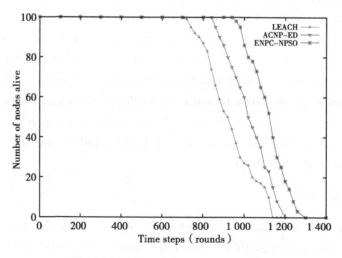

Fig. 1 Number of nodes alive over time

The total data messages received at the base station is shown in Fig. 2. The plot clearly indicates the effectiveness of the proposed protocol in delivering more data messages than LEACH and ACNP – ED. Our protocol offers improvement in data delivery by factors of 49% over LEACH, 30% over ACNP-ED. The reason for this is because the proposed protocol can take advantage of higher energy nodes as a CH by considering the remaining energy of the CH and balance the energy consumption. Hence, more data messages are sent to the base station, and it is unlikely that the CH will run out of energy before the steady state phase ends. In contrast LEACH do not take into account the energy of a node when selecting the CH, and may select the CH with insufficient energy to remain alive during the data transfer phase, and though AC-NP-ED consider the energy of a node and the distance to the BS, it's still randomly select the CH. While our protocol partition the network evenly, consider the nodes residual energy and the intra-cluster distance during the CH selection.

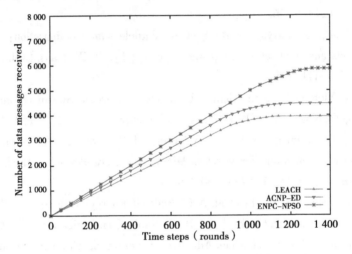

Fig. 2　Total amount of packets received at BS over time

5　CONCLUSION

In this paper we have presented an Efficient Node Partition Clustering protocol using Niching Particle Swarm Optimization (ENPC-NPSO). We partition the network effectively and adaptively, consider the nodes residual energy and the intra-cluster distance and using NPSO to select the best solution as the CH. Results from the simulations indicate that the proposed protocol using NPSO algorithm gives a higher network lifespan, delivers more data to the base station compared to LEACH and ACNP – ED. Furthermore, the proposed protocol produces better clustering by evenly allocating the CHs throughout the sensor network area. Our future work includes the multi-hop routing among the cluster heads and the implementation of NPSO in hardware to further improve energy efficiency.

REFERENCES

[1] J. Yick, B. Mukherjee, D. Ghosal. Wireless sensor network survey [J]. Computer Networks, 2008, 52 (12): 2292-2330.

[2] A. Abbasi, M. Younis. A survey on clustering algorithms for wireless sensor networks [J]. Computer Communications, 2007, 30 (14-15): 2826-2841.

[3] R. Kulkarni, G. Venayagamoorthy. Particle Swarm Optimization in Wireless Sensor Networks: A Brief Survey [J]. IEEE Trans. Systems, Man & Cybernetics, 2011, 41 (2): 262-267.

[4] R. Xu, J. Xu. A Comparison Study of Validity Indices on Swarm-Intelligence-Based Clustering [J]. IEEE Trans. Systems, Man & Cybernetics, 2012, 42 (4): 1243-1256.

[5] Y. Valle, G. Venayagamoorthy, et al.. Particle swarm optimization: Basic concepts, variants and applications in power systems [J]. IEEE Trans. Evol. Comput., 2008, 12 (2): 171-195.

[6] X. Li. Niching Without Niching Parameters: Particle Swarm Optimization Using a Ring Topology [J]. IEEE Trans. Evol. Comput., 14 (1): 150-169.

[7] W. B. Heinzelman, A. P. Chandrakasan, H. Balakrishnan. An Application-Specific Protocol Architecture for Wireless Microsensor Networks [J]. IEEE Trans. Wireless Commun., 2002, 1 (4): 660-770.

[8] S. Muruganathan D. Ma et al. A Centralized Energy-Efficient Routing Protocol for Wireless Sensor Networks [J]. IEEE Radio Communications, 2005, 43: 8-13.

[9] R. Wang, G. Liu, H. Zhao. Adaptive clustering algorithm based on node partition for wireless sensor networks [J]. Journal of Dalian Maritime University, 2008, 34 (1): 45-48.

[10] J. Tillet, R. Rao, F. Sahin. Cluster-head identification in ad hoc sensor networks using particle swarm optimization [R]. IEEE International Conference on Personal Wireless Communications, 2002.

[11] N. Latiff, C. Tsimenidis, B. Sharif. Energy-aware clustering for wireless sensor networks using particle swarm optimization [R]. In Proc IEEE PIMRC, 2007.

[12] T. Rappaport. Wireless Communications: Principles & Practice [M]. Prentice-Hall: USA, 1996.

[13] J. Kennedy, R. Eberhart. Particle swarm optimization [R]. In Proc. IEEE Int. Conf. Neural Netw., 1995.

[14] K. Pahlavan, A. Levesque, Wireless Information Networks [M]. Wiley: USA, 1995.

[15] T. Blackwell. A Study of Collapse in Bare Bones Particle Swarm Optimization [J]. IEEE Trans. Evolutionary Computation, 2012, 16 (3): 354-372.

An energy distance aware clustering protocol with Dual Cluster Heads using Niching Particle Swarm Optimization

Energy efficient utilization is an important criteria and factor that affects the design of wireless sensor networks (WSNs). In this literature, we propose an energy distance aware clustering protocol with Dual Cluster Heads using Niching Particle Swarm Optimization (DCH-NPSO). The protocol selects two cluster heads in each cluster, the Master Cluster Head (MCH) and the Slave Cluster Head (SCH), and the selection needs to consider the network state information carefully and deliberately. Simulation results show that the protocol we proposed can balance the energy dissipation and extend the network lifetime effectively.

1 INTRODUCTION

The application of WSNs becomes more wide in recent years; there are a large number of typical applications, such as environment surveillance and exploration, target tracking, and volcano activity monitoring[1]. The energy of wireless sensor nodes is usually supplied by dry/lithium batteries, and, in some dangerous situations, the replacement or recharge of batteries is impossible. So, energy-efficient utilization is an important issue for WSNs, while clustering is as an energy efficient technique and is often used to prolong the sensor network's lifetime[2].

There are several optimization problems in the whole information transmission process, WSN issues, such as clustering, localization, data fusion, and deployment that are often formulated as optimization problems[3]. Traditional mathematical optimization methods need tremendous computation efforts, and with the problem scale increase, the computation grows up exponentially. As to implement on tiny wireless sensor nodes, an optimization method that needs moderate and even minimal computation efforts yet generates better results is needed. Bioinspired optimization methods are computation efficient in contrast to traditional mathematical methods[4]; the authors[5] have demonstrated and compared the general mathematical formulation for this kind of problems. In this paper, we take advantage of bioinspired optimization to resolve clustering formation.

Particle swarm optimization (PSO) is a bioinspired optimization, and it is simple, effective, and popular optimization method[6]. It has been used to address WSN issues such as clustering, deployment, localization, and date aggregation[3]. Niching is also inspired by nature[7]; different species must compete to survive by taking on different roles. Niching is a significant technique for multimodal optimization, and within the framework of PSO, some niching methods have been proposed and developed[8,9].

Several clustering protocols have been proposed for WSNs, LEACH[10] is a famous clustering protocol and selects nodes with a probability to act as cluster heads periodically. BCDCP[11] makes use of the base station to select clusters and routing path and perform randomized shift of cluster heads.

Particle swarm optimization has been used to resolve WSN issues such as clustering, localization, deployment, and data fusion[3]. A protocol using PSO[12] aimed to minimize the intracluster distance and optimize the energy dissipation of the network. A clustering particle swarm optimization[13] has been presented for dynamic optimization and employed a hierarchical clustering method to locate multiple optima for dynamic optimization problems.

In this literature, we propose an energy distance aware clustering protocol with Dual Cluster Heads using Niching Particle Swarm Optimization (DCH-NPSO). The protocol generates two cluster heads in each cluster using Niching PSO, a Master Cluster Head (MCH) and a Slave Cluster Head (SCH). The MCH receives and aggregates the data that have been collected from its member nodes and then sends it to the SCH; the SCH transmits the aggregated data to the sink directly. In our designed scheme, MCH does not communicate with the sink directly, that can balance the energy consumption. The simulation results show that the protocol can balance the network loads, minimize the energy consumptions, and prolong the network lifespan of WSNs.

2 SYSTEM MODEL

2.1 Network Model

We assume a number of N wireless sensor nodes distributed among an $M \times M$ squared field randomly. Sensed data are gathered in a periodical way, andwe define each period as a round. Each round contains three phase, a set-up phase uses DCH-NPSO to part the network into clusters and selects MCH and SCH using Niching PSO, a steady phase when data collected and aggregated by the MCH, and a finish phase followed when data is transmitted to the BS (Base Station, also called sink node); the steady phase should be longer than other phases to minimize the energy dissipation. The MCH is used for data collection and data aggregation. The aggregated data are sent to the SCHs, and then the SCHs transmit it to the sink node directly and get some information through a piggyback manner to save more energy. The partial structure of the network in a round is as shown in Fig. 1.

We make assumptions about the network model as follows.

(1) All nodes are equipped with small batteries, so each node has initial energy allocated and the energy is constrained.

(2) A BS situates far away from the sensing field, and it has no energy constraints.

(3) Each node has its location information. There are a number of existing works on local-

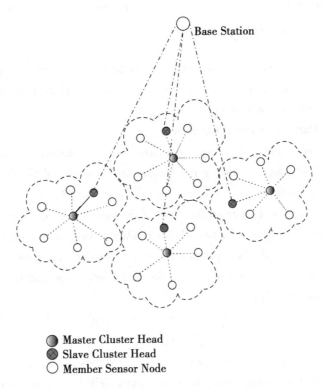

- ● Master Cluster Head
- ◉ Slave Cluster Head
- ○ Member Sensor Node

Fig. 1 DCH-NPSO partial network architecture

ization in WSNs[14-16]. For example, we use the solution proposed in[16] to obtain the localization.

(4) All nodes are equipped with power control units to adjust the transmit power according the distance.

(5) All nodes have no mobility.

2.2 Radio Model

The radio model we adopt is as used in[10]. In this model, the transmitter dissipates energy to drive the amplifier and the radio components, and the receiver dissipates energy to drive the radio components. The transmitter has power control ability to properly set the amplifier according to the distance and can be turned off to avoid receiving unintended sensing data, to minimize the energy consumption. The energy dissipation for transmit an-vbit packet over distance d is

$$E_{Tx}(l,d) = E_{elec}(d) + E_{amp}(l,d) = \begin{cases} l \times E_{Tx} + l \times \varepsilon_{fs} \times d^2, & d \leq d_o \\ l \times E_{Tx} + l \times \varepsilon_{mp} \times d^4, & d > d_o \end{cases} \quad (1)$$

Where d_0 is the threshold distance[17] and is given via $d_o = \sqrt{\varepsilon_{fs}/\varepsilon_{mp}}$. The energy dissipation that the receiver requires to receive an l-bit message is $E_{Rx}(l) = l \times E_{elec}$. E_{elec} elec is the energy consumed per bit to run the receiver or the transmitter circuit, ε_{fs} and ε_{mp} depend on the transmitter amplifier model, the energy for data aggregation E_{DA} is set to $5nj/bit/message$[18].

3 DCH-NPSO PROTOCOL

Our proposed protocol takes full advantage of Niching PSO for clustering, takes the network states information into account deliberately, and selects SCH to balance the energy dissipation to improve the energy efficiency.

3.1 Niching Particle Swarm Optimization

Particle swarm optimization is a bionic optimization that is inspired by nature such as bird flocking and fish school, and it was first put forward by Kennedy and Eberhart[19]. In a classic PSO, a number of m particles fly at a certain speed, respectively, in a D-dimension space, and each particle has a fitness value on the basis of the fitness function which is defined as related to the problems. Theparticles fly towards the position defined as personal best position and global best position (i.e., the best position refers to a position that gives the best fitness value). During each iteration, the particles' velocity and position are updated based on the following equations, respectively:

$$v_{id}^{k+1} = \omega v_{id}^k + c_1 \xi (p_{id}^k - x_{id}^k) + c_2 \eta (p_{gd}^k - x_{id}^k) \quad (2)$$

$$x_{id}^{k+1} = x_{id}^k + v_{id}^{k+1} \quad (3)$$

Where v_{id} is particle i's velocity in d dimension, k stands for the current iteration, x_{id} is the particle i's position in d dimension, ω is the inertia weight, deciding how the particle inherit present velocity, c_1 and c_2 are acceleration coefficient (or called learning factor) which are positive constants normally, ξ and η are stochastic numbers between 0 and 1, p_{id} is the particle's best position, and p_{gd} is the global best position[20,21].

When the issue target is to find multiple global optima, niching methods are valuable and efficient. Niching PSO searched in parallel, so the chance of getting trapped into a local optimum is reduced[7,9]. There has been a lot of existing work on niching methods[7-9]; in this paper, we use a ring topology niching method as[7], and each particle interacts with its immediate neighbors only.

3.2 The Fitness Function of NPSO

The fitness function is closely correlated with the features of the problem and determines the property of the optimal solution directly. So, the design of the fitness function is very important. Based on[3,12], we consider the conditions that affect the system performance and define the fitness function deliberately and comprehensively. At first, we should take the residual energy of the sensor node into account, because if the node becomes the MCH, it dissipates more energy than the member nodes. Then, the distance between the MCH and its member nodes should be considered as the MCH takes charge of gathering data from its member nodes. At last, we should take the distance between the SCH and the BS into account as SCH sends the aggregated data to the BS directly. We define the fitness function $f(i)$ of node i based on these considerations:

$$f(i) = \alpha f_1(i) + \beta f_2(i) + \gamma f_3(i) \tag{4}$$

$$f_1(i) = \sum_{i=1}^{N} E(n_i) \Big/ \sum_{k=1}^{k_{opt}} E(CH_{p,k})$$

$$f_2(i) = \sum_{k=1}^{k_{opt}} \sum_{\forall n_i \in C_{p,k}} d(n_i, CH_{p,k}) / |C_{p,k}|$$

$$f_3(i) = \frac{d(CH_{p,k}, BS)}{k_{opt}} \sum_{k=1}^{k_{opt}} d(CH_{p,k}, BS)$$

$$0 \leq \alpha, \beta, \gamma \leq 1, \quad \alpha + \beta + \gamma = 1$$

Where n_i is the sensor node i, $E(n_i)$ is the residual energy of sensor nodes i, $CH_{p,k}$ is the nodes that belong to cluster C_k of particle p, N is the number of node alive in the network, and k_{opt} opt is the optimal number of clusters in the network[10,12], so $f_1(i)$ takes the network's energy efficiency into account and select the nodes that has more residual energy as cluster head. $|C_{p,k}|$ is the number of nodes that belong to cluster C_k of particle p, $d(n_i, CH_{p,k})$ is the Euclidean distance between them, so $f_2(i)$ minimizes the intracluster distance between the CH and its member nodes, and $f_3(i)$ minimizes the distance between the CHs and the BS. The fitness function $f(i)$ has the maximum of the residual energy and the minimum of the distance between the nodes among the cluster and the distance between the CH and the BS. So, based on the above fitness function, we choose the node that has the maximum fitness value as MCH and the suboptimal solution as SCH, that is the optimal selections during that round.

3.3 DCH-NPSO Setup

The proposed protocol runs in periodic way, and we use Niching PSO and the fitness function(4) to select the optimal set of CHs and their member nodes. The sensed data transmissions begin after the clusters have been selected; each MCH gathers and aggregates the data collected from its members. The aggregated data are sent to the SCH, and the SCH transmits the data to the BS directly. The flowchart of DCH-NPSO is shown in Fig. 2.

The MCH sets up a TDMA schedule for its member nodes to avoid collection during data transmission, allowing the radio units to be turned off at all times except the transmission period for each member sensor. And we use a fixed spreading code and CSMA (Carrier Sense Multiple Access), a similar approach used in[22] during data transmission to minimize and balance the energy dissipation. Once the MCH finishes gathering data from its member sensors, the MCH performs data aggregation and sends the aggregated data to the SCH and then the SCH sends the aggregated data to the BS directly.

4 SIMULATION

We perform the simulations to evaluate the proposed protocol's performance. We assumethat in a 100m×100 m-squared sensing area, 100 sensor nodes are distributed random-

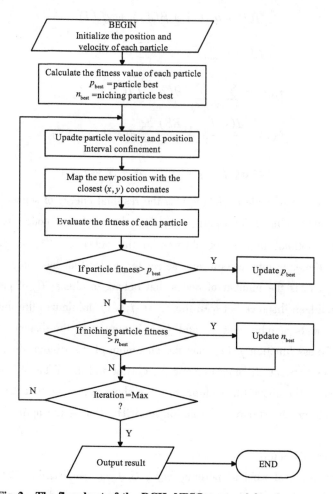

Fig. 2 The flowchart of the DCH-NPSO protocol for cluster setup

ly, and the BS is situated at (50,175) m, the datamessage length is 75 bytes with 15-byte packet header. The simulation parameters are given in Table 1, in which the radio parameters are as described in Section 2 and particle swarm optimization is as described in Section 3, and we use ring topology as in[7] to formniching. The performances of our proposed protocol are compared with clustering protocol LEACH, LEACH-C[10], and PSO-C[12].

We use the number of rounds to measure the number of nodes alive during the simulation; it as shown in Table 2. We can see that, as for LEACH, the first node is "dead" (i.e., the node has consumed all its initial energy) in 725 rounds, LEACH-C in 843 rounds and PSO-C in 954 rounds, while DCH-PSO in 1134 rounds, and our proposed protocol prolongs the network lifetime about 410 rounds compared to LEACH. The reason is that LEACH and LEACH-C do not take the residual energy of sensor nodes into account and select cluster head randomly, although PSO - C can prolong the network lifetime compared to LEACH and LEACH-C; it still does not balance the energy consumption. Our protocol DCH-NPSO takes the residual energy into account and selects SCH in each cluster, which balances and minimizesthe energy consumption.

Fig. 3 shows the total data messages received at the BS, the plot clearly indicates the effectiveness of the proposed protocol that delivers more data messages than LEACH, LEACH-C, and PSO-C, because the proposed protocol takes advantage of higher energy sensors such as MCHs by considering the residual energy of the nodes and selects SCH to balance the energy dissipation. Thus, more data messages are transmitted to the BS. In contrast, LEACH does not consider the residual energy of a sensor node when choosing the MCH and may choose the MCH with insufficient energy to stay alive during the data transmission phase. Although PSO-C takes the node's residual energy and the intracluster distance into consideration, it does not take actions to balance the energy consumption, while our protocol divides the network into clusters evenly, takes the nodes residual energy and the intracluster distance into account during the MCH selection, and selects SCH to balance the energy dissipation.

Table 1 Simulation parameters for DCH-NPSO

Parameter	Value
E_{elec}	$50nj/bit$
ε_{fs}	$10pj/bit/m^2$
ε_{mp}	$0.0013pj/bit/m^4$
E_{DA}	$5nj/bit/message$
c_1	2.5~0.5
c_2	0.5~2.5
ω	0.8~0.3

Table 2 Network lifespan comparison

Protocol	First node dead	Half nodes dead	All nodes dead
LEACH	725	1 053	1 418
LEACH-C	843	1 136	1 580
PSO-C	954	1 308	1 636
DCH-NPSO	1 134	1 545	1 788

As can be seen from Fig. 3 and Table 2, the protocol we proposed can prolong the network lifetime and transmit more sensed data to the BS and thus improves the system robustness and enhances the system scalability. LEACH, LEACH-C, and PSO-C, on the other hand, either generate an uneven distribution of cluster heads throughout the network or do not balance the energy dissipation. Our protocol produces performs even clustering distribution and selects SCHs to balance the energy consumption.

5 CONCLUSION

In this literature, we have presented an energy distance aware clustering protocol with

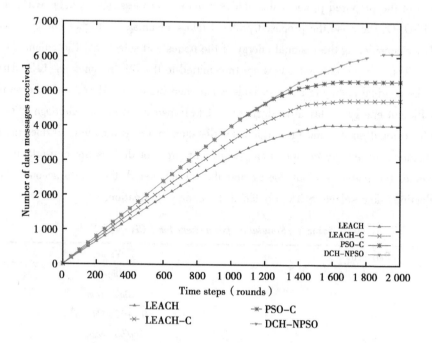

Fig. 3 The flowchart of the DCH-NPSO protocol for cluster setup

Dual Cluster Heads using Niching Particle Swarm Optimization (DCH-NPSO). We uses Niching PSO to select MCH and SCH in a cluster, taking the factors such as residual energy and the distance into account to balance and minimize the energy consumption. Results from the simulations show that the proposed protocol using Niching PSO prolongs the network lifetime and deliversmore data to the BS compared to LEACH, LEACH-C, and PSOC. Our future work includes the multihop routing among the network and explores rational parameters of Niching PSO to further improve the energy efficiency and the system performance.

6 CONFLICT OF INTERESTS

The authors declare that there is no conflict of interests regarding the publication of this paper.

REFERENCES

[1] J. Yick, B. Mukherjee, D. Ghosal. Wireless sensor network survey [J]. Computer Networks, 2008, 52 (12): 2292-2330.

[2] A. A. Abbasi, M. Younis. A survey on clustering algorithms for wireless sensor net-

works [J]. Computer Communications, 2007, 30 (14-15): 2826-2841.

[3] R. V. Kulkarni, G. K. Venayagamoorthy. Particle swarm optimization in wireless-sensor networks: a brief survey [J]. IEEE Transactions on Systems, Man and Cybernetics Part C: Applications and Reviews, 2011, 41 (2): 262-267.

[4] R. Xu, J. Xu, D. C. Wunsch. A comparison study of validity indices on swarm-intelligence-based clustering [J]. IEEE Transactions on Systems, Man, and Cybernetics, Part B: Cybernetics, 2012, 42 (4): 1243-1256.

[5] N. M. Abdul Latiff, C. C. Tsimenidis, B. S. Sharif. Performance comparison of optimization algorithms for clustering in wireless sensor networks [R]. in Proceedings of the IEEE International Conference on Mobile Adhoc and Sensor Systems (MASS'07), 2007.

[6] Y. del Valle, G. K. Venayagamoorthy, S. Mohagheghi, et al. Particle swarm optimization: basic concepts, variants and applications in power systems [J]. IEEE Transactions on Evolutionary Computation, 2008, 12 (2): 171-195.

[7] X. Li. Niching without niching parameters: particle swarm optimization using a ring topology [J]. IEEE Transactions on Evolutionary Computation, 2010, 14 (1): 150-169.

[8] A. E. R. Brits and F. van den Bergh. A niching particle swarm optimizer [R]. in Proceedings of the Asia-Pacific Conference on Simulated Evolution and Learning (SEAL'02), 2002.

[9] A. P. Engelbrecht, B. S. Masiye, G. Pampard. Niching ability of basic particle swarm optimization algorithms [R]. in Proceedings of the IEEE Swarm Intelligence Symposium (SIS'05), 2005.

[10] W. B. Heinzelman, A. P. Chandrakasan, H. Balakrishnan. An application-specific protocol architecture for wireless microsensor networks [J]. IEEE Transactions on Wireless Communications, 2002, 1 (4): 660-670.

[11] S. D. Muruganathan, D. C. F. Ma, R. I. Bhasin, et al. A centralized energy-efficient routing protocol for wireless sensor networks [J]. IEEE Communications Magazine, 2005, 43 (3): S8-S13.

[12] N. M. A. Latiff, C. C. Tsimenidis, B. S. Sharif. Energy aware clustering for wireless sensor networks using particle swarm optimization [R]. n Proceedings of the 18th Annual IEEE International Symposium on Personal, Indoor and Mobile Radio Communications (PIMRC'07), 2007.

[13] S. X. Yang, C. H. Li. A clustering particle swarm optimizer for locating and tracking multiple optima in dynamic environments [J]. IEEE Transactions on Evolutionary Computation, 2010, 14 (6): 959-974.

[14] D. Niculescu, B. Nath. Ad hoc positioning system (APS) using AOA [R]. in

Proceedings of the IEEE 22nd Annual Joint Conference of the IEEE Computer and Communications Societies (INFOCOM'03), 2003.

[15] T. He, C. Huang, B. M. Blum J. A. Stankovic, and T. Abdelzaher. Range-freelocalization schemes for large scale sensor networks [R]. in Proceedings of the 9th Annual International Conference on Mobile Computing and Networking (MobiCom'03), 2003.

[16] P. Kumar, A. Chaturvedi, M. Kulkarni. Geographical location based hierarchical routing strategy for wireless sensor networks [R]. in Proceedings of the International Conference on Devices, Circuits and Systems (ICDCS'12), 2012.

[17] T. Rappaport. Wireless Communications: Principles & Practice [M]. Prentice-Hall, Upper Saddle River, NJ, USA, 1996.

[18] A. Wang, W. R. Heinzelman, A. P. Chandrakasan. Energyscalable protocols for battery-operated microsensor networks [R]. in Proceedings of the IEEE Workshop on Signal Processing Systems (SiPS'99), 1999.

[19] J. Kennedy, R. Eberhart. Particle swarm optimization [R]. in Proceedings of the IEEE International Conference on Neural Networks, 1995.

[20] Y. Shi, R. Eberhart. A modified particle swarm optimizer [R]. in Proceedings of the IEEE International Conference on Evolutionary Computation, 1998.

[21] T. Blackwell. A study of collapse in bare bones particle swarm optimization [J]. IEEE Transactions on Evolutionary Computation, 2012, 16 (3): 354-372.

[22] K. Pahlavan, A. Levesque. Wireless Information Networks [M]. JohnWiley & Sons, New York, NY, USA, 1995.

An Energy Efficient Clustering protocol based on Niching Particle Swarm Optimization

According to the energy constraints characteristics of Wireless Sensor Networks, how to optimize clustering, reduce the node energy consumption and balance the network energy dissipation is an main target, we proposes an Energy Efficient Clustering protocol based on Niching Particle Swarm Optimization (NPSO-EEC), the algorithm considers the factors such as the node's residual energy and neighbor nodes' status, etc. The simulation results show that the proposed protocol can balance the nodes energy consumption effectively, reduce the sensor nodes' death rate, and prolong the network lifetime.

1 INTRODUCTION

Wireless Sensor networks (WSNs) are consisted by tens of thousands of low power consumption micro sensor nodes, which are cheap and deployed in the monitoring areas, such as in military, environmental monitoring, and other fields, especially suitable for deployment in the bad environment and the remote places that unfavorable to be reached[1]. Sensor nodes usually adopt micro batteries to offer limited energy supply, it is difficult to supply energy when the batteries are run out after the deployment, the first goal in the design of a WSN is the energy efficient.

WSNs adopt clustering protocol that can improve the sensor network scalability and prolong the network lifetime[2]. The main task of the protocol is to select a set of nodes that act as cluster heads efficiently, the cluster heads are not only distributed uniformly in the network but also reduce the energy consumption within the clusters, but that is an NP hard problem. If we adopt the swarm intelligence algorithm, such as Genetic Algorithms (GA or GAs)[4], Particle Swarm Optimization (PSO)[5-6], we can solve the problem effectively.

Particle Swarm Optimization (PSO) was used to optimize the selection of cluster heads in[9-10], it was based on[11] to solve clustering problems, but it often falls into local optimum, while Niching Particle Swarm Optimization (NPSO) can effectively avoid falling into local optimum. In this paper, we propose an Energy Efficient Clustering protocol based on Niching Particle Swarm Optimization (NPSO-EEC), the protocol considers the factors such as node residual energy and thestate of the neighbor nodes, the protocol not only optimize the selection of cluster heads but also balance the energy consumption, so as to prolong the network lifetime and improve the scalability and robustness of the system.

2 NICHING PARTICLE SWARM OPTIMIZATION

Particle Swarm Optimization (PSO) is an algorithm that simulate birds flying, it has the characteristics such as simple, ease to implement, does not need gradient information and has less parameters, etc. So it achieves good performance in resolving problems of discrete and continuous optimization problems, and has good robustness and generality, it has get widely attention and application since it has been proposed in 1995[13].

In the nature evolution, niching phenomenon are generally exist, the biology always tend to live together with which are similar to their shape and feature, mating and propagate progenies. Inspired by this, we introduce niching phenomenon into particle swarm optimization, the application practice has proved that the niching method has increase the speed of convergence and avoid premature.

Particle swarm optimization algorithm is composed of m particle fly at a certain speed in a D-dimension space, during the flight each particle consider its own and the neighbors' optimal fitness value position, based on this information, the particle changes its states and position. The particle's position and speed changes according to the following equations[14]:

$$v_{iD}^{k+1} = \omega v_{iD}^k + c_1 \xi (p_{iD}^k - x_{iD}^k) + c_2 \eta (p_{gD}^k - x_{iD}^k) \quad (1)$$

$$x_{iD}^{k+1} = x_{iD}^k + v_{iD}^{k+1} \quad (2)$$

Where ω is inertia weight, it decide how much the particle inherit its current velocity, c_1 and c_2 are accelerated coefficient factor (learning factor) and are positive constant number, the accelerated coefficient factor has self-learning ability and learn from the excellent individuals in the swarm. ξ and η are the pseudo-random number that uniform distributed in the range 0-1.

3 NPSO-EEC CLUSTERING PROTOCOL

The fitness functions of NPSO-EEC. The determination of the fitness value function is closely related with the problem domain, and determines the algorithm's performance directly. Based on[4,9,10], we consider various factors that may influence the system's performance, and do many experiences to explore the optimal fitness function and parameters. At first we should consider the node's residual energy, that because the cluster head consume more energy than normal nodes, and the cluster head is responsible for the data collection within the cluster, so the minimal Euclidean distance between the cluster head and other nodes in the cluster, the best performance we will get.

Based on the above consideration, we adopt the fitness function as follows:

$$f(i) = \alpha f_1(i) + (1 - \alpha) f_2(i) \quad (3)$$

$$f_1(i) = E(i) / \frac{1}{m} \sum_{i=1}^{m} E(i)$$

$$f_2(i) = \frac{1}{N}\sum_{k=1}^{N}d(i,k) / \frac{1}{m}\sum_{k=1}^{m}d(i,k)$$

Where $E(i)$ is i th node's residual energy, m stands for the number of living nodes among the cluster, N is the number of node alive among the network, $d(i,k)$ is the Euclidean distance between the node i and k. So $f_1(i)$ is the cluster head's residual energy in proportion to the average energy, $f_2(i)$ stands for the nodes' average Euclidean distance in the network with ratio to the average Euclidean distance in the cluster, we can adjust the proportion of the fitness function through α. We can see from the above, if a node has more residual energy and the average distance is smaller, the fitness value is optimal, we select such node as cluster head to balance the energy consumption and prolong the network lifetime.

NPSO-EEC Protocol. NPSO-EEC protocol is performed periodically, first the protocol parts the network into clusters preliminary, and use NPSO-EEC protocol to select the optimal cluster head and send the cluster information message to the cluster.

1. Initial clustering

We adopt clustering algorithm such as LEACH to divide the network into cluster preliminary, the cluster head during that stage we called assistant cluster head.

2. Collect nodes information

Each node in the cluster sends its states information such as residual energy and location information to the assistant cluster head.

3. Select optimal cluster head using NPSO-EEC

This is the core step, we use Formula (3) to optimize the cluster heads selection.

4. Decide the optimal cluster head

The assistant cluster head nodes send cluster information to the node that has been selected as cluster head.

NPSO-EEC protocol considers the node's residual energy, the state information of its neighbor nodes to optimize cluster head selection, that balance the energy consumption and avoid occur "blind node" frequently.

4 SIMULATION

We do experiments using Matlab, let's assume there are 100 nodes distributed in 150m× 150m area, the base station locates at ($x = 75$, $y = 200$). In order to estimate the performance of the protocol, we compare it with clustering protocol such as LEACH[11], PSO-C[9]. The parameters we use in PSO-C and NPSO-EEC are shown in Table 1[11,12].

Table 1 Simulation Parameters

Parameter	Value
E_o	0.5J

(continuation)

Parameter	Value
E_{Tx}, E_{Rx}	$50nJ/bit$
E_{DA}	$5nJ/bit/message$
ε_{fs}	$10pJ/bit/m^2$
ε_{mp}	$0.0013pJ/bit/m^4$
m	20
c_1, c_2	2
ω	$0.9 \sim 0.4$
x_{max}, v_{max}	100

The nodes alive in the network are as shown in Fig. 1, the clustering protocol NPSO-EEC we proposed is significantly better than LEACH and PSO-C. That is because in LEACH the energy consumption is not balanced and the cluster heads are selected randomly, PSO-C adopts particle swarm optimization and gets trapped into local optimum sometimes. While our protocol balance the energy consumption in the cluster, optimize the cluster head selection and avoid appear "blind node" early, thus prolong the network lifetime.

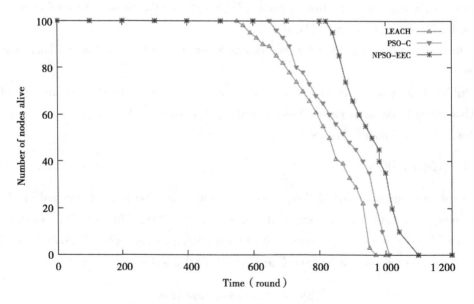

Fig. 1　Number of nodes alive in the network

5 CONCLUSION

In this paper we proposed an Energy Efficient Clustering protocol based on Niching Particle Swarm Optimization, the simulation results show that the protocol has good performance, strong robustness and better expansibility. The protocol prolongs the network lifetime and gets good performance according to optimize the cluster head selection, balance the energy consumption. We do some work in this area, and the parameter selection and realize in hardware is the next step we will do.

REFERENCES

[1] Yick J, Mukherjee B, Ghosal D. Wireless sensor network survey [J]. Computer Networks, 2008, 52 (12): 2292-2330.

[2] Abbasi A, Younis M. A survey on clustering algorithms for wireless sensor networks [J]. Computer Communications, 2007, 30 (14-15): 2826-2841.

[3] Bollobas B. Random Graphs [M]. Academic Press, 1985.

[4] Wang D W, Yung K L, Ip W H. A Heuristic Genetic Algorithm for Subcontractor Selection in a Global Manufacturing Environment [J]. IEEE Trans. SMC Part-C, 2001, 31 (2): 189-198.

[5] Kulkarni R, Venayagamoorthy G. Particle Swarm Optimization in Wireless Sensor Networks: A Brief Survey [J]. IEEE Trans. Systems, Man & Cybernetics, 2011, 41 (2): 262-267.

[6] Li Xd. Niching Without Niching Parameters: Particle Swarm Optimization Using a Ring Topology [J]. IEEE Trans Evol. Comput., 2010, 14 (1): 150-169.

[7] Latiff N, Tsimenidis C, Sharif B. Performance comparison of optimization algorithms for clustering in wireless sensor networks [R]. Proc IEEE Int. Conf. Mobile Ad Hoc Sens. Syst., 2007.

[8] Xu R, Xu J. A Comparison Study of Validity Indices on Swarm-Intelligence-Based Clustering [J]. IEEE Trans. Systems, Man & Cybernetics, 2012, 42 (4): 1243-1256.

[9] Latiff N, Tsimenidis C, Sharif B. Energy-aware clustering for wireless sensor networks using particle swarm optimization [R]. Proc IEEE PIMRC, 2007.

[10] Liang Y, Yu H B, Zeng P. Optimization of clusterbased routing protocols in wireless sensor network using PSO [J]. Control & Decision, 2006, 21 (4): 453-456.

[11] Heinzelman W B, Chandrakasan A P, Balakrishnan H. An Application-Specific Protocol Architecture for Wireless Microsensor Networks [J]. IEEE Trans. Wireless Commun., 2002, 1 (4): 660-770.

[12] Rappaport T. Wireless Communications: Principles & Practice [M]. New Jersey: Prentice-Hall, 1996.

[13] Kennedy J, Eberhart R. Particle swarm optimization [R]. Proc IEEE Int. Conf. Neural Netw., 1995.

[14] Li C, Yang S, Nguyen T. A Self-Learning Particle Swarm Optimizer for Global Optimization Problems [J]. IEEE Trans. Systems, Man & Cybernetics, 2012, 42 (3): 627-646.

Energy-aware Clustering Protocol with Dual Cluster Heads using Niching Particle Swarm Optimization for Wireless Sensor Networks

Energy efficiency is one of the most important design criteria for wireless sensor networks (WSNs). In this literature, we propose an energy-aware clustering protocol with Dual Cluster Heads using Niching Particle Swarm Optimization (DCH-NPSO). The protocol generates two cluster heads in each cluster, the selection of the Master Cluster Head (MCH) and the Slave Cluster Head (SCH) needs to consider the network state information. Simulation results show the protocol can balance the energy consumption and extends the network lifetime effectively.

1 INTRODUCTION

In recent years, WSNs have been applied in many applications, such as circumstance exploration, seismic activity monitoring, target tracking and environment surveillance[1] etc. Sensor nodes are usually powered by small batteries, and in some danger circumstances, the recharge or replacement of batteries are impossible. So energy efficiency is one of the important issues for WSNs, while clustering is one of the energy-efficient techniques, it is often coupled with data fusions to prolong the sensor network's lifetime[2].

Clustering is often formulated as optimization problems[3], traditional optimization methods need enormous computation efforts and grow up exponentially with the problem scale increase. As to implement on a small sensor nodes, an optimization method that requires appropriate even minimal computation efforts and yet produces better results is needed. Bio-inspired optimization methods are computation efficient in contrast to traditional methods[4].

Particle Swarm Optimization (PSO) is one of the bio-inspired optimization methods, which is a simple, popular and effective optimization algorithm[5]. It has been applied to address WSN issues such as deployment, localization, clustering and data fusion[3]. It also has many good qualities, such as ease of implementation, high speed of convergence and computational efficiency etc. Niching is also inspired by nature[6], In natural ecosystems, individual species must compete to survive by taking on different roles. Niching is an important technique for multimodal optimization, now within the framework of PSO, several niching methods have been developed[7][8][9].

There are some clustering protocols that have been proposed for wireless sensor networks. LEACH[10] is a well-known cluster based protocol that runs periodically and selects nodes with probability to become cluster heads. In HEED[11], the nodes' probability of becoming a cluster head are depend on its residual energies, and the nodes that are not covered by any clusters double their probability to become a cluster head. DCCG[12] algorithm consists of two levels of clustering: local clustering and global clustering.

The particle swarm optimization has been applied to address WSN issues such as clustering, deployment, localization, and data aggregation[3][13][14]. A protocol using PSO has been proposed in[15], it has the objective of minimizing the intra-cluster distance and optimizing the energy consumption of the network. The authors present a clustering particle swarm optimization[16] for dynamic optimization problem, and employ a hierarchical clustering method to locate and track multiple peaks for dynamic optimization problems.

In this literature, we propose a Dual Cluster Heads clustering protocol using Niching Particle Swarm Optimization (DCH-NPSO). The protocol generates two cluster heads using PSO in each cluster, a Master Cluster Head (MCH) and a Slave Cluster Head (SCH). The MCH receives and aggregates the data from its member nodes, and sends to the SCH; the SCH transmits the aggregated data to the sink directly. In our designed scheme, MCH is not communicated with the sink directly, that can save and balance the energy consumption. Our simulation results show that the protocol can balance the network loads, minimize the energy dissipations and prolong the lifetime of wireless sensor networks.

2 SYSTEM MODEL

2.1 Network model

We assume in a $M \times M$ squared field, a total of N sensors distributed randomly among the field. Sensed data are gathered in a periodical manner, and each period is defined as a round. Each round consists three phase, a set-up phase uses DCH-NPSO protocol to format clusters, selects MCH and SCH using Niching PSO, then a steady phase when data gathered by the MCH, at last a finish phase when data transmitted to the BS. The MCH is used for the date collection and date aggregation. The aggregated data are sent to the SCH, the SCH transmits aggregated data to the sink directly and get some information through a piggyback manner to save more energy. The partial structure of the network in a round is as showed in Figure 1, the steady phase should be longer than others to minimize the energy consumption.

We make assumptions about the network model as follows:

All nodes are energy constrained and with initial energy allocated.

A base station (BS, also called sink node) locates far away from the sensing field.

All nodes are equipped with power control devices to adjust the transmit power according to the distance.

Each node knows his location information.

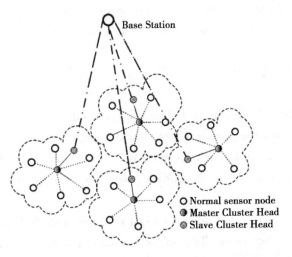

Fig. 1 DCH-NPSO partial network architecture

2.2 Radio model

We adopt the radio model as used in[10]. In this model, the transmitter consumes energy to run the radio components and the amplifier, the receiver consumes energy to run the radio components. The transmitter has power control ability that can consume minimum energy by properly setting the amplifier according the distance and can be turned off to avoid receiving unintended data. The energy consumption for transmit an l-bit packet over distance d is:

$$E_{Tx}(l, d) = E_{elec}(d) + E_{amp}(l, d) \\ = \begin{cases} l \times E_{Tx} + l \times \varepsilon_{fs} \times d^2, & d \leq d_o \\ l \times E_{Tx} + l \times \varepsilon_{mp} \times d^4, & d > d_o \end{cases} \quad (1)$$

Where d_0 is the threshold distance[17] and is given by $d_o = \sqrt{\varepsilon_{fs}/\varepsilon_{mp}}$. The energy consumption that a receiver need to receive an l-bit message is: $E_{Rx}(l) = l \times E_{elec}$ is the energy dissipated per bit to run the transmitter or the receiver circuit, ε_{fs} and ε_{mp} depend on the transmitter amplifier model, the energy for data aggregation is set as $E_{DA} = 5\text{nJ/bit/message}$[18].

3 DCH-NPSO PROTOCOL

Our protocol takes advantage of Niching PSO for clustering, considers the network states information deliberately, and selects SCH to balance the energy dissipation to improve energy efficiency.

3.1 Niching particle swarm optimization

Particle Swarm Optimization is a bionic optimization that is inspired by nature such as bird flocking and fish school, and it was first proposed by Kennedy and Eberhart[19]. In a standard PSO, a number of m particles fly at a certain speed in a D-dimension space, and each particle has a fitness value according the fitness function which is defined related with the problems. The particles fly towards a region defined as personal best position and global best position (i.e., the best position refers to the position that gives the best fitness value). In each iteration, the particles' velocity and position are updated based on the equations respectively as follows:

$$v_{id}^{k+1} = \omega v_{id}^k + c_1\xi(p_{id}^k - x_{id}^k) + c_2\eta(p_{nd}^k - x_{id}^k) \tag{2}$$

$$x_{id}^{k+1} = x_{id}^k + v_{id}^{k+1} \tag{3}$$

Where v_{id} is particle i's velocity in d dimension, x_{id} is the particle position, ω is the inertia weight, it decides how the particle inherit present velocity, c_1 and c_2 are learning factor (or called acceleration coefficient) which are positive constants normally, ξ and η are random numbers between 0 and 1, p_{id} is the particle's best position, p_{gd} is the global best position[20][21].

Niching methods are valuable when the target is to find a single/ multiple global optima, and a niching PSO searches in parallel, so the chance of getting trapped in a local optima is reduced[7][9]. There has been a lot of existing work on niching methods[7][8][9], as in[7], a PSO using a ring topology, each particle interacts only with its immediate neighbors.

3.2 The fitness function of NPSO

The fitness function is closely related with the characteristics of the problem, and determines the property of the optimal solution directly. So it is important to design the fitness function, based on[3][15], we take into account the conditions that influence the system performance and define the fitness function comprehensively. First, we should consider the residual energy of the node, because if it becomes the MCH, it dissipates more energy than the member nodes. Then the distance between the MCH and its member nodes should be take into account as the MCH responsible for collecting data from its member nodes. At last, the distance between the SCH and the BS also should be considered due to the SCH sends the aggregated data to the BS directly. We define the fitness function $f(i)$ of node i based on these considerations.

$$f(i) = \alpha f_1(i) + \beta f_2(i) + \gamma f_3(i) \tag{4}$$

$$f_1(i) = E(i) / \frac{1}{m}\sum_{i=1}^{m} E(i)$$

$$f_2(i) = \frac{1}{N}\sum_{k=1}^{N} d(i,k) / \frac{1}{m}\sum_{i=1}^{m} d(i,k)$$

$$f_3(i) = \frac{D(i)}{d_{max} - d_{toBS}}$$

$$0 \leqslant \alpha, \beta, \gamma \leqslant 1, \alpha + \beta + \gamma = 1$$

Function $f_1(i)$ is the ratio of node i's residual energy compare to the average energy within the cluster, $E(i)$ is the energy reside in node i, m is the number of member nodes within the cluster. Function $f_2(i)$ is the ratio of average distance among the network to the average distance within the cluster to node i, N is the number of nodes alive in the network, $d(i, k)$ is the Euclidean distance between node i and node k. d_{max} is the maximum distance between node i and the BS, d_{toBS} is the average distance between the nodes and the BS, and $D(i) = d_{max} - d_{toBS}(i)$. α, β, γ are the weight of each sub functions. The fitness function that defined above has simultaneously maximum the residual energy and minimum the distance among the cluster and the distance between the CH and the BS. So based on the fitness function, we select the node that has the maximum value as MCH, and the sub optimal solution as SCH, it is the optimum selections during that round.

3.3 DCH-NPSO setup

The protocol we presented executes in period, first we can use such as LEACH to partition the network. Each cluster use the fitness function (4) to choose MCH and SCH respectively. The data transmissions begins after clusters have been formed, each MCH receives and aggregates the data from its members, the aggregated data are sent to the SCH, then the SCH transmits aggregation data to the BS directly. We use a fixed spreading code and CSMA (Carrier Sense Multiple Access), a similar approach that used in[22] during data transmission to balance and minimize the energy consumption. The flowchart of DCH-NPSO is shown in Figure 2, and the detailed steps are as follows:

(1) Initialize particle swarm (Randomly initialize position and velocity of each particle).

(2) Calculate the fitness of each particle using Formula (4).

(3) Find the personal and niching best for each particle. The personal best is the current position of the particle, and the niching best is the position of the particles in the niches (or subspaces) that have experienced and have maximum fitness. There has been a lot of existing work on niching methods[7][8][9]. For example, we can use the solution presented in[7] to achieve the niching method, which is simply and effective.

(4) Update each particle's position and velocity using Formula (2) and (3).

(5) Map the new updated position with the closest coordinate.

(6) Repeat steps 2) to 5) until the maximum number of iterations is reached. Select the global best as MCH, and the sub optimal as SCH. The initial cluster head transmits the information that contains the MCH and SCH ID to all nodes in the cluster.

The MCH sets up a TDMA schedule for its members to avoid collisions among data messages, allowing the radio devices of each member to be turned off at all times except during their transmission time. Once the cluster head finishes receiving data from its entire members at the end of each round, the cluster head performs data aggregation and sends the aggregated data to the SCH, the SCH sends the aggregated data to the sink.

4 SIMULATION

We run the simulations to evaluate the performance of the proposed protocol, we assume there are 150 nodes randomly distributed in a 150m×150m network area with the BS located at (75, 200), the data message size is 175 bytes with the packet header is 25 bytes long. The

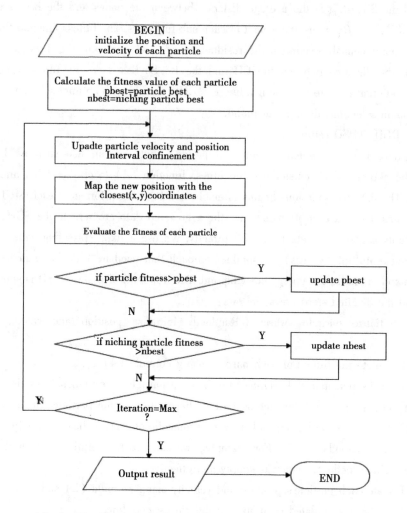

Fig. 2 The flowchart of the DCH-NPSO protocol for cluster setup

simulation parameters are given in Table 1, in which the radio parameters are as described in section II and particle swarm optimization parameters are as described in section III, we use the ring topology as in[7] to form niching. The performances of our proposed protocol are compared with clustering protocol LEACH[10] and PSO-C[15].

Table 1 Simulation parameters for DCH-NPSO

Parameter	Value
Network coverage	(0, 0) ~ (150, 150) m
Sink location	(75, 200) m
E_{elec}	$50 nJ / bit$
ε_{fs}	$10\ pJ/bit/m^2$

(continuation)

Parameter	Value
ε_{mp}	$0.0013 pJ/bit/m^4$
E_{DA}	$5 nJ/bit/message$
c_1	$2.5 \sim 0.5$
c_2	$0.5 \sim 2.5$
ω	$0.9 \sim 0.4$

The number of nodes alive over the simulation is measured by the number of rounds as shown in Table 2. We can see that as to LEACH, the first node "dead" (i. e., the node has lost its all initial energy) in 758 round, while PSO-C in 986 round and DCH-PSO in 1094 round, the protocol we proposed prolong the network lifetime about 350 rounds compare to LEACH. That because LEACH is not an energy-efficient algorithm which doesn't consider the residual energy of sensor nodes, although PSO-C can prolong the stability period and lifetime of networks compare to LEACH, it is still does not balance the energy consumption. While our proposed protocol considers the residual energy and selects a slave cluster head in each cluster, that balances the energy consumption and minimizes the energy dissipations.

Table 2　Network lifespan comparison

Protocol	First node dead	Half nodes dead	All nodes dead
LEACH	758	1 108	1 426
LEACH-C	864	1 276	1 534
PSO-C	986	1 418	1 682
DCH-NPSO	1 094	1 625	1 790

The total data messages received at the base station is shown in Figure 3. The plot clearly indicates the effectiveness of the proposed protocol in delivering more data messages than LEACH, LEACH-C and PSO-C. The reason is that the proposed protocol can take advantage of higher energy nodes as a CH by considering the remaining energy of the nodes and select SCH to balance the energy consumption. Hence, more data messages are sent to the base station, and it is unlikely that the CH will run out of energy before the steady state phase ends. In contrast LEACH does not take into account the energy of a node when selecting the CH, and may selects the CH with insufficient energy to remain alive during the data transfer phase, and though PSO-C considers the energy of a node and the intracluster distance, it does not balance the energy dissipation. While our protocol partitions the network evenly, considers the nodes residual energy and the intra-cluster distance during the CH selection, and selects SCH to balance the energy consumption.

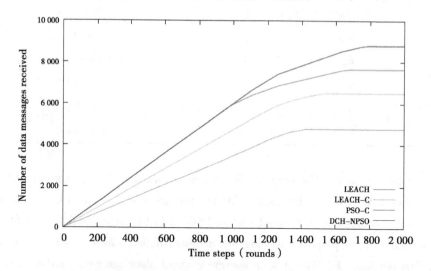

Fig. 3 The flowchart of the DCH-NPSO protocol for cluster setup

As can be seen from the simulation results, the proposed protocol can prolong the network lifetime and transmit more sensing data to the BS, thus enhance the system robustness and scalability. LEACH, LEACH-C and PSO-C on the other hand, either produce an uneven distribution of cluster heads throughout the sensing field or do not balance the energy consumption and consume more energy by cluster heads. Our proposed protocol produces even clustering and selects SCHs to balance the energy consumption.

5 CONCLUSION

In this paper we have presented an energy-aware Clustering Protocol with Dual Cluster Heads using Niching Particle Swarm Optimization (DCH-NSPO). We clustering the network consider the nodes residual energy, the intra-cluster distance, and using NPSO to select MCH and SCH in a cluster to balance the energy consumption. Results from the simulations indicate that the proposed protocol using Niching PSO gives a higher network lifetime, delivers more data to the base station compared to LEACH, LEACH-C and PSO-C. Furthermore, the proposed protocol produces better clustering by evenly allocating the CHs throughout the sensor network area. Our future work includes the multi-hop routing among the cluster heads and explores reasonable parameters of Niching PSO to further improveenergy efficiency.

REFERENCES

[1] J. Yick, B. Mukherjee, D. Ghosal. Wireless sensor network survey [J]. Computer Networks, 2008, 52 (12): 2292-2330.

[2] A. A. Abbasi, M. Younis. A survey on clustering algorithms for wireless sensor networks [J]. Computer Communications, 2007, 30 (14-15): 2826-2841.

[3] R. V. Kulkarni, G. K. Venayagamoorthy. Particle Swarm Optimization in Wireless Sensor Networks: A Brief Survey [J]. IEEE Trans. Systems, Man, and Cybernetics, 2011, 41 (2): 262-267.

[4] R. Xu, J. Xu, D. C. Wunsch. A Comparison Study of Validity Indices on Swarm-Intelligence-Based Clustering [J]. IEEE Trans. Systems, Man, and Cybernetics, 2012, 42 (4): 1243-1256.

[5] Y. Valle, G. K. Venayagamoorthy, S. Mohagheghi, et al. Particle swarm optimization: Basic concepts, variants and applications in power systems [J]. IEEE Trans. Evol. Comput., 2008, 12 (2): 171-195.

[6] A. E. R. Brits, F. van den Bergh. A niching particle swarm optimizer [C]. Proc. Asia-Pacif. Conf. Simul. Evol. Learn. (SEAL2002), 2002.

[7] X. Li. Niching Without Niching Parameters: Particle Swarm Optimization Using a Ring Topology [J]. IEEE Trans. Evol. Comput., 2010, 14 (1): 150-169.

[8] S. Bird, X. Li. Adaptively choosing niching parameters in a PSO [R]. Proc. Genet. Evol. Comput. Conf. (GECCO'06), 2006.

[9] A. Engelbrecht, B. Masiye, G. Pampara. Niching ability of basic particle swarm optimization algorithms [R]. Proc. IEEE Swarm Intell. Symp., 2005.

[10] W. B. Heinzelman, A. P. Chandrakasan, H. Balakrishnan. An Application-Specific Protocol Architecture for Wireless Microsensor Networks [J]. IEEE Trans. Wireless Commun., 2002, 1 (4): 660-770.

[11] O. Younis, S. Fahmy. HEED: A Hybrid, Energy-Efficient, Distributed Clustering Approach for Ad Hoc Sensor Networks [J]. IEEE Trans. Mobile Comput., 2004, 3 (4): 366-379.

[12] J. Huang, J. Zhang. Distributed Dual Cluster Algorithm Based on Grid for Sensor Streams [J]. Journal of JDCTA, AICIT, 2010, 4 (9): 225-233.

[13] M. Soliman, G. Tan. Conditional Sensor Deployment Using Evolutionary Algorithms [J]. Journal of JCIT, AICIT, 2010, 5 (2): 146-154.

[14] M. Romoozil, H. Ebrahimpour-komleh. A Positioning Method in Wireless Sensor Networks Using Genetic Algorithms [J]. Journal of JDCTA, AICIT, 2010, 4 (9): 174-179.

[15] N. M. Abdul Latiff, C. C. Tsimenidis, B. S. Sharif. Energy-aware clustering for wireless sensor networks using particle swarm optimization [R]. Proc. IEEE PIMRC'07, 2007.

[16] S. X. Yang, C. H. Li. A Clustering Particle Swarm Optimizer for Locating and Tracking Multiple Optima in Dynamic Environments [J]. IEEE Trans. Evol. Comput., 2010, 14 (6): 959-974.

[17] T. Rappaport. Wireless Communications: Principles & Practice [R]. NJ: Prentice-Hall, USA, 1996.

[18] A. Wang, W. Heinzelman, A. Chandrakasan. Energy-scalable protocols for battery-operated microsensor networks [R]. in Proc. IEEE SiPS' 99, 1999.

[19] J. Kennedy, R. Eberhart. Particle swarm optimization [R]. Proc. IEEE Int. Conf. Neural Netw., 1995.

[20] Y. Shi, R. Eberhart. A modified particle swarm optimization [R]. IEEE Int. Congress on Evolutionary Computation, 1998.

[21] T. Blackwell. A Study of Collapse in Bare Bones Particle Swarm Optimization [J]. IEEE Trans. Evol. Comput., 2012, 16 (3): 354-372.

[22] K. Pahlavan, A. Levesque. Wireless Information Networks [M]. New York: Wiley, USA, 1995.

Solar-powered Wireless Sensor Network's Energy Gathering Technology

We analyze the solar-powered wireless sensor network's energy gathering techniques, aiming to prolong the lifetime of wireless sensor network. We summarize wireless sensor network node's energy autonomy system, its characteristics in detail and new technology adopts, provides some suggestions and new ideas in the design and research of solar-powered wireless sensor networks.

1 INTRODUCTION

Wireless sensor networks have widely applications in military, agriculture and environment areas[1-2], the sensor nodes are usually powered by battery, and the nodes often place in danger areas, such as enemy control areas, radiation fields and volcano, etc. It's very difficult to replace and even unable to replace or recharge the energy supply system, so the energy provide and management is one of the research problems in wireless sensor network studies. Although we can reduce the energy consumption of the nodes, optimize the network structure and adjust the node's cycle time to reduce the nodes' or the network's energy dissipation[3-4], we can't solve the energy supply problem permanently. If we utilize energy collection techniques to gather energy supply from the surroundings, such as through solar energy to get a steady energy supply, we can resolve this problem in a new direction.

Energy gathering technology gets energy from light, mechanical operation, the differential of temperature and vibrations, etc. The light, especially the solar energy has got widely applied, that is because its techniques is relatively mature and the density of its distribution. In our daily life, high-performance power solar energy system has got widely used, but it's not suitable for apply in wireless sensor network according to the installation, the cost and the efficiency. The small applications using solar energy supply has also been studied, its main methods is to recharge the batteries, that method has prolong the lifetime of the batteries, but there has main shortcoming due to the battery charge and recharge frequently, that results in the decline in the performance. While through improving energy efficiency and increase multi-stage energy storage to extend the life time of battery, the power supply system has certain pertinence, the system need to specific design for each applications, this is because each system has different applications environment and features.

And by improving energy efficiency, as well as increase the multi-stage energy storage ways to extend the service life of the battery, the power supply system has certain pertinence, each system needs specific design for each application, this is because each system and system features of the different application environment.

We have an overview over wireless sensor network's node energy autonomy system, analyze the characteristics of each system and new techniques they have adopted, provide some methods and suggestions to the design of solar-powered wireless sensor network.

2 THE STRUCTURE OF SOLAR ENERGY GATHERING NODE

The structure of solar energy gathering node used in wireless sensor networks is as shown in Fig. 1[5-6]. It is constituted by four modules, the solar energy gathering module, energy management module, energy storage module and energy consumption module. The solar energy gathering module transforms the solar energy through energy conversion; energy management module is responsible for the management and control of battery charge and recharge, and monitor the circuit; energy storage module complete the storage of solar energy gathering, and provide energy supply, we usually use capacitor or battery as the energy storage module; energy consumption module stands for the sum energy consumption including the energy consumed in communication, computing and storage, etc.

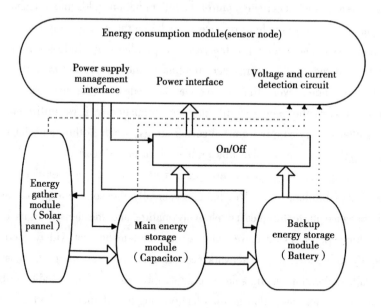

Fig. 1 The structure of solar energy gathering node

3 SOLAR ENERGY GATHERING SYSTEM

Heliomote[7], Fleck[8] and ZebraNet[9], these three systems' use the diode to connect with the rechargeable batteries on the solar panels, when charge the battery, the solar panel's voltage is 0.7V greater than the battery voltage. When the solar panels voltage is lower than the battery voltage, charge can't work thus waste energy, and the system adopt NiMH rechargeable battery, the charge and recharge time are limited, with these restrictions, the wireless sensor nodes can't use for a long time.

Prometheus[5] and Trio[10] solve the battery disabled problem that the battery charge and recharge frequently, the system adopts a secondary energy storage structure, equipped with main energy storage module (super capacitor) and backup energy storage module (lithium battery). Solar panels store the energy in super capacitor first, and use it first, the backup energy storage start to work when the capacitor's voltage is low. The design reduces the battery charge and charge time, prolong its lifetime, but due to the capacitor charging lacks effective control, the energy utilization rate is still not high.

Everlast[11] and eShare[12] adopt capacitor as the energy storage module, the system design a maximum power point tracking (MPPT) circuit to improve energy utilization. The shortcoming of the system is the stored energy is weak, and the maximum power point tracking (MPPT) circuits need to be real-time controlled by micro-controller, that add software cost, and if the sunshine is not enough, the energy storage may run out, these cause the system's reliability is not high.

Ambimax[13] can use a variety of environment energy, such as solar and wind energy, and design hardware maximum power point tracking (MPPT) circuit to maximize the energy gathered from the environment. The system uses capacitor as energy storage module to resolve the battery aging problem, but due to the system is complex and the hardware's overheads, this scheme is not suitable for the application in the low cost and low price demand wireless sensor network.

Duracap[14] adopts capacitor as the energy storage module, designs and achieves the maximum requirements of different solar power supply. The system contains "cold" start module to solve the initial status with no electricity that the system can't be started. The system has the advantages that according to different applications, it provides different interfaces, and maximizes the utilization of energy in the environment. The disadvantage is that the charge capacity is limited and energy management is not reasonable, that lead to energy depletion and limit the energy consumption demand and applications.

EasiSolar[15] system adopts adaptive solar maximum power output of panels (MPPT) circuit to maximize the collection of solar energy. The system uses two levels energy storage structure, the rechargeable battery's low leakage and high energy density, wireless charge and recharge characteristics of super capacitor, combined with efficient energy management method and efficient charge and discharge management to make full use of environmental energy reasonably.

SHE solar energy collection suite is a development kit that can transfer light energy into electric power, the kit is made by Texas Instruments (TI) and can be applied in many areas, such as agriculture, transportation, industry and commerce, it can satisfy wireless sensor network's demand for alternative energy. The size of SHE suite is as a credit card, it combines Cymbet company's EnterChip (thin-film technology), CC2500 transceiver, TI MSP430 microprocessor and development tools, collector can use in a variety of lighting environment, and the energy storage, processing and transmission can work efficiently.

4 THE CHARACTERISTICS OF SOLAR ENERGY COLLECTION SYSTEM

Single energy storage module structure. There have two types of energy storage module, capacitor and battery. The capacitor can store relative small energy and the leakage current is large, that leads to the lack and waste of energy; Rechargeable batteries have limited lifetime and failure problems due to frequent charging and recharging.

Energy efficiency is not high. The energy collection, storage, management need energy and consume some energy, some system use solar power supply directly, at that time, the energy can be used only when the solar panels voltage is higher than the capacitor voltage. Due to these reasons, the energy utilization rate is not high.

Simple energy management method. The energy management method is relatively simple and inefficient in most system, so the environment energy can't be effectively used, this is due to the energy distribution, storage and management is not reasonable.

Certain specific in energy supply system. Each system has its specific energy requirement and application environment, so the design of power supply system is different for each application.

5 CONCLUSION

Wireless sensor network is widely used, and the energy supply and management is one of the main researches. In this paper, we analyze the current solar gathering system, summarize the characteristics of the solar gathering system and the techniques they have adopted, provide some suggestions and ideas in the design and study about the solar gathering techniques for wireless sensor network.

REFERENCES

[1] Cui L, Ju H L, Miao Y, et al. Overview of wireless sensor network [J]. Journal of Computer Research and Development, 2005, 42 (1): 163-174.

[2] Wang B Y. Review on internet of things [J]. Journal of Electronic Measurement and Instrument, 2009, 23 (12): 1-7.

[3] Ren H J, Dai X H, Wang Z, et al. System design of nodes of wireless sensor networks [J]. Journal of Electronic Measurement and Instrument, 2006, 20 (6): 31-35.

[4] Mikko K, Jukka S, Timo D, et al. Energy-efficient reservation-based medium access control protocol for wireless sensor networks [J]. Journal of EURASIP Journal on Wireless Communications and Networking, 2010, 57: 1-22.

[5] Jiang X, Polastre J, Culler D. Perpetual environmentally powered sensor networks [R]. 4th International Conference on Information Processing in Sensor Networks, California, USA, 2005.

[6] Jeong J, Jiang X F, Culler D. Design and analysis of micro-solar power systems for wireless sensor networks [R]. 5th International Conference on Networked Sensing System, Kanazawa, Japan, 2008.

[7] Raghunathan V, Kansal A, Hsu J, et al. Design consideration for solar energy harvesting wireless embedded systems [R]. 4th International Conference on Information Processing in Sensor Networks, California, USA, 2005.

[8] Corke P, Valencia P, Sikka P, et al. Long-duration solar-powered wireless sensor networks [R]. 4th Workshop on Embedded Networked Sensors, Cork, Ireland, 2007.

[9] Zhang P, Sadler C M, Lyon S A, et al. Hardware design experiences in ZebraNet [R]. 2nd International Conference on Embedded Networked Sensor Systems, Maryland, USA, 2004.

[10] Dutta P, Hui J, Jeong J, et al. Trio: Enabling sustainable and scalable outdoor wireless sensor network deployments [R]. 5th International Conference on Information Processing in Sensor Networks, Nashville, TN, USA, 2006.

[11] Simjee F, Chou P H. Everlast: Long-life, supercapacitor operated wireless sensor node [R]. International Symposium on Low Power Electronics and Design, Tegernsee, Germany, 2006.

[12] Zhu T, Gu Y, He H, et al. Share: A capacitor-driven energy storage and sharing network for long-term operation [C]. 8th ACM Conference on Embedded Networked Sensor Systems, Zurich, Switzerland, 2010.

[13] Park C, Chou P H. AmbiMax: Efficient, autonomous energy harvesting system for multiple-supply wireless sensor nodes [C]. 3th Annual IEEE Communications Society Conference on Sensor, Mesh and Ad Hoc Communications and Networks, Reston, VA, USA, 2006.

[14] Chen Ch Y, Chou P H. DuraCap: A supercapacitorbased, power-bootstrapping, maximum power point tracking energy-harvesting system [C]. International Symposium on Low Power Electronics and Design, Austin, TX, USA, 2010.

[15] Zhang Jingjing, Zhao Ze, Chen Haiming, et al. EasiSolar: Design and implementation of a high-efficiency solar energy-harvesting sensor node system [J]. Chinese Journal of Scientific Instrument, 2012, 33 (9): 1952-1960.

Virtual Area Partition Clustering protocol with Assistant Cluster Head

Most existing clustering algorithms are suitable for homogeneous wireless sensor networks, but in practice there exists more heterogeneous scenarios. In this paper, a Virtual Area Partition Clustering protocol with Assistant Cluster Head (VAPC-ACH) is proposed for homogeneous and heterogeneous wireless sensor networks. The simulation results show that VAP-ACH can reduce and balance energy dissipation, thus prolong the sensor network lifetime and improve the system performance.

1 INTRODUCTION

Wireless sensor networks are application-specified networks and have a great number of applications, such as environment monitoring, military tracking and disaster rescuing, etc. So we must consider the factors such as the object, application environment, hardware, cost and other system constraints[1][2].

A wireless sensor network consists of a number of tiny battery powered multifunctioning devices with limited energy. Once deployed, the sensor nodes are often unreachable to users, and the replacement or recharge of batteries are impossible, so energy efficient is the vital factor for WSNs. Clustering is one of the energy-saving techniques that extending the sensor network's lifespan[3].

In this paper, we present a Virtual Area Partition Clustering protocol with Assistant Cluster Head (VAPC-ACH), which is an energy-efficient clustering protocol. The protocol obeys the idea of energy efficient to obtain good performance in terms of system lifespan and data delivery. We use virtual area partition clustering protocol that guarantees the mean value of the number of cluster heads is optimal in each round and select an Assistant Cluster Head (ACH) if needed in each cluster to balance the energy consumption. It also enhances the running rate of clustering for it has a characteristic of high speed parallel. Our simulation results and analysis show that VAPC-ACH can achieve better performance in system lifespan and data delivery by virtual area partition clustering and assistant cluster head that balancing the energy consumption evenly in the networks.

The outline of the paper is organized as follows. Section II surveys the related work. Section III describes the network and energy consumption models. Section IV presents VAPC-ACH protocol in detail. Section V provides the simulation results and followed by the conclusion in Section VI.

2 RELATED WORK

A WSN networked a large number of low-powered nodes together, traditional techniques such as direct transmissions and Minimum Transmission Energy (MTE) routing protocol have to be avoided due to their deficiency, an adaptive clustering scheme called Low-Energy Adaptive Clustering Hierarchy (LEACH) is proposed in[4] that randomly rotating the role of cluster head among all the nodes in the network. SEP[5] protocol consider node's heterogeneity, and based on weighted election probabilities of each node to become cluster head according to the remaining energy in each node, and yields longer stability. Another clustering-based protocol PEGASIS[6] which prolong the network lifetime by using local collaboration among sensor nodes. BCDCP[7] utilizes the base station to set up clusters and routing paths, perform randomized rotation of cluster heads. HEED[8] considers a hybrid of energy and communication cost that makes the energy consumption is more evenly. VAP-E[9] based on node partition to balance the loads. BEES[10] is a lightweight bio-inspired backbone construction protocol, it can help mitigate many of the typical challenges in sensor networks by allowing the development of simpler network protocols. The author proposed protocol[11] to select a minimum number of sensors to achieve full k-coverage of a field while guaranteeing connectivity between them and get better performance.

3 NETWORK AND ENERGY CONSUMPTION MODELS

We suppose there are N nodes distributed randomly in an $M \times M$ squared region with the sink node (also called the Base Station, BS for simplicity) located far away from the region. We use heterogeneous network model in this paper to strengthen practicability of wireless sensor networks, the energy of each node is randomly distributed in $[E_0, \alpha E_0]$, where α is a constant and $\alpha > 1$, and E_0 is the minimal energy of sensor nodes.

We assume a sensor network model that has properties as follows:

- A base station is located far away from the sensing field, and all sensor nodes have no mobility.
- The sensor nodes are energy limited with initial energy allocated.
- The nodes are equipped with power equipment to vary their transmitted power.
- Every node has location information. There are some localization algorithms have been proposed before, e.g.,[12][13], so we assume all nodes have known their locations.

The clustering techniques are adopted in wireless sensor networks, which divide all sensor nodes into two types respectively, member nodes and cluster heads. The cluster heads consume more energy, and charge to manage their member nodes, aggregate the data and then transmit to the BS; while member nodes only send sensing data to their cluster heads. We suppose that every member node in clusters will send l-bit message to its cluster head in each round. Each cluster has a CH which collects data from its members, aggregates and sends to the BS.

A typical WSN node is constituted of four major components: a sensor unit, a data processor unit, a power supply unit and a wireless communication unit that consists of transceiver/receiver circuit, antenna and amplifier[12][14][15]. Although in a sensor node, the energy is dissipated in all of the components except power supply unit, we mainly consider the energy consumed related to the wireless communication unit since the goal of this paper is to present an energy efficient protocol to prolong the network lifespan, and the energy consumed by the cluster head nodes during the data aggregation is also take into account.

We use a simple energy consumption model as used in[4]. In this model, a receiver consumes energy to run radio units, a transmitter consumes energy to run radio units and power amplifier. Thus, if we transmit an l-bit message through the distance d, the energy consumption that a transmitter need is:

$$E_{Tx}(l, d) = E_{elec}(d) + E_{amp}(l, d)$$
$$= \begin{cases} l \times E_{Tx} + l \times \varepsilon_{fs} \times d^2, & d \leq d_o \\ l \times E_{Tx} + l \times \varepsilon_{mp} \times d^4, & d > d_o \end{cases} \quad (1)$$

Where d_0 is the threshold distance[13] and is given by $d_o = \sqrt{\varepsilon_{fs}/\varepsilon_{mp}}$. The energy consumption that a receiver need to receive an l-bit message is: $E_{Rx}(l) = l \times E_{elec}$.

The radios have power control ability that can consume minimum energy by properly setting the power amplifier according the distance and can be turned off to avoid receiving unintended data. E_{elec} is the energy dissipated per bit to run the transmitter or the receiver circuit, ε_{fs} and ε_{mp} depend on the transmitter amplifier model, the energy for data aggregation is set as E_{DA} = 5nJ/bit/message[16].

4 VAPC-ACH PROTOCOL

A. Optimum number of clusters

We present an efficient clustering protocol to promote the existing clustering protocols' performance. Use the energy consumption model as described above, the amount energy consumed by all nodes during a round is:

$$E_r = l\left(NE_{DA} + 2NE_{elec} + k\varepsilon_{mp}d_{toBS}^4 + N\varepsilon_{fs}\frac{1}{2\pi}\frac{M^2}{k^2}\right)$$

Where and k is the number of clusters, d_{toBS} is the Euclidian distance between the BS and the cluster head, and l is the number of bits in each message, other parameters are as described above.

In order to minimize the energy consumption, we can calculate the optimal value of k from E_r to minimize the energy consumption. We set the derivative of the continuous function when $k>0$ about E_r with respect to k to zero $E'_r = l\left(\varepsilon_{mp}d_{toBS}^4 - \frac{N}{2\pi}\varepsilon_{fs}\frac{M^2}{k^2}\right) = 0$

So we get the optimum number of clusters, $k_{opt} = \dfrac{\sqrt{N}}{\sqrt{2\pi}} \sqrt{\dfrac{\varepsilon_{fs}}{\varepsilon_{mp}}} \dfrac{M}{d_{toBS}^2}$.

B. Virtual Area Partition Clustering

In a given sensor network, when k_{opt} is determined, we perform virtual area partition clustering on the networks. We part the network into k_{opt} partitions by $2\pi/k_{opt}$ in origin of network region center. It's not that the sensor nodes are divided, but the network is parted and the sensor nodes belong to its partition.

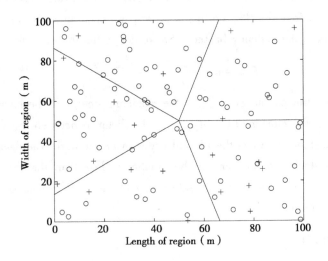

Fig. 1 Virtual area partition clustering with $k_{opt} = 5$

As shown in Fig. 1, a 100m× 100m region with 100 nodes distributed randomly, we use virtual area partition clustering that divided the network by $k_{opt} = 5$.

Most energy efficient clustering protocols[4][8][17-20] consider the residual energy as an important factor that decide which node to be cluster head in a cluster, the more energy a node has, the more opportunity it is to become a cluster head. That can prolong the network lifetime by balancing the energy consumption contrast to random select node as cluster head. However, from Formula (1) we can see that the distance is also a factor that affects the energy dissipation. We select node that has more energy and the distance between the node and the BS is minimal as cluster head, that can save more energy. So based on the consideration above, we define the following probability to decide which node to be cluster head.

$$p(i) = p_{opt} \left[(1 + \lambda) \dfrac{E_i(r)}{\bar{E}(r)} - \lambda \dfrac{d_{toBS}(i)}{\overline{d_{toBS}}} \right] \qquad (2)$$

Where p_{opt} is the ratio of optimal cluster heads, and $p_{opt} = k_{opt}/N$, $E_i(r)$ is node i's residual energy during round r, $\bar{E}(r)$ is the average energy of the alive nodes, $d_{toBS}(i)$ is the distance between node i and the BS, $\overline{d_{toBS}}$ is the average distance between nodes and the BS.

Obviously, we select the node that has greater $p(i)$ as cluster head, that can prolong the network lifetime due to the node has more energy and the distance between the node and the BS is minimal. We calculate the average energy of the nodes through a piggyback manner, that need not extra cost and make the energy consumption evenly among all nodes.

C. Assistant Cluster Heads

After the cluster head has been selected, it's time to decide whether to select an ACH in the cluster or not. We consider from the following several aspects, the residual energy of the CH, the distance between the CH and the BS. Generally if cluster head has little residual energy, the distance is far from the BS, on this occasion, we need an ACH to balance the energy dissipation and improve the system robustness. So we define the threshold function as:

$$T_{ACH} = c \cdot \frac{E_{max} - E_{re}}{E_{max}} \cdot \frac{d_{(CH, BS)}}{d_{avg}} \tag{3}$$

Where c is a control parameter; E_{max} is the maximum energy of the cluster head; E_{re} is the residual energy of the cluster head; $d_{(CH,BS)}$ is the distance between the CH and the BS; d_{avg} is the average distance between the BS and nodes in the network. The decision is made by the CH choosing a stochastic number, if the number is less than the threshold T_{ACH}, the cluster need to select a node as ACH, we select an node as ACH to balance the energy consumption during that round.

5 SIMULATION

We evaluated the performance of the proposed protocol, we run the simulations for 100 nodes randomly distributed in a 100m×100m network area with the BS located at ($x=50$, $y=180$), the energy of each node is randomly distributed in $[E_0, \alpha E_0]$, where $E_0 = 0.5J$ and $\alpha = 2$. For simplicity, we assume 20% nodes have αE_0 energy, the other nodes have E_0 energy. The data message size was 100 bytes with the packet header was 15 bytes long. The simulation parameters of the energy consumption model are described in[4][5]. The performance of our protocol was compared with the clustering protocols for wireless sensor networks LEACH[4] and VAP-E[9].

Table 1 Simulation Parameters

Parameter	Value
Network coverage	(0, 0) ~ (100, 100) m
Sink location	(50, 180) m
E_0	0.5J
E_{elec}	50nJ/bit
ε_{fs}	10pJ/bit/m^2
ε_{mp}	0.0013pJ/bit/m^4

(continuation)

Parameter	Value
E_{DA}	$5nJ/bit/message$
α	2
λ	0.3
Data message size	100 bytes
Data message header	15 bytes

The number of nodes alive over the simulation is measured by the number of rounds as shown in TABLE II. We can see that as to LEACH, the first node "dead" (i.e., all nodes have lost their all initial energy) in 697 round, while VAP-E in 838 round and VAPC-ACH in 968 round, the protocol we proposed prolong the network lifetime about 300 rounds compare to LEACH. That because LEACH is not an energy-efficient algorithm which doesn't consider the residual energy of sensor nodes, although VAP-E can prolong the stability period and lifetime of networks compare to LEACH, it is still does not balance the energy consumption. While ourproposed protocol considers the residual energy and selects assistant cluster heads to balance the energy consumption.

Table 2　Network Lifetime Comparation

Protocols	First node dead	Half nodes dead	All nodes dead
LEACH	697	936	1 196
VAP-E	838	984	1 227
VAPC-ACH	968	1 167	1 293

The total data messages received at the BS is shown in Fig. 2. It clearly indicates the effectiveness of the proposed protocol in delivering more data messages than LEACH and VAP-E. Our protocol offers improvement in data delivery by factors of 63% over LEACH and 34% over VAP-E. The reason for this is because the proposed protocol can take advantage of higher energy nodes as a CH by considering the remaining energy of the CH candidates and balance the energy consumption by using assistant cluster heads to balance the energy consumption. In contrast LEACH does not take into account the energy of a node when selecting the CH, and may select the CH with insufficient energy to remain alive during the data transfer phase, and VAP-E does not balance the energy dissipation. While our protocol considers the nodes'residual energy and select ACHs to communicate with the BS during the data transfer phase adaptively.

As can be seen from the simulation results, the proposed protocol can balance the energy consumption and cluster heads are located at the place that have more energy, and select assistant cluster heads to balance the energy dissipation if in need. LEACH and VAP-E on the other

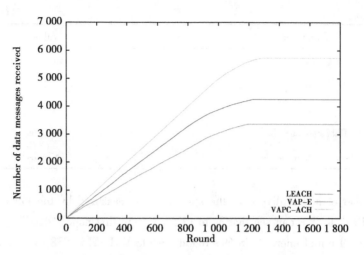

Fig. 2 Total data message received at the BS

hand, either produce an uneven distribution of cluster heads throughout the sensor field or do not balance the energy consumption and consume more energy by cluster heads. Our proposed protocolproduce even partitions and selects assistant cluster heads to balance the energy consumption if in need.

6 CONCLUSION

In this paper we have presented a Virtual Area Partition Clustering Protocol with Assistant Cluster Heads (VAPC-ACH) for wireless sensor networks. We have defined a threshold function that takes into account the residual energy and the distance between the cluster head and the BS to decide whether we need an assistant cluster head in a cluster. Results from the simulations indicate that the proposed protocol gives a higher network lifespan and the energy consumption is minimal compared to LEACH and VAP-E. Furthermore, the proposed protocol selects the optimal nodes as cluster heads throughout the sensor network area and uses assistant cluster head to balance the energy consumption. Our future work includes exploring more detailed parameters and factors that affect system performance to further improve energy efficiency, system robustness and scalability.

REFERENCES

[1] J.Yick, B.Mukherjee, D.Ghosal. Wireless sensor network survey [J]. Computer Networks, 2008, 52 (12): 2292-2330.

[2] I.F.Akyildiz, W.Su, Y.Sankarasubramaniam, et al. Wireless sensor networks: A survey [J]. Computer Networks, 2002, 38 (4): 393-422.

[3] A.A.Abbasi, M.Younis. A survey on clustering algorithms for wireless sensor net-

works [J]. Computer Communications, 2007, 30 (14-15): 2826-2841.

[4] W.B.Heinzelman, A.P.Chandrakasan, H.Balakrishnan. An Application-Specific Protocol Architecture for Wireless Microsensor Networks [J]. IEEE Trans.Wireless Communications, 2002, 1 (4): 660-770.

[5] Smaragdakis, G., Matta, I., AND Bestavros.A.SEP: A stable election protocol for clustered heterogeneous wireless sensor networks [R]. In Second International Workshop on Sensor and Actor Network Protocols and Applications, 2004.

[6] S.Lindsey, C.Raghavendra, K.M.Sivalingam. Data Gathering Algorithms in Sensor Networks using Energy Metrics [J]. IEEE Trans.Parallel and Distributed Systems, 2002, 13 (9): 924-935.

[7] S.D.Muruganathan, D.C.F.MA, et al. A Centralized Energy-Efficient Routing Protocol for Wireless Sensor Networks [R]. IEEE Radio Communications, 2005.

[8] O.Younis, S.Fahmy. HEED: A Hybrid, Energy-Efficient, Distributed Clustering Approach for Ad Hoc Sensor Networks [J]. IEEE Trans. Mobile Computing, 2004, 3 (4): 366-379.

[9] R.Wang, G.Liu, C.Zheng. A Clustering Algorithm based on Virtual Area Partition for Heterogeneous Wireless Sensor Networks [R]. In Proc.IEEE Mechatronics & Automation, 2007.

[10] AbdelSalam H.S., Olariu, S.BEES: Bio-inspired backbone Selection in Wireless Sensor Networks [J]. IEEE Trans. Parallel and Distributed Systems, 2012, 23 (1): 44-51.

[11] Ammari H.M., Das, S.K..Centralized and Clustered k-Coverage Protocols for Wireless Sensor Networks [J]. IEEE Trans. Computers, 2012, 61 (1): 118-133.

[12] V.Raghunathan et al. Energy-Aware Wireless Microsensor Networks [J]. IEEE Signal Processing Magazine, 2002, 1 (2): 40-50.

[13] T.Rappaport. Wireless Communications: Principles & Practice [M]. Englewood Cliffs, NJ: Prentice-Hall, 1996.

[14] S.Reddy, C.R.Murthy. Dual-Stage Power Management Algorithms for Energy Harvesting Sensors [J]. IEEE Trans. wireless communications, 2012, 11 (4): 1434-1445.

[15] A.B.da Cunha, D.C.da Silva. Behavioral Model of Alkaline Batteries for Wireless Sensor Networks [J]. IEEE Latin America Transactions, 2012, 10 (1): 1295-1304.

[16] A.Wang, W.Heinzelman, A.Chandrakasan. Energy-scalable protocols for battery-operated microsensor networks [R]. In Proc. 1999 IEEE Workshop Signal Processing Systems (SiPS'99), 1999.

[17] H.M.Ammari, S.K.Das. Centralized and Clustered k-Coverage Protocols for Wireless

Sensor Networks [J]. IEEE Trans.Computers, 2012, 61 (1): 118-133.

[18] J.Lee, W.L.Cheng. Fuzzy-Logic-Based Clustering Approach for Wireless Sensor Networks Using Energy Predication [J]. IEEE Sensors Journal, 2012, 12 (9): 2891-2897.

[19] Z.Huang, H.Okada, K.Kobayashi et al. A study on cluster lifetime in multi-hop wireless sensor networks with cooperative MISO scheme [J]. Journal of Communications and Networks, 2012, 14 (4): 443-450.

[20] X.Gao, Y.Vanq, D.Zhou. Coverage of communication-based sensor nodes deployed location and energy efficient clustering algorithm in WSN [J]. Journal of Systems Engineering and Electronics, 2012, 21 (4): 698-704.

Wireless Sensor Networks' Application Research Progress

Based on the summary of Wireless Sensor Networks' characteristics, we analyze the Wireless Sensor Networks' applications and its development trends, and put forward the factors that restrict the development of Wireless Sensor Networks. In this paper, we analyze the Wireless Sensor Networks development opportunities and challenges in the future, and provide some new ideas and enlightenment for the comprehensive development and widely used.

1 INTRODUCTION

Wireless Sensor Networks (WSNs) has many characteristics such as low cost, high precision, easy to operate, etc. It is consisted of a number of cheap sensor nodes that is deployed in the monitoring areas, these nodes has sensing, measurement, communication and computing ability[1-2]. Due to wireless sensor networks have irreplaceable advantages, it has widely used in many applications, such as military, environmental monitoring, industrial and agricultural production, medical health, building detection, space exploration and other fields, in these area, wireless sensor networks has made great values. In recent years, the concept of Internet of things has been put forward, and has closely relationship with wireless sensor networks. As the support technology of the Internet of things, it undertakes the sensing and data transmission tasks, and there are some factors that restrict its development, so wireless sensor networks face challenges and opportunities.

Many research institutes at home and abroad have carried out technology and application researches on wireless sensor networks. We focus on wireless sensor networks' application technology, their developing trends and summarize the restricted factors.

2 WIRELESS SENSOR NETWORK NODE'S STRUCTURE AND CHARACTERISTICS

Wireless sensor node's structure. The structure of wireless sensor node is shown in Fig. 1, it is consisted of 4 units, sensing module (A/D conversion, sensor), data storage, control and processing module (storage components, micro-processor unit), wireless communication module (wireless transceiver components, network protocol stack), and power supply module (lithium battery or dry cell), this is the basic functional units of wireless sensor network. Sensing module is responsible for collecting the sensing data within the scope of monitoring area, and transforms the information according to the user needs; data storage con-

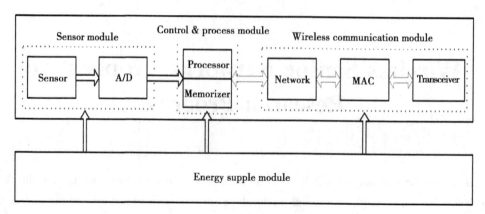

Fig. 1 Wireless sensor network node's structure

trol module takes charge of coordinate and control behaviors of the sensor nodes, and store the collected data information; wireless communication module takes mainly charge of information interaction between wireless sensor nodes, sensing data collection and the control and management commands receive/send; power supply module is batteries commonly, and provide the energy that the nodes needed[3-4].

The characteristics of wireless sensor network. The characteristics of wireless sensor network are as follows[5-6]:

(1) Nodes with limited resources. Node miniaturization leads to limited battery energy supply, and the limited energy only satisfies limited node resources.

(2) Self-organization. Wireless sensor network nodes have self-organization functions, the nodes can communicate with each other, and they don't need to rely on any support infrastructure.

(3) Multi-hop routing. Sensor node is limited by power, energy saving technology and communication distance, they can communicate directly in a limited area. If a node communicates with other nodes out of its communication scope, they must route through neighbor nodes.

(4) Large number of nodes deployed. Wireless sensor network deploys a large number of sensor nodes in a certain area to maintain the network's fault tolerance and anti-destroying ability.

3 THE APPLICATION STATUS OF WIRELESS SENSOR NETWORK

Wireless sensor network has wide application field and prospects, its development and application will bring huge influence in human life and production. As the brilliant application prospect of wireless sensor network, many countries carried out researches and applications in military, industry and academic area[7-13].

The applications in military field. As wireless sensor network has the characteristic of low cost, rapid deployment, intensive distribution and high fault-tolerant, etc., that makes it very suitable for application in military conditions, such as reconnaissance enemy condition, invasion detection, monitoring enemy's army deployment, material transformation and the determination of biological attacks, etc. Typical application is to spread a large number of micro sensor nodes in the broad battlefield area use airplane, these nodes form network using Ad-hoc method, collect a number of information in the battle field, then classify and send the information to the headquarters, these information are tactical support to make the decision.

The applications in environment monitor. People have paid more and more attention to the environment today, the environment monitor has widely application, such as air pollution detection, water pollution detection, indoor environment detection, plant growth environment, precision agriculture detection, etc. Wireless sensor network can finish the tasks that traditional system can't complete in environment detections.

Australian scientists detect the distribution of the toad in north Australia using wireless sensor networks, as the toads' sounds are unique, it is very effective method that use sound as the detection characters. Researchers use signal process on the sensor node, then the data that send to the control center are minimal, finally the distribution of the toad is got through analysis.

Intel has built the world's first wireless vineyard in Oregon, sensor nodes are distributed in every corner of the yard, and detect the oil temperature, humidity and harmful things in the areas every minute to ensure the grape grow in a health environment. The researchers have found that subtle variations in climate change can affect the wine quality greatly. We can grasp the relationship of the grape wine quality with the sunshine, temperature and humidity of the environment where the grape grow up according to the data and analysis we have got for years.

The applications in Healthcare area. University of California has proposed a human health monitoring platform CustMed that based on wireless sensor network, the platform adopts sensor nodes that can be wear and can detect pressure, the reaction and temperature, etc. The nodes called "dot-mote" are developed by university of California, Berkeley, and produced by Crossbow Co., we can examine our body's status through a pocket PC easily.

Intel research center is working on home care system based on wireless sensor networks, the system is one part of the "deal with an aged society technology project". The system places sensor nodes in the shoes, furniture and household appliances, help the older and the handicapped to help themselves, the system can sense the actions of the older and record it, and give attention or warning when need, provide safety to the older and help them by medical staff and social workers.

Building detection applications. Wireless sensor network can be used in the detection of the buildings status, it has low cost and can solve the traditional wire layout's shortcoming such as wire aged and easy to be damaged.

The University of Southern California has carried out a building detection system called

"NETSHM" using wireless sensor network, the system can also locate the damage position. The system was deployed in "The Four Seasons" building, at Los Angeles.

Fujitsu announced they have a study on how to install earthquake detection sensor nodes in the buildings with American Xerox company in Dec. 2004, the want to provide better warning mechanism to the residents.

4 THE APPLICATION RESTRICTING FACTORS

In the design of wireless sensor network, we should prolong the network lifetime and use limited resources. The lifetime of single sensor nodes is not the critical factors, how to maintain the network's connectivity and prolong the network lifetime is the main target. Currently, there has several factors that constrains wireless sensor network.

The cost. This is the problem we should consider first in wireless sensor network applications. In wireless sensor network applications, we usually need a large number of micro sensors, the cost is the primary problem we should consider. With the development of SOC (System On Chip), the cost will gradually reduce in a single node with communication function and processing function integrated on a chip.

Nodes and reliable communication. Wireless sensor network is a self-organized networks, the communication between nodes can be reached directly, so we must use routing technology sometimes. So we should choose the appropriate node connection mode, network topology and routing protocol according to the different applications is the core problem.

Energy supply. Sensor nodes use battery to provide energy supply, we should provide stable power energy supply to ensure the system running. The battery energy shortage problem is outstanding due to the restrictions of the sensor nodes' size. The solutions to resolve the problem includes research high performance battery, reduce sensor node's energy consumption and self-energy collection technology, etc.

Multi-sensor information fusion. With the development of sensing technology and the application area expanding, sensor network has gradually diversified, and the data usually comes from different nodes or different sensors in a node, that bring a challenge to the data management and transmission work.

5 CONCLUSION AND PROSPECT

Wireless sensor network has extremely broad application prospect, it has great application values in military, environment monitoring, agriculture production detect, health care and other new emerging areas such as smart home, intelligent transportation, space exploration, etc. With the development of the Internet of things, wireless sensor network will integrated into every aspect of our lives, and will get deep into every corner of human industrial and agricultural production and even in the forefront of space exploration.

But we should clearly realize that the technology and application of wireless sensor network is far from mature, and there still many factors that restrict the development of wireless sensor network. These restrict factors are both challenges and opportunities, we should grasp these opportunity and intensify research and development, lead and promote the development of the industry.

REFERENCES

[1] Jennifer Yick, Biswanath Mukherjee, Dipak Ghosal.Wireless sensor network survey [J]. Computer Networks, 2008, 52 (12): 2292-2330.

[2] Akyildiz I, Su W, Sankarasubramaniam Y, et al.Wireless sensor networks: a survey [J]. Computer Networks (Elsevier) Journal, 2002, 38 (4): 393-422.

[3] Mauri Kuorilehto, Marko Hannikainen, Timo D.Hamalainen.A Survey of Application Distribution in Wireless Sensor Networks [J]. EURASIP Journal on Wireless Communications and Networking.2005, 5 (5): 774-788.

[4] Zoran.Bojkovic, Bojan.Bakmaz.A survey on wireless sensor networks deployment [J]. World Scientific and Engineering Academy and Society, 2008, 7 (12): 1172-1181.

[5] Dietrich I., Dressler F.On the lifetime of wireless sensor networks [J]. Trans.Sen. Netw.2009, 5 (1): 5-39.

[6] Kohvakka M., Suhonen J., Kuorilehto M., et al.Energy-efficient neighbor discovery protocol for mobile wireless sensor networks [J]. Ad Hoc Networks, 2009 (2): 24-41.

[7] Li D, Wong KD, Hu YH, et al.Detection, classification, and tracking of targets [J]. IEEE Signal Processing Mag, 2002, 19: 17-29.

[8] He T, Krishnamurthy S, Stankovic JA, et al.An energy-efficient surveillance system using wireless sensor networks [R]. MobiSys'04, Boston, MA, 2004.

[9] http://www.coe.berkeley.Edu/labnotes/0701brainybuildings.html.

[10] Kintner Meyer M, Brambley M R.Wireless Sensors: How Cost-Effective Are They in Commercial Buildings.PNNL-SA-36839.

[11] WernerAllen G, Johnson J, Ruiz M, et al.Monitoring Volcanic Eruptions with a Wireless Sensor Network [R]. In Second European Workshop on Wireless Sensor Networks, Istanbul, Turkey, 2005.

[12] HuW, Tran VN, Bulusu N, et al.The Design and Evaluation of a Hybrid Sensor Network for Cane toad Monitoring [R]. In Proceedings of Information Processing in Sensor Networks (IPSN 2005/SPOTS 2005), Los Angeles.

[13] Noury N, Herve T, Rialle V, et al.Monitoring behavior in home using a smart fall sensor [R]. In: Proceedings of the IEEE EMBS Special Topic Conference on Micro technologies in Medicine and Biology, Lyon: IEEE Computer Society, 2000.

第四篇

项目资料

茶园水肥一体自适应调控模式研究

一、项目背景与总体思路

茶叶是我国重要的经济作物，也是世界三大无酒精饮料作物之一。山东省是江北产茶大省，目前茶园面积已达38.30万亩，产量1.89万吨，茶叶产值达到31.5亿元。随着"山东茶"市场份额的不断扩大，山东省茶园面积还在不断拓展，茶产业发展的规模效益不断显现。根据《山东省茶叶发展规划（2014—2020）》目标，到2020年，全省茶园面积将发展到50万亩，茶叶产量3万吨，干毛茶总产值将达60亿元以上。

然而，由于山东地处我国茶区最北端，属于纬度最高的产茶区，干旱缺水、肥料利用率低等问题一直困扰着山东省茶产业的发展。水分与肥料的高效低耗、资源节约和环境友好是茶叶优质高产与可持续发展的重点课题之一。我们以茶叶优质高效安全生产为目标，以设施茶园水分与肥料高效利用、环境与食品安全和资源节约为重点，研究开发感知采集影响茶树生长指标的物联网设备，针对茶树在不同生长期开展环境因子、茶树蒸腾速率和土壤指标之间的相关关系进行研究，建立各指标间的动态模型，研究水肥一体按需精量自适应调控模式，使水肥利用率显著提高，建立茶叶优质高产示范基地，加强技术培训，提高技术转化率，实现茶叶优质高效安全生产与可持续发展。

关于物联网技术在茶园方面的应用研究，目前的文献主要集中在有关物联网技术在农业上的应用，柳平增等（2011）研究了物联网的农业生产过程智能控制系统；戴起伟等（2012）研究了面向现代设施农业应用的物联网技术模式；郭理等（2014）研究了物联网的农业生产过程智能控制架构；刘佳海等（2014）研究了物联网在农产品流通信息化应用；周静怡等（2104）研究了面向物联网的农田数据传输中间件；冷波等（2014）研究了基于物联网技术的智慧茶园控制技术；美国、英国、日本、以色列等农业现代化程度较高国家也进行了物联网技术在农业方面的研究，但是茶园物联网技术应用方面的研究较少，结合茶树自身及生长环境采用物联网技术进行水肥一体自适应调控研究尚属空白，未见有关茶园水肥一体自适应调控模式方面的研究报道。

在茶树环境—蒸腾—土壤指标模型研究方面，Maga'n et al.（2008）研究了水肥一体灌溉条件下，营养液中盐度含量对番茄果实和品质的影响；Gallardo et al.（2009）研究了水肥一体栽培条件下，番茄蒸腾、排水、氮素吸收的模型；Gallardo et al.（2009）将作物生长模型和作物蒸腾模型有机组合模拟了温室番茄的氮素吸收和氮素流失量；国内孙霞等（2004）结合基质栽培中营养液的pH值、浓度等参数具有时变、非线性的特点，采用自适应模糊控制器对营养液的调配供液进行控制；刘佳（2008）利用pH值和EC值调节营养液；曾祥国等（2012）研究了3种营养液配方对草莓植株生长、光合特性、产量及品质的影响，以筛选出最适宜草莓生长的营养液配方；但国内外未见有茶树环境—蒸腾—土壤指标动态关系模型方面的研究，在该动态模型基础上的茶园水肥一体自适应调控专家决策系统研究目前尚属空白。

在茶园水肥一体自适应灌溉系统研究方面，于舜章等（2009）构建了设施黄瓜水肥一体化滴灌技术，与畦灌冲肥做了比较；李家念等（2012）研制了柑橘园水肥一体化滴灌自动控制装置，该装置即可进行水肥一体滴灌，又可进行清水滴灌；袁洪波等（2014）研究了日光温室水肥一体灌溉循环系统，设计了温室水肥一体灌溉循环利用系统，构建回收管路，收集过量的水肥并实现循环利用；以上水肥一体系统仅用于作物水肥滴灌，未与物联网智能感知装备结合，不能按照茶树环境—蒸腾—土壤指标动态模型进行水肥一体自适应调控精量供给。我们提出的基于物联网的设施茶园水肥一体自适应调控模式及系列技术应用在国内均属新技术，填补了该领域的空白。

另随着人们对食品安全的日益关注，劳动力成本的增加，以及国外对农产品溯源制度的严格要求，我们研究开发的环境智能感知装备与技术将逐步为农业资源、环境、生产管理系统和农产品消费者采用，省去了农产品生长环境现场手工监控的烦琐，节省人力物力的投入，有效降低监控成本，提高工作效率和监控智能化水平，有利于实现农业集约化生产，满足现代农业要求；应用水肥一体自适应调控系统，省工节水，且水肥自适应同步管理调控，杂草生长少，灌水施肥与中耕除草节省大量劳力，减少水肥的流失与浪费，提高水分和肥料利用率，使茶树处于适宜生长环境中，提高茶叶的产量。茶园水肥一体自适应调控技术是节约资源、保护环境、高产、高效四位一体的农业关键技术，从节约成本与提高产量两个方面将产生巨大的经济效益，因此极具推广价值。

为此，该项目以茶叶优质高产为目标，以肥料与水分的高效低耗、资源节约和环境友好为研究重点，针对生产中存在的大水漫灌、化学肥料过量施用等问题，研究开发了感知茶树生长环境指标的物联网设备，针对茶树在不同生长期开展环境因子、茶树蒸腾速率和土壤指标之间的相关关系进行研究，建立各指标间的动态模型，研究水肥一体按需精量自适应调控模式，使水肥利用率显著提高。

我们研究应用的茶园环境信息感知装备与水肥一体自适应调控系统具有良好的先进性，属于低成本新技术，能提高农业产业的技术含量，为设施农业的智能管理打下良好的基础，同时可以有效提高农业产业各环节的工作效率，促进现代农业向低能耗、高效率的方向发展。在山东省茶叶产区青岛、日照、临沂、泰安、威海等地建立了2处茶园水肥一体化示范基地，并通过加强技术培训，提高技术转化率，带动周边大力发展茶园水肥一体化技术，产生了广泛的社会效益和生态效益。

二、项目主要创新点

创新点一：解析了温室小气候环境因子对茶树蒸腾速率的影响关系，对基于修正彭曼-蒙特斯方程（M-P-M方程）的茶树蒸腾速率进行修正和优化，确立了茶树蒸腾速率计算模型。

1. 作物系数 K_c 的确定

根据FAO推荐，在充分供水条件下作物需水量计算公式如下：

$$ET_c = K_c \cdot ET_0$$

式中，ET_c 为作物需水量（充分供水条件下作物腾发量，mm）；K_c 为作物系数。

FAO-56 作物系数表中列出了不同作物 Kcini、Kcmid、Kcend 的典型值，根据 Allen 研究，K_c 的变化主要由其具体的作物特征来决定，受气象条件的影响有限。作物系数最重要的影响因素是作物自身的生理生态指标，例如作物种类、品种、生育阶段以及冠层状况等。而近年来，我国学者研究发现，温室作物系数比大田环境下有变小的趋势，我们在综合研究分析茶树特征、茶园实验基地气象信息的基础上，最终确定茶树生长阶段的作物系数为 0.95。

2. 温室茶树蒸腾蒸发模型 ET_0（Tea）确定

由于温室小气候环境与露天环境的水热运移模式有很大不同，Boulard、Demrati 等提出利用温室能量平衡和 Penman-Monteith 方程（P-M 方程），推导出基于室内气象数据的温室作物蒸腾量计算模型；王健、陈新明等从温室内总辐射和风速因子入手，提出适于参考作物腾发量的 P-M 温室修正式；刘浩等建立了包含气象数据、叶面积指数和冠层高度为主要参数的日光温室番茄蒸腾量估算模型。但是针对于茶树这一经济作物，其在温室内的蒸腾蒸发模型的研究目前还相对较少。我们采用多元线性回归分析等方法，解析了温室小气候环境因子对茶树蒸腾速率影响关系，提出了适于北方地区温室茶树腾发量的计算模型 ET_0（Tea）。

在 P-M 公式和 P-M 温室修正式中假设作物高度是 $h_c = 0.12$m，而实际作物高度是一个时间变量，所以在计算不同作物腾发量时不能忽略作物实际高度这一参数。为了提高计算温室实际作物腾发量的精度，本文选取温室茶树作为实际作物进行试验，在公式中引入作物高度 h_t 这一参数。结合 P-M 方程和 P-M 温室修正式，推导出适于温室茶树的 ET0（Tea）计算方法：

$$ET_0(\text{Tea}) = \frac{0.408\Delta(R_n - G) + \gamma \dfrac{39661(e_a - e_d)}{(T + 273)\left[\ln\left(\dfrac{Z - 0.64h_t}{0.13h_t}\right)\right]^2}}{\Delta + \gamma + \dfrac{15\gamma}{\left[\ln\left(\dfrac{Z - 0.64h_t}{0.13h_t}\right)\right]^2}}$$

式中，h_t 为茶树冠层高度（m）；其他参数意义同前。

根据汪小旵等的方法计算 e_a、e_d、Δ：

$$e_a = 0.6107\exp[17.4T/(239 + T)]$$

$$\Delta = 4158.6 e_a T/(T + 239)^2$$

$$RH = \frac{e_d}{e_a} \cdot 100\%$$

式中，RH 为温室内空气相对湿度（%）。

根据强小嫚的研究可知土壤热通量（G）和地表净辐射通量（R_n）具有很好的线性关系。

白天：$G = 0.1R_n$

夜晚：$G = 0.5R_n$

3. 作物腾发量的逐日变化与日变化

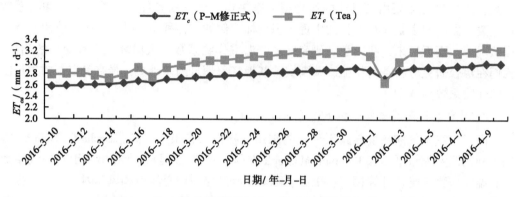

图1 ET_c（P-M修正式）及 ET_c（Tea）的逐日变化图

利用P-M温室修正式和温室茶树蒸腾蒸发模型ET_0（Tea）对茶园温室2016年3月10日至4月10日的作物腾发量进行逐日统计，其变化过程如图1所示。

从图1可以看出，2种方法计算出的ET_c值具有相同的变化趋势，且总体上看，3月至4月茶树腾发量随气温的变暖呈现升高趋势；P-M温室修正式的计算值几乎均小于温室茶树蒸腾蒸发模型的计算值，这是由于P-M温室修正式中参考作物为苜蓿，在计算式中假设植株高度$h_c=0.12m$，远低于'黄山种'茶树，且总叶数量和总叶面积同样小于'黄山种'茶树；在3月13日、3月17日和4月2日的蒸腾蒸发量均出现不同程度的下降，推测其原因是这三天的天气状况均是阴天，且3月17日伴有雾出现，4月2日阴转小雨，太阳辐射遭到不同程度的削弱。

为了检验温室茶树蒸腾蒸发模型的计算精度，选取试验时间段内具有代表性的2天，3月13日（阴天）和4月5日（晴天）进行分析，由于清晨温室内湿气较大，作物腾发基本从9：00左右开始，所以选择当天10：00—17：00（7h）的实测数据绘制图2，利用P-M温室修正式和温室茶树蒸腾蒸发模型ET_0（Tea）对作物ETc进行计算，两种方法的计算结果与实测值进行对比分析。

从图2（a）可以看出ET_c（P-M修正式）、ET_c（Tea）和ET_{ca}这三者的变化趋势基本一致，而ET_c（Tea）和ET_{ca}的数值更为接近，阴天条件下，作物腾发从10：00一直增大，13：00左右达到峰值，之后开始平缓下降，16：00后下降显著；但总的来说试验时间段内蒸腾蒸发量的变化幅度不是很大。从图2（b）可以看出，晴天条件下ET_c（P-M）、ET_c（Tea）和ET_{ca}这三者的变化趋势也基本一致，但试验时间段内腾发量明显高于阴天，ET_c（Tea）和ET_c重合度也更好，峰值出现在13：00左右。

4. 基于手机的远程温湿度监控系统及方法

传统的监测系统和方法多采用铺设线缆的方式进行数据采集，工程量大、成本高，智能化程度低、误差大，因此现有技术还有待于改进和发展。我们针对现有技术的上述缺陷，提供一种基于手机的远程温湿度监控系统及方法，旨在解决现有技术智能化程度

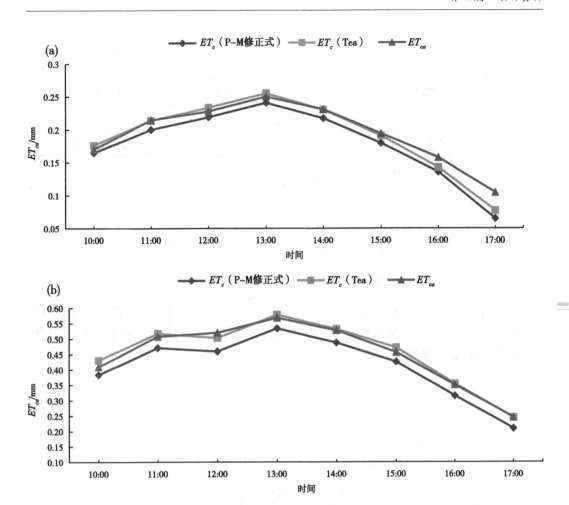

图2 阴天（a）和晴天（b）ET_c（P-M修正式）、ET_c（Tea）及ET_{ca}的变化图

低、误差大的问题。

发明公开了一种基于手机的远程温湿度监控系统及方法，包括若干个温度传感器和湿度传感器、无线监测节点、田间控制站、喷雾单元、喷雾采集单元、加热单元、加热采集单元、服务器以及手机终端，温度传感器和湿度传感器连接无线监测节点，喷雾采集单元和加热采集单元分别连接喷雾单元和加热单元，并将采集的数据也发送到无线监测节点，田间控制站收集无线监测节点发送的数据并通过无线网络发送给服务器，手机终端根据服务器发送的温湿度数据远程控制喷雾单元和加热单元。通过将采集到的温湿度数据发送到手机终端，从而实现远程监控，智能化程度高，精度高。系统结构示意图如图3所示。

创新点二： 提出了一种无参数小生境粒子群优化算法；优化了环境信息智能感知节点拓扑结构、节点间信息传输方式，在此基础上，研究开发了低成本、高效的设施茶园环境信息智能感知节点和网关节点（图4）。

1. 提出了一种无参数小生境粒子群优化算法

粒子群算法可描述如下：在 D 维搜索空间中，由 m 个粒子组成的一粒子群体，群

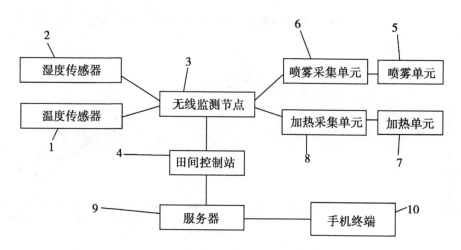

图3　基于手机的远程温湿度监控系统

体中的每个粒子以一定的方向和速度飞行，每个粒子在飞行搜索的过程中，考虑群体（邻域）内其他粒子的历史最优点和自身曾经搜索到的历史最优点，在此基础上进行位置、速度等状态的变化。

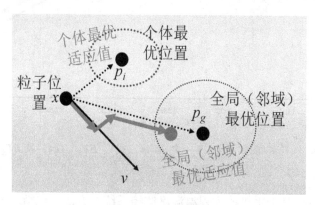

图4　粒子群算法

小生境受大自然的启示，来自生物学的一个概念。在自然生态系统中，物种个体为了生存竞争必须承担不同的角色，不同的物种依据环境进化成不同小生境（子群体），而不是仅进化成一个群落中毫不相干的个体。换句话说，生物在进化过程中，通常会与自己相同的物种生活在一起，共同繁衍生息。

小生境都是在某一特定的区域中，比如北极熊不能生活在热带，大熊猫不能生活在北极等，把里面包含的思想提炼出来，运用到智能优化中去，保持群体内物种的多样性，利于优化算法探索能力与开发能力的平衡，提高算法的效率。

小生境技术将来源于大自然中的小生境概念应用到各种智能优化算法中去，在智能优化算法中，每一代划分为若干子群体，每个子群体中选出一些最优适应值的个体，作为子类的优秀代表组成群，然后在种群中，或者不同的种群之间进行个体的变异、杂交等产生新一代个体群，同时可辅助分享、排挤、选择等机制协助完成任务。

目前众多小生境构造方法存在严重的问题，算法的性能严重依赖于小生境参数的设置，这对于普通用户来说比较困难。在 fitness sharing 中的共享参数 σ_{share}、species conserving GA（SCGA）中的种群距离 σ_s、clearing 中的距离测量 σ_{clear}、speciation-based PSO（SPSO）中的种群半径 r_s 等，均定义了不同的距离量度参数，Shir 等试图采用固定小生境半径、Bird 和 Li 等试图降低小生境半径对结果的敏感性，但是参数依然存在或者引入了新的参数，我们提出了一种采用环形结构的小生境构造方法，该方法不引入任何参数，相比以上小生境的构造方法，该方法具有明显的优势，且性能优良。

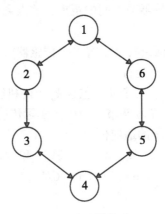

图 5　小生境构成

我们利用左右邻居节点的构成小生境，如图 5 所示，该小生境的构造方法不需要参数，且取得了较好的性能，但是我们认为仅依靠左右节点序号的邻居信息，在粒子的社会经验的学习与传播方面过于局限，为此我们将左右邻居进行重新定义，定义左邻居为节点左侧半径范围内的任意节点，右邻居为节点右侧半径范围内的任意节点，我们定义的左右邻居节点能增加一定的干扰性，使粒子在飞行时受社会经验吸引更大或者逃离局部最优的可能加大，从而增加求解的质量。

根据以上的描述，算法的实现过程可以用以下的伪代码来表示：

```
begin
粒子群初始化；
While（not met termination criterion）{
    for（i=1；i<=Population Size；i++）
    if（fit(x⃗ᵢ) > fit(p⃗ᵢ)）
    x⃗ᵢ -> p⃗ᵢ；
    for（i=1；i<=Population Size；i++）
    neighborhoodBest(p⃗ₗ, p⃗ᵢ, p⃗ᵣ) -> p⃗ₙ,ᵢ；
    for（i=1；i<=Population Size；i++）{
        运用公式（4-1）、（4-2）对粒子进行更新
    }
}
```

End

2. 优化了环境信息智能感知节点拓扑结构

设计了一种基于小生境粒子群优化的自适应分簇协议（an Adaptive Clustering Protocol using Niching Particle Swarm Optimization，ACP-NPSO），我们充分利用小生境粒子群算法的优良特性，充分考虑影响能耗的因素，根据网络中节点的具体情况进行分簇，进一步均衡了能量消耗，延长了系统的生存周期。

我们利用基于小生境粒子群优化的自适应分簇协议 ACP-NPSO 进行分簇的时候，充分考虑网络的状态，比如传感器节点间的距离、节点剩余能量、以及节点与汇聚节点的距离等因素。

在如图 6 所示无线传感器网络模型中有 N 个传感器节点，我们利用公式计算出最优簇头数量 k_o，然后按照如下步骤运行：

（1）随机初始化 m 个粒子，每个粒子包含 k_o 个随机簇头信息
（2）将每个粒子包含的 k_o 个簇头节点对应至实际传感器节点
（3）利用公式计算每个粒子的适应值

针对存活节点 $n_i(i=1,2,3...,N)$，计算 n_i 和所有簇头节点之间的距离

若满足 $d(n_i,CH_{p,k}) = \min_{\forall k=1,2,...,k_o}\{d(n_i,CH_{p,k})\}$，将 n_i 分配到簇头 $CH_{p,k}$ 对应的簇中

运用公式计算粒子的适应值

（4）寻找个体和小生境内粒子最优适应值。个体最优指的是该粒子运行过程中曾经的最优适应值；小生境最优是在我们构造的小生境内粒子的曾经（历史）最优适应值。并对最优适应值进行更新。
（5）运用公式更新粒子的位置与速度
（6）限制粒子位置和速度的更新
（7）重复（2）至（6）直到达到终止条件或者达到最大迭代次数

算法的流程图如图 7 所示。

3. 实验结果分析

我们评估 ACP-NPSO 的性能，对仿真实验结果进行分析，并与 LEACH、PSO-C 分簇协议进行比较。

从图 8 中我们可以看出 LEACH 所形成的分簇数量过多，这是由于簇头选择的随机性造成的，因为在 LEACH 中每个传感器节点根据一定的概率决定是否成为簇头，具有很强的随机性。PSO-C 所形成的分簇较 LEACH 所形成的分簇有所改进，但与 ACP-NPSO 相比较则存在大小不一的情况。这是因为 ACP-NPSO 充分考虑了网络的状态来决定最优簇头数量，然后根据节点的剩余能量、节点间的距离以及节点与基站的距离来决定最优簇头的选择，因此形成的分簇分布均匀，保持最优簇头数量，降低网络的能量消耗，提高系统的性能。

无线传感器网络中存活节点的数量随时间变化如图 9 所示。从图 9 中我们可以看到，我们提出的 ACP-NPSO 能明显延长网络的生存时间，且首个节点耗尽能量的时候也明显延迟，这是由于我们提出的分簇协议根据网络的状态决定最优簇头数量，然后运

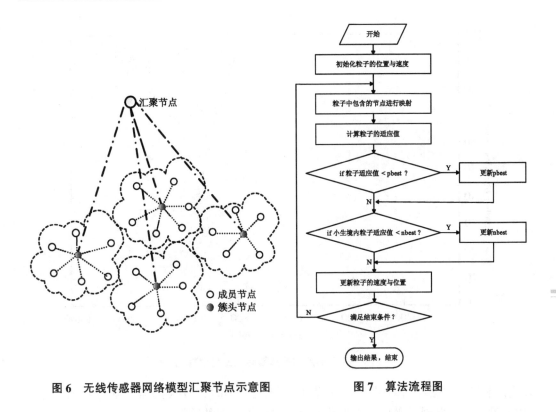

图6 无线传感器网络模型汇聚节点示意图　　图7 算法流程图

图8 LEACH 分簇数量示意图

用小生境粒子群算法进行最优簇头的选择，且簇头的选择充分考虑了节点的剩余能量、节点间距离等因素，使簇头分布更加合理，从而减少能量消耗。而 LEACH 协议每个节点根据一定的概率决定是否成为簇头，具有较强的随机性，且簇头选择未考虑节点剩余能量以及节点间的距离等因素，从而簇头数量具有较强的随机性，且分簇不均匀；PSO-C协议采用粒子群优化分簇选择，但是簇头数量未根据网络状态动态变化，且未考虑节点与基站间的距离影响，从而导致能量消耗较高。我们提出的 ACP-NPSO 协议能避免以上缺点，从而能明显延长网络的生存时间。

4. 开发了低成本、高效的设施茶园环境信息智能感知节点

设备定时自动采集环境信息（环境温度、环境湿度、光照强度、土壤湿度等），并传输至物联网云平台。环境智能感知节点与装备采用太阳能电池板供电，锂电池存储的

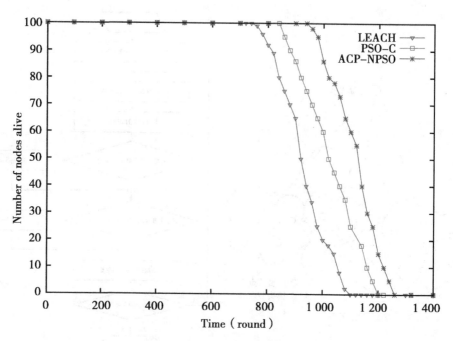

图 9 存活节点的数量随时间变化示意图

方式，采集的数据定时自动（目前设置为每隔 1 小时，可根据需要随意设置）通过 GPRS 发送至物联网云平台，为大数据分析提供了一定的数据。

5. 基于运动匹配的视频自动标注方法及自动标注系统

在现实生活中时常会遇到需要对实时拍摄的视频进行标注，如对特定目标进行实时监护视频拍摄的同时对视频内特定目标进行标注，在大型游乐场所、景区实时拍摄的游客游览、纪念视频进行标注以便游客检索、提取，但是现有的视频标注多是对已有视频、根据视频内容进行训练、匹配再进行标注，目前尚没有利用目标运动状态对视频内容进行自动语义描述的方法。针对上述问题，本发明提供了一种基于运动匹配的视频自动标注方法，其能对实时拍摄的视频自动同步实时标注。为此，本发明还提供了一种基于运动匹配的视频自动标注系统（图10）。

本发明提供了一种基于运动匹配的视频自动标注方法，其能对实时拍摄的视频自动同步实时标注；其特征在于：采集运动目标的目标运动视频，同时采集运动目标运动时附带的运动传感器模块的传感器数据，再分别对目标运动视频、对传感器运动数据分别进行视频运动特征、传感器运动特征的提取，然后将视频运动特征、传感器运动特征进行特征匹配分析，最后根据所述特征匹配分析结果对目标运动视频进行视频标注。一种基于运动匹配的视频自动标注系统，其包括用于采集运动目标的传感器运动数据的运动传感器模块；视频采集模块；运动匹配模块；其中，运动匹配模块与运动传感器模块、视频采集模块之间均通过无线通信模块进行数据通信。该方法的流程图如图 11 所示。

6. 一种计算机通讯设备用电缆及其制备方法

计算机的通信都是采用电缆传输，现有的通信电缆都是设置在户外，对电缆的耐气

图 10 视频自动标注示意图

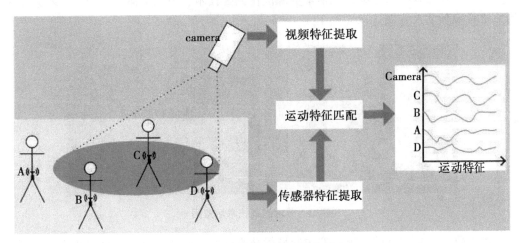

图 11 视频自动标注流程图

候性有很高的要求，但是现有的电缆耐气候性还达不到人们的预期，通信电缆的阻燃问题也是需要引起关注的问题，电缆在燃烧时释放出大量的烟雾和有毒、腐蚀性气体，是火灾中的危险因素，在火灾中妨碍人们的安全撤离和灭火工作，如何解决上述技术问题成为本领域技术人员努力的方向。

针对以上问题，项目组发明了一种计算机通讯设备用电缆及其制备方法，本发明原料来源广泛，制备成本低，本发明制备的成品耐寒性能、耐热性能和阻燃性能好，不会在户外低温条件下冻裂，保证通信线路的稳定运行，避免了通信电缆因外界环境因素造成伤害而影响通信线路运行的现象。

创新点三：确立了多参数耦合条件下的关系模型的方法和策略，建立了茶园灌溉模式，开发了茶园水肥一体智能决策专家系统；发明设计了土壤多参数测量装置及方法、土壤养分智能化原位监测系统、基于手机的远程温湿度监控系统及方法、一种基于运动匹配的视频自动标注方法及自动标注系统。

1. 提出了茶树水肥按需供给自适应调控模式，并开发了专家系统

研究确立了多参数耦合条件下的关系模型的方法和策略，建立最优调控模式数学模型，得出茶树不同生育阶段水肥一体供应量、供应时间和供应次数等多种模式的控制策略，提出茶树水肥按需供给自适应调控模式。

通过该项目研究，共建立了3种茶园灌溉模式：①微喷带喷灌与滴灌模式；②茶树基部滴灌模式；③茶树顶部微喷模式。这三种模式中，①②两种模式可以同时实现滴灌施肥。以上3种灌溉模式均可在茶园水肥一体智能决策专家系统的调控下运行。通过研究发现，利用上述措施，可以实现节水22%~25%，节肥23%~23%，茶叶产量比对照提高18%~25%（详细数据参照5、产量统计）。

节水灌溉技术是比传统的灌溉技术明显节约用水和高效用水的灌水方法、措施和制度等的总称。灌溉用水从水源到田间，到被作物吸收、形成产量，主要包括水资源调配、输配水、田间灌水和作物吸收四个环节。在各个环节采取相应的节水措施，组成一个完整的节水灌溉技术体系，包括水资源优化调配技术、节水灌溉工程技术、农艺及生物节水技术和节水管理技术（图12）。

图12　节水灌溉技术示意图

国外研究者早就开始寻求农作物的最优灌溉制度。从20世纪60年代开始，研究者就已对作物水模型进行了大量的研究，以其与产量的关系给出规划模型，用以寻求供水条件下农作物的最优灌溉制度。线性规划（Linear Programming，简称LP），是处理线性目标函数和线性约束的一种较为成熟的方法。Ghahraman和Sepaskhah用NLP模型求解了不同效益费用比时冬小麦的缺水灌溉问题。Flin和Musgrave（1967）论证了DP法应用于有限灌溉水量时求解最优灌溉制度的可行性，并以阶段初的可供水量为状态变量

建立了相应的 DP 模型。

我们以茶园应用为例，开发了设施作物灌水量控制专家系统，在简介茶叶相关知识及茶树的耗水规律，茶树适时喷灌的技术指标的基础上，设计完成了设施作物灌水量控制专家系统。

农业物联网信息管理系统以创新产业核心技术、探索产业技术集成应用模式、培育相关新兴产业为合作重点，开展技术交流、服务、项目合作，全面覆盖了农业物联网领域的纵向创新链条与横向产业领域。

探索产业技术集成应用模式。开展物联网技术在农业资源利用、农业生态环境监测、特色农牧业精细生产管理、农机作业导航与调度、农产品/食品安全溯源等领域的集成应用，构建中国智慧农业系统。

建立农业物联网研发基地、物联网产业化中试基地，建立物联网核心技术产品生产线，培育和壮大一批农业物联网相关企业，促进物联网技术产业化。我们设计完成了农业物联网信息管理系统。

2. 发明设计了土壤多参数测量装置及方法

现有的土壤多参数测量装置功能单一，数据传输的可靠性低，很难完成更大范围的网络覆盖，信息很难实现共享，最终监测系统之间形成了若干个"信息孤岛"，缺乏规范的土壤参数数据发布平台，导致数据的利用效率低。我们针对现有技术的上述缺陷，提供一种土壤多参数测量装置及方法，旨在解决现有技术智能化程度低、误差大的问题。

本发明包括若干个温度传感器、湿度传感器、盐分传感器、pH 值传感器、无线监测节点、田间控制站、喷雾单元、喷雾采集单元、加热单元、加热采集单元、服务器以及手机终端，温度传感器、湿度传感器、盐分传感器和 pH 值传感器连接无线监测节点，喷雾采集单元和加热采集单元分别连接喷雾单元和加热单元，并将采集的数据也发送到无线监测节点，田间控制站收集无线监测节点发送的数据并通过无线网络发送给服务器，手机终端根据服务器发送的温湿度数据远程控制喷雾单元和加热单元。通过将采集到的数据发送到手机终端，从而实现远程监控，智能化程度高，精度高。本发明结构示意图如图 13 所示。

3. 土壤养分智能化原位监测系统

目前国内外针对土壤养分监测主要有实验室监测，但是实验室监测需要将样本带回到实验室中进行监测，在运输过程中土壤内部的水分以及各种养分的含量可能因为外界环境变化的影响导致监测的数据不精确，如果是在原地进行监测则会浪费大量的人力资源，从而增加成本。针对现有技术的不足，发明了一种土壤养分智能化原位监测系统，具备智能化监测成本较低等优点，解决了土壤监测过程中采集数据不准确以及成本较高的问题。

发明一种土壤养分智能化原位监测系统，包括外壳，所述外壳的侧面开设有通孔，所述通孔的内部套装有信息采集探头，所述信息采集探头贯穿通孔且延伸至外壳的内部，所述信息采集探头延伸至外壳内部的外表面套装有弹簧，所述弹簧延伸至外壳内部的一端固定连接有防脱块，所述外壳内部的固定连接有置物板，所述置物板的轴心处套

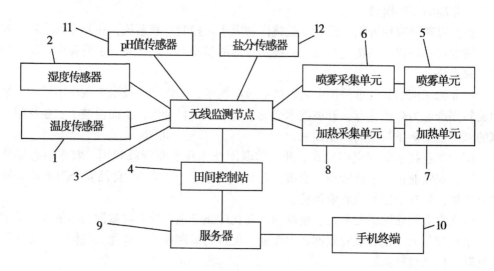

图 13　土壤多参数测量装置结构示意图

装有传动轴，所述传动轴的一端固定连接有导向板，所述传动轴远离导向板的一端与电机的输出端固定连接，所述电机的顶部固定连接有保护板。该土壤养分智能化原位监测系统，通过设置了一体成型的带有锥形底的外壳，能够使得外壳可以方便快捷的插入土壤，从而进行较好的固定与采集工作。发明的系统外壳剖面示意图如图 14 所示。

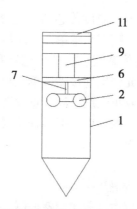

图 14　土壤养分智能化原位监测系统外壳剖面

创新点四：集成基于物联网的茶园水肥一体化技术与应用，在物联网设备的自动调控下，对茶树进行水肥一体化管控，建立了茶园水肥一体化技术体系。提出了水肥一体化技术规程，制定了山东省地方标准《茶园水肥一体化操作技术规程》（DB37/T 2898—2016），并制定了企业技术标准 2 项。

1. 物联网技术
2. 水肥耦合技术
3. 水肥一体化技术规程
4. 有机肥+水肥一体化技术

(1) 针对不同品种、不同季节、不同制茶种类，研制不同的茶树专用水溶性肥料及其配方。

(2) 茶园专用水肥一体化技术模式创新：研究茶园水肥一体化高效低耗综合利用理论与技术，创建"1管2行"水肥供给模式，实现"低压供肥、高压供水"。

(3) 针对北方茶园土壤pH值普遍偏高的现状，开展水肥一体化对茶园土壤养分变化与茶树吸收规律的研究，填补该领域研究空白。

5. 建立了茶园水肥一体化技术体系，并制定了企业技术标准

通过对水肥一体化技术研究，建立了茶园水肥一体化技术体系，并制定了山东省地方标准《茶园水肥一体化操作技术规程》（DB37/T 2898—2016）；制定了企业技术标准2个，《农业环境监控管理系统》（Q/GQCY 002—2017）、《智慧茶园水肥一体管理系统》（Q/GQCY 001—2017）。

水肥一体化试验产量统计信息（表1至表12）。

表1 27个处理8次茶叶干重

处理	I	II	III	IV	V	VI	VII	VIII	合计
W1N1S1	12.20	15.00	8.00	19.80	25.62	22.00	14.80	11.00	128.42
W1N1S2	12.60	15.00	14.80	30.20	31.00	18.00	13.80	10.80	146.20
W1N1S3	14.80	20.80	14.40	36.60	34.12	25.00	18.20	10.20	174.12
W1N2S1	14.80	18.60	17.20	38.80	40.08	21.80	16.60	8.80	176.68
W1N2S2	15.20	17.00	15.20	39.20	43.20	22.00	21.40	11.00	184.20
W1N2S3	18.00	14.80	17.00	35.80	35.82	24.20	17.00	8.40	171.02
W1N3S1	15.60	11.60	12.00	25.20	34.69	19.20	11.60	15.20	145.09
W1N3S2	14.60	13.40	19.60	24.20	48.30	15.20	25.60	13.40	174.30
W1N3S3	12.00	14.00	8.20	25.20	17.11	13.00	15.20	13.60	118.31
W2N1S1	13.60	27.80	12.20	42.80	26.75	27.80	19.00	6.20	176.15
W2N1S2	11.80	18.40	16.80	43.60	25.05	26.00	19.80	14.40	175.85
W2N1S3	13.80	22.80	33.96	80.20	29.02	42.20	23.00	8.20	253.18
W2N2S1	15.40	22.60	28.40	28.00	29.30	38.60	18.50	8.00	188.80
W2N2S2	11.40	17.40	17.20	22.40	21.37	31.80	15.00	11.80	148.37
W2N2S3	12.60	24.40	16.60	33.40	32.71	35.40	17.60	8.80	181.91
W2N3S1	16.00	22.00	21.60	26.80	34.12	35.60	16.80	11.00	183.92
W2N3S2	15.80	20.20	19.40	35.20	57.65	34.00	17.20	11.80	211.25
W2N3S3	15.80	19.60	27.40	57.80	17.40	24.60	26.20	13.00	201.80
W3N1S1	11.60	29.80	29.00	41.60	31.29	42.00	31.20	8.60	225.09
W3N1S2	12.20	31.60	35.60	55.80	37.53	37.60	37.20	8.80	256.33

(续表)

处理	I	II	III	IV	V	VI	VII	VIII	合计
W3N1S3	11.20	24.40	33.00	58.40	29.87	29.00	33.60	8.20	227.67
W3N2S1	16.40	27.40	24.00	47.60	37.81	34.00	29.60	9.20	226.01
W3N2S2	10.60	19.00	22.80	40.00	30.72	41.20	24.00	8.00	196.32
W3N2S3	13.60	23.60	26.40	76.40	22.78	22.20	16.60	14.40	215.98
W3N3S1	12.20	19.00	20.60	36.20	15.41	25.40	24.60	12.20	165.61
W3N3S2	17.60	19.20	24.00	50.00	51.13	27.40	20.40	12.80	222.53
W3N3S3	15.60	26.60	24.00	53.20	38.09	36.60	20.00	17.00	231.09

表2　27个处理8次茶叶干重的方差分析表

变异来源	平方和	df	均方	F值	显著水平
区组	18 413.12	7	2 630.45	48.36	0.00
W	2 095.30	2	1 047.65	19.26	0.00
N	86.07	2	43.04	0.79	0.45
S	179.92	2	89.96	1.65	0.19
W×N	509.86	4	127.46	2.34	0.06
W×S	234.64	4	58.66	1.08	0.37
N×S	504.55	4	126.14	2.32	0.06
W×N×S	530.96	8	66.37	1.22	0.29
误差	9 900.37	182	54.40		
总和	32 454.80	215			

表3　3个灌水处理茶叶干重间的SSR检验

处理	均值	5%显著水平	1%极显著水平
W3	27.31	a	A
W2	23.91	b	B
W1	19.70	c	C

表4　3个N肥处理茶叶干重间的SSR检验

处理	均值	5%显著水平	1%极显著水平
N1	24.49	a	A
N2	23.46	a	A
N3	22.97	a	A

表5 3个施肥处理茶叶干重间的SSR检验

处理	均值	5%显著水平	1%极显著水平
S3	24.65	a	A
S2	23.82	a	A
S1	22.44	a	A

表6 8次茶叶干重（X1~X8）与总干重（Y）间的相关系数

相关系数	X1	X2	X3	X4	X5	X6	X7	X8	Y	显著水平P
X1	1.000	-0.122	-0.016	0.034	0.531	-0.102	-0.218	0.162	0.151	0.452
X2	-0.122	1.000	0.636	0.531	0.019	0.709	0.616	-0.393	0.750	0.000
X3	-0.016	0.636	1.000	0.702	0.066	0.638	0.737	-0.225	0.883	0.000
X4	0.034	0.531	0.702	1.000	-0.064	0.324	0.457	-0.048	0.790	0.000
X5	0.531	0.019	0.066	-0.064	1.000	0.100	0.057	-0.012	0.335	0.087
X6	-0.102	0.709	0.638	0.324	0.100	1.000	0.413	-0.432	0.671	0.000
X7	-0.218	0.616	0.737	0.457	0.057	0.413	1.000	-0.318	0.693	0.000
X8	0.162	-0.393	-0.225	-0.048	-0.012	-0.432	-0.318	1.000	-0.201	0.315
Y	0.151	0.750	0.883	0.790	0.335	0.671	0.693	-0.201	1.000	0.000

表7 8次茶叶干重（X1~X8）与总干重（Y）间的通径系数

因子	直接	→X1	→X2	→X3	→X4	→X5	→X6	→X7	→X8
X1	0.058		-0.018	-0.003	0.015	0.150	-0.024	-0.039	0.012
X2	0.146	-0.007		0.132	0.228	0.005	0.165	0.109	-0.030
X3	0.208	-0.001	0.093		0.302	0.019	0.149	0.131	-0.017
X4	0.430	0.002	0.078	0.146		-0.018	0.075	0.081	-0.004
X5	0.283	0.031	0.003	0.014	-0.028		0.023	0.010	-0.001
X6	0.233	-0.006	0.103	0.133	0.139	0.028		0.073	-0.032
X7	0.177	-0.013	0.090	0.153	0.196	0.016	0.096		-0.024
X8	0.075	0.009	-0.057	-0.047	-0.021	-0.003	-0.101	-0.056	

表8 27个处理8次茶叶平均干重（g）

处理	Ⅰ	Ⅱ	Ⅲ	Ⅳ	Ⅴ	Ⅵ	Ⅶ	Ⅷ	合计
W1N1S1	0.81	2.14	1.33	2.20	2.33	2.44	2.47	1.10	14.83
W1N1S2	0.84	2.14	2.47	3.36	2.82	2.00	2.30	1.08	17.00
W1N1S3	0.99	2.97	2.40	4.07	3.10	2.78	3.03	1.02	20.36
W1N2S1	0.99	2.66	2.87	4.31	3.64	2.42	2.77	0.88	20.53

(续表)

处理	I	II	III	IV	V	VI	VII	VIII	合计
W1N2S2	1.01	2.43	2.53	4.36	3.93	2.44	3.57	1.10	21.37
W1N2S3	1.20	2.11	2.83	3.98	3.26	2.69	2.83	0.84	19.74
W1N3S1	1.04	1.66	2.00	2.80	3.15	2.13	1.93	1.52	16.24
W1N3S2	0.97	1.91	3.27	2.69	4.39	1.69	4.27	1.34	20.53
W1N3S3	0.80	2.00	1.37	2.80	1.56	1.44	2.53	1.36	13.86
W2N1S1	0.91	3.97	2.03	4.76	2.43	3.09	3.17	0.62	20.97
W2N1S2	0.79	2.63	2.80	4.84	2.28	2.89	3.30	1.44	20.97
W2N1S3	0.92	3.26	5.66	8.91	2.64	4.69	3.83	0.82	30.73
W2N2S1	1.03	3.23	4.73	3.11	2.66	4.29	3.08	0.80	22.94
W2N2S2	0.76	2.49	2.87	2.49	1.94	3.53	2.50	1.18	17.76
W2N2S3	0.84	3.54	2.77	3.71	2.97	3.93	2.93	0.88	21.58
W2N3S1	1.07	3.14	3.60	2.98	3.10	3.96	2.80	1.10	21.74
W2N3S2	1.05	2.89	3.23	3.91	5.24	3.78	2.87	1.18	24.15
W2N3S3	1.05	2.80	4.57	6.42	1.58	2.73	4.37	1.30	24.82
W3N1S1	0.77	4.26	4.83	4.62	2.84	4.67	5.20	0.86	28.06
W3N1S2	0.81	4.51	5.93	6.20	3.41	4.18	6.20	0.88	32.13
W3N1S3	0.75	3.49	5.50	6.49	2.72	3.22	5.60	0.82	28.58
W3N2S1	1.09	3.91	4.00	5.29	3.44	3.78	4.93	0.92	27.36
W3N2S2	0.71	2.71	3.80	4.44	2.79	4.58	4.00	0.80	23.84
W3N2S3	0.91	3.37	4.40	8.49	2.07	2.47	2.77	1.44	25.91
W3N3S1	0.81	2.71	3.43	4.02	1.40	2.82	4.10	1.22	20.53
W3N3S2	1.17	2.74	4.00	5.56	4.65	3.04	3.40	1.28	25.84
W3N3S3	1.04	3.80	4.00	5.91	3.46	4.07	3.33	1.70	27.31

表9 27个处理8次茶叶平均干重的方差分析表

变异来源	平方和	df	均方	F值	显著水平
区组	284.42	7	40.63	54.95	0.00
W	39.26	2	19.63	26.55	0.00
N	2.51	2	1.26	1.70	0.19
S	2.71	2	1.35	1.83	0.16
W×N	8.23	4	2.06	2.78	0.03

(续表)

变异来源	平方和	df	均方	F值	显著水平
W×S	4.17	4	1.04	1.41	0.23
N×S	6.96	4	1.74	2.35	0.06
W×N×S	6.72	8	0.84	1.14	0.34
误差	134.58	182	0.74		
总和	489.56	215			

表10　3个灌水处理茶叶平均干重间的SSR检验

处理	均值	5%显著水平	1%极显著水平
W3	3.33	a	A
W2	2.86	b	B
W1	2.28	c	C

表11　3个N肥处理茶叶平均干重间的SSR检验

处理	均值	5%显著水平	1%极显著水平
N1	2.97	a	A
N2	2.79	a	A
N3	2.71	a	A

表12　3个施肥处理茶叶平均干重间的SSR检验

处理	均值	5%显著水平	1%极显著水平
S3	2.96	a	A
S2	2.83	a	A
S1	2.68	a	A

农村土地管理信息系统

一、项目背景

中国必须依法维护农民的土地承包经营权，增强集体经济实力，增加农民在土地增值收入中的比重。基于此，我们应积极推进农村土地管理的信息化建设，在现代化建设和全面建设小康社会进程中发挥越来越重要的作用。2012年，农业部发布了"全国农村管理信息化发展规划（2013—2020）"，强调了重点建设土地流转信息共享服务和监管平台，逐步实现土地流转信息，在线服务和在线监管的在线分配。农业部"2013年农村经营管理要点"还提到，要制定全国农村土地承包信息化建设规划，指导农村土地承包经营信息系统的建立，同时，农村土地权的登记和认证正式纳入基层工作。2013年，中央一号文件要求五年内基本完成农村土地承包经营权的登记和认证，妥善解决农民承包地面积不准确、面积不清的问题。近年来，国家不断出台新文件，规范农村土地经营权的管理和指导。

目前，通过加快农村产权改革步伐，我们将积极引导农民，农业产业化龙头企业，新的企业实体等。关注重点农业产业，以具体项目为载体，通过股权、租赁、分包、转让等多种方式进行土地流转。在土地集中集约化、适度规模经营等方面进行积极有效的探索，取得了良好的成效。

以青岛为例，2012年年底青岛市土地流转面积为75万亩，占耕地总面积的12%，其中中等规模经营面积超过100亩规模的有20万亩。根据规划，到2018年，城市中的农村土地管理总面积占耕地面积的比例可达35%以上。通过土地承包经营权转让实现土地集约经营是农村土地的必然趋势，交易量的持续增长，合同管理权转让的管理要求也越来越高。因此，强调建立土地承包经营权管理信息系统的必要性。

目前在土地经营权流转工作确权颁证、交易服务、纠纷处理等方面有了良好的成效，不少农户通过土地流转实现了自己的事业抱负。但是，在实际应用中，还存在服务对象范围小、管理系统不规范、效率低等问题。因此急需建立规范、高效、广泛应用的土地流转管理平台，改善管理，使更多的农民和土地承包商获得更方便，更优质的服务，加快城乡统筹协调发展的速度。

全国各城市在计划中提出加快推进和建立市、县、乡转让农村土地承包经营权的市场体系。加强与农村产权交流的合作与交流，做好农村土地承包经营权交易的服务、监督和指导。积极创建条件，建立覆盖全区县的农村土地承包经营权管理信息系统。

二、项目意义

农村土地承包经营权信息管理信息系统依托传统农村土地信息系统的运作方式，依托信息应用软件的优势，为土地流转、确权登记、参谋分析、行政监管、纠纷仲裁等相关业务提供技术支撑，实现农村土地工作管理的信息化，为涉及农村土地的问题提供全面的信息处理机制。

（一）经济意义

充分发挥平台媒体的引导和支撑作用，信息的丰富性和全面性，传播的便利性和时空性，沟通的开放性和互动性，促进农村土地市场健康发展，实现农村土地的优化配置和流通，建立完整的农村土地承包经营权监督管理制度。将农业生产转变和提升到机械化、集约化、现代化和高效率的方式，解放农村生产力，振兴农村生产要素，它将带来巨大的经济和社会效益。

（二）政治意义

农村土地问题关系农民的切身利益，社会的稳定与和谐，国家的长治久安。平台通过推动信息化在行政管理部门、农民群体中的普及作用，促成上级部门调动管理和基层参与的统筹部署，落实中共中央对农村经济社会发展的要求，为农民朋友土地流转提供良好的服务，把农村土地所有权确权到每一位农民，逐步形成农村土地承包经营制度，产权明晰，权力明确，流通畅通，权益保护，分配合理。它对农村社会稳定、经济社会稳定发展、党群关系和谐发展具有重要的积极意义。

（三）社会意义

平台建设以高效的沟通机制，提高农村土地管理利用水平，强化农村土地物权意识，在现代化、工业化、城镇化进程中农民的利益得到维护，解决农村土地流转、权利确认、所有权等问题，为构建统一的城乡农村土地市场，实现城乡发展，推进农村基层建设提供动力。

农村土地管理信息系统主要以农村土地承包经营权信息管理为重点，以当前农村土地权登记工作为契机。集中部署农村土地信息化应用软件系统为突破口，全力打造一个覆盖市、县（区）、乡（镇、街道）、村四级应用，为农村土地承包信息管理系统提供标准化的权证管理和多种转移方式。它将在平台上发布土地信息和相关信息，为各级农村土地工作提供全面的信息支持。为确保确权成果应用的效果，该系统包括农村承包土地信息管理系统，合同管理权纠纷仲裁系统，土地基础信息管理系统和合同管理权交易系统的应用平台。实现综合服务，为农村土地安全成果创建新的应用模型，可根据不同地域、不同时期、不同客户群的特点，立足于个性化的建设机制，实现一站式服务。

三、系统总体设计

（一）整体系统分层管理原则

土地管理信息系统围绕承包经营权进行管理，总体原则是"整体规划、分层管理"，其中"整体规划"是指统一规划、统一设计开发信息管理系统；"分层管理"是指按市、县（区）、乡（镇，街）和村的四级实现。"市级监管、区级处理、镇级采集、村级查询"的分层次管理。

系统的构建分为两个阶段，即确定信息化管理阶段和管理权的管理流程（图1）。

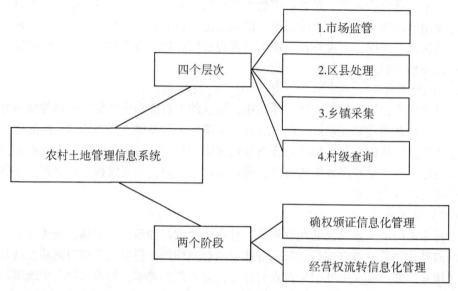

图1 系统分层管理架构

（二）以"市，县（区），乡（镇，街），村"四级应用模式为框架

农村土地管理系统框架包括市政监督、县（区）级管理、乡（镇，街）级入口、村级查询四级应用模式。市级监督主要是市级农业主管部门，如农业局（或农业委员会）；县（区）级管理主要是县（区）级农业局；镇级进入主要是土地流转管理，主要管理部门是乡镇经济管理审计中心；村级应用主要功能是实现对本村承包土地的信息查询（图2）。

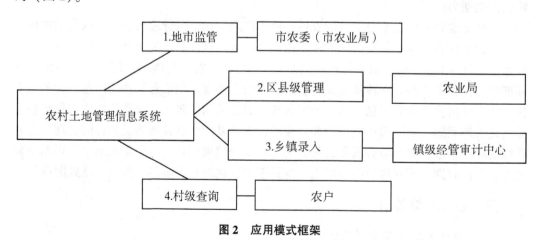

图2 应用模式框架

农村土地管理制度着重于农村土地承包经营权的管理，是一个综合的应用平台。各级农业管理部门通过平台进行土地流转工作，并根据自身职能处理相关业务。通过该平台实现业务处理，提高工作效率，工作更加透明化智能化。各级各类土地流转数据可进

行自动汇总，供市级相关部门管理领导查询、统计、分析等，提供决策依据。

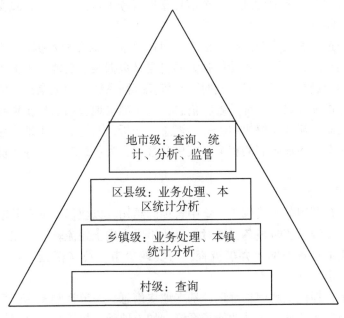

图3　农业管理部门权限

与土地经营权流转工作相关的各级农业主管部门，在该平台上享有不同的使用权限（图3），具体权限如下。

地市级管理部门：对所辖区县内的数据进行统计、分析、监管，从中挖掘出有用有价值的信息，为农村土地承包经营权管理提供宏观的指导管理，对土地确权、流转、承包、合同签订、纠纷仲裁等方面起到协调、公证、监督作用。

区（县）级管理部门：针对各区县所属的乡镇数据进行汇总、存储，对各乡镇信息进行统一管理，集中存储，统一上报。系统可按照农村土地的类别、时间、流转方式等进行分类汇总，能够查看各种查询条件下的具体信息，达到信息共享，便于查询管理的目的。

在乡（镇、街道）级管理部门，主要是对辖区内村级数据进行录入和采集，确保基础数据采集工作的权威性、系统性、全面性，集中建立一个准确及时、综合性高、信息量大、实时性强的数据采集平台，提供一个沟通顺畅的信息化渠道，方便各级用户使用。

（三）以业务管理能力为支撑

农村土地信息管理系统着重于农村土地承包经营权的管理，涵盖城市、县（区）、乡镇（城镇，街道）和村庄。系统通过应用层和表现层的融合内网与外网应用系统的无缝贯通，通过一体化方式为各级政府提供农村土地管理的便捷应用方式。

1. 外网应用平台

作为网站门户平台，我们设置"首页、需求信息、转出信息、政策文件、经验分享、流转规程、合同模板、流转动态、新闻报道、土地流转大厅、网上仲裁"等模块，平台根据不同的属性提供在线发布功能。该网站作为根基农村土地管理部门的门户来使用，是集合信息发布、土地管理、一个信息管理服务平台，集成了在线交易、行政审查和经验共享等功能。

外网平台通过对相关资源的整合汇总，对内网业务的授权数据进行实时查询抽取。同时出于安全性考虑，系统设置功能强大的自定义数据共享功能，根据各级用户的实际需求可对内网用户进行适当公开。同时平台依托外网系统在数据信息上的传播优势，为广大农民群众提供涉及农村土地的各种信息，及时准确地发布农村土地政策法规，规范指导农村土地交易、流转及解决纠纷等，汇总、分析、对比农村土地实用信息，自动选择、统计、发布全国各地政府的相关政策信息数据，为各级政府管理人员、农经系统提供决策参考。

2. 内网应用平台

内网管理平台利用分布式应用，集中管理的结构，以数据传输和共享管理技术为核心，实现全面、逐级、实时报备，通过信息化手段完成土地流转、确权、仲裁等系列工作，建立农村土地管理的网上办事流程，为实现"市，县（区），乡（镇，街），村"四级应用模式奠定基础。

内网管理平台提供土地信息登记、基本地块信息、情节视频图像信息、土地承包经营权证书、土地档案信息、各种功能子系统，如土地承包权转让、合同管理、仲裁争议管理、流通信息发布和统计分析，它涵盖了土地承包经营权管理的关键内容。其中，土地经营权转让管理职能模块提供转移、分包、租赁、持股、借贷、交换、灵活互动等多种土地流转方式，以满足不同管理的需要（图4）。

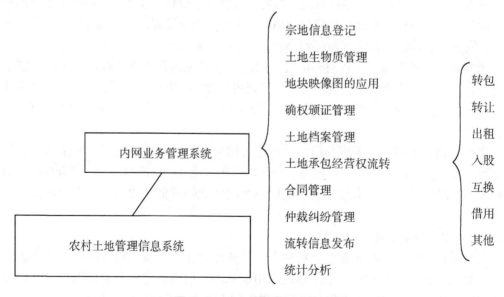

图4 Intranet 服务管理系统功能

（四）以先进的技术手段为保证

农村土地承包经营权信息管理采用一系列先进技术作为项目建设的保障。所有用户无需安装任何软件即可使用浏览器管理 LAN 或 Internet 上的操作。针对数量众多的用户，我们拟采用 Oracle 大型数据库作为存储手段，确保用户的服务质量。管理平台可结合 Web 地图来直观显示每个地块的地理位置，实现地块相关信息的查询等各种功能。管理平台还提供定制工作流程的能力，以快速满足每个用户的个性化需求；另管理平台的移动应用将明显提升用户的操作体验，提供更加便携方便的操作（图 5）。

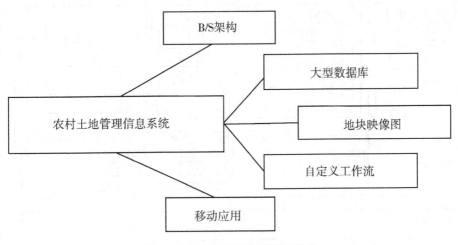

图 5　农村土地管理信息系统技术

四、技术解决方案

（一）基于 SOA 的技术架构

面向服务的体系结构 SOA 可以说是领先的集成应用程序体系结构。通过业务服务的概念，提供 IT 的基本应用功能，这些服务可以自由集成，相互理解和安排。同时，我们可以随时适应未来和新的需求，例如，在 SOA 框架下，使用标准接口封装项目信息检索，各种指标的查询，用户权限认证等，它作为服务发布，然后作为服务部署在集成了系统数据和功能的平台上。任何想要访问另一个应用程序的应用程序都可以通过服务发现或服务表示来确定所访问服务的属性和调用格式，实现标准化应用程序之间的协作，同时满足不同应用程序系统之间的松散耦合原则，以避免由于单方面程序而导致的内部或系而对另一方应用造成冲击。

利用已建立的 SOA 架构，可以实现城市、区域和县公司的信息服务以及各种业务系统，可以是以服务形式发布的 IT 资产，以松散耦合的方式共享，并且可以快速集成各种服务。开发出多种组合应用，实现"整合与开发"效果，实现快速，及时的业务需求响应（图 6）。

基于 SOA 的应用系统可以轻松地与其他应用系统交互，从而消除信息孤岛现象。

终端用户界面集成

应用程序连接

流程集成

信息集成

构建集成开发模型

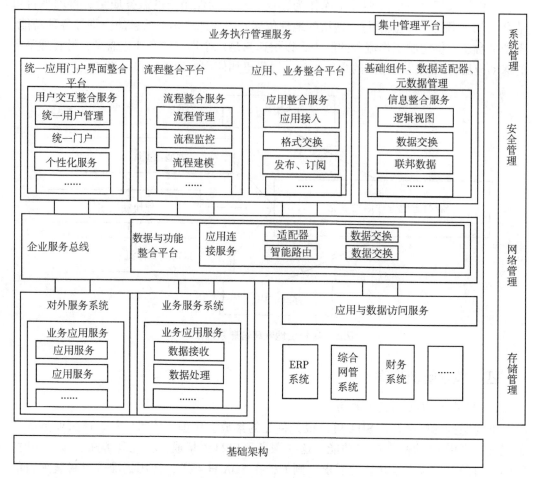

图 6 基于 SOA 技术架构

（二）系统架构

农村土地管理信息系统采用 B/S 架构，客户端通过 Web 模式进行业务操作，如图 7 所示。操作终端可以分为两类，定制终端和通用浏览器终端。定制终端需要安装特殊的客户端软件，以完成系统的管理，查询和统计等操作；通用浏览器终端只需要安装通用浏览器软件，主要执行日常操作和查询统计分析功能。

（三）应用集成

应用程序集成解决方案可以多个级别和多种形式呈现。企业应用程序集成的级别取决于多种因素，包括业务类别、系统大小、预算、项目复杂性或应用程序集成。目前，农村土地管理信息系统和其他系统应用的集成解决方案可以采用以下一种或多种方法。数据同步的导入和导出方法，数据同步的集成应用程序引擎，集成业务处理的工作流技术，使用数据仓库分析重要的业务数据并与中间件技术集成，下面对这五种方式分别进行介绍。

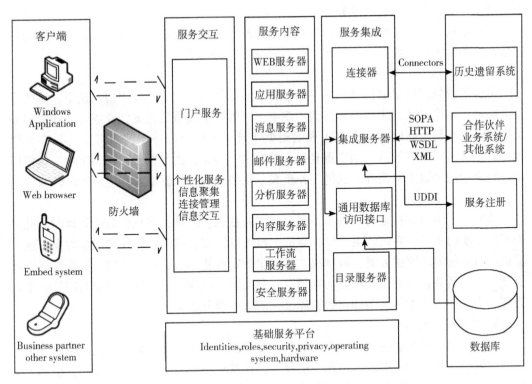

图 7　农村土地管理信息系统架构

1. 导入导出方式

土地转让管理平台提供了强大的数据交换工具，可通过数据交换向导进行配置，以交换任何支持的格式和常规数据。实现与其他相关信息系统的数据交换。土地流转管理系统的报告，清单和文件可以根据用户的需要直接转换为标准文件。

导入和导出时，可以定期自动操作或手动操作。手动操作可以根据需要对导入和导出的数据进行适当的修正，便于灵活的数据交换。

土地转让管理平台提供修改、控制、存储、审核和分析数据的界面。

2. 应用集成引擎

HUB-SPOKE（Wheel Hub）架构是解决应用程序集成问题的最常用方法，通常与"发布—订阅引擎"结合使用。轮毂包括适配器（AAPTER，也称为协议转换器）和集线器（HUB，也称为数据路由器或发布引擎）。

适配器负责历史系统的接口，将原始形式的信息（例如 CICS，TCP/IP，甚至字符终端信息）转换为集线器可以读取的通用规范；适配器还可用于将数据格式转换为通用格式，如 ASCII 或 UNICODE，适配器的另一个最重要的任务是转换数据结构。

数据路由器或集线器将数据发送到目标系统，确保数据在传输过程中不会丢失。集线器提供路由表，该表提供用于标识和映射目标系统和源系统的规则，实现发布—订阅引擎以及对请求的智能过滤。源系统（发布者）发布数据和更新，而目标系统（订阅者）请求新数据或特殊类型的数据更新。集线器根据规则自动将数据传输到他感兴趣的位置，并执行相应的管理任务。

该方法确保一个系统中的数据更新可以反映在另一个系统中，并且可以连接新的订户系统而不影响发布者系统的操作，且不需要太多的人工干预。

3. 工作流技术

工作负载处理的集成涉及识别，监视和更改业务流程，通常使用模式模型。该技术将流经系统的信息和组织边界映射到图像中，以创建集成解决方案。业务处理通常很复杂，并且不容易重新绘制为人工生成的模型工具。工作流程技术擅长有序的程序化过程，其显著特征是有效执行预定重复任务的要求。但是，大多数组织在多个部门和组织中都有复杂的多级业务流程，这种工作负载系统很难满足数据处理和应用程序同步。

4. 数据仓库

数据仓库可以在指定时间内从不同系统捕获数据。但是，在执行数据处理时，此方法不支持实时事务和系统集成。数据集成工具可以读取、移动、转换和加载数据，通常提供定时或批处理解决方案，以解决大量数据传输或初始加载仓库问题，数据集成工具仅允许在加载数据或直接读取时在物理层复制数据。这样，当数据在特定时间从一个地方移动到另一个地方时，不提供数据库级别的增值业务逻辑。

5. 中间件技术

中间件是各种企业应用程序集成解决方案提供商为应用程序系统提供的集成方法。

数据访问技术：不更改应用程序层程序，提供对不同数据源的直接访问，例如 ODBC、JDBC、ADO 和其他标准数据接口。

基于消息的中间件技术（MOM）：MOM 提供了一种异步信息传送机制。MOM 中间件在所有需要集成的应用程序系统中安装 MOM 的 API 调用程序，以便将消息传递到中间件进行处理，优点是发送方和接收方不必在线等待。

远程过程调用（Remote Procedure Call）：使用客户端/服务器模型，双方需要在线同步，等待收件人回复以继续工作，因此等待将减慢，但降低错误的可能性。

服务技术（Web Service）：Web 服务已成为系统集成和异构系统的首选技术。通过集成信息标准访问，可以使用 Web 服务技术连接和部署内部和合作伙伴信息以及所需信息，提供服务的系统以 Web 服务的形式发布业务流程，调用者需要根据协议调用所需的服务，并且每个服务都需要。

农村土地管理信息系统涉及多个业务平台的集成，包括业务集成和数据集成，各地部署的环境条件和应用场景存在重大差异。单一集成方法不能适应所有情况，需要根据实际情况采用不同的集成方法，或者混合使用各种集成技术来实现系统集成。

（四）系统安全

计算机系统的许多安全问题是由构成系统的某些环境或缺陷引起的，并且可以通过技术手段来改进，这包括基于系统硬件平台的安全措施（如热备份、数据灾难恢复、数据备份、关键组件冗余、网络操作系统选择、网络防火墙部署和设置、网络防病毒、入侵检测、系统管理、安全评估等，以及基于应用的安全措施（如访问控制、安全认证、日志审计等）。

系统安全性主要从以下几个方面考虑：数据备份、系统容灾、系统容错、安全操作和维护、安全监控以及安全事件响应和恢复。

　　一般而言，信息平台的不安全因素是由计算机系统和人为因素引起的。实施安全系统需要采取管理措施，以最大限度地减少人为因素的破坏性，并通过可行的技术手段减少自然因素对系统的危害。

测土配方施肥专家系统

一、项目背景

测土配方施肥专家系统的主要服务对象是农民、肥料经销人员、农业管理人员、农业专家等。系统集成农业相关部门的土壤养分检测数据和专家的测土配方施肥模型，对农民的地块进行建档、配方施肥模型建立、农户生产过程进行管理和服务，针对不同的用户，能提供相应的服务。对于特定的用户，系统可以为其建立账户，提供针对农户地块的测土配方施肥指导意见，为其建立施肥日志；对于肥料经销人员，主要是帮助农户进行管理；对于农业管理人员，该系统可以提供土地空间数据库管理和土壤养分信息管理，便于管理者维护和更新数据管理和维护功能，如绘图属性数据。同时系统可与政府部门的其他系统对接，根据农户地块档案信息对农户地块的信息数据进行管理；对于农业专家，可通过系统平台查看农户的施肥情况（日志），根据农户的地块档案等信息和种植情况，给出有价值的指导。

随着网络技术和智能手机的在农村的普及应用，系统提供了多终端支持（PC、触摸屏、智能手机、PAD等），用户体验感更强，并在操作的易用性、简便性、便捷性上下功夫，针对农业朋友的操作习惯、方式等均进行了详细的调研，做了专门针对性的设计。

二、项目意义

土壤测试和配方施肥作为精准农业和智能农业的重要组成部分，党和国家高度重视土壤测试和配方施肥的应用和推广。自2010年以来，农业部特别提出，"进村到田"的土壤测试和配方施肥项目应该作为一项关键任务。决定自20世纪80年代以来，在我国开展土壤测试和施肥的综合研究和开发，以及土壤测试和配方施肥信息技术的研究和建设。随着计算机等信息技术与手段的普通，我们国家目前取得了较好的成绩与效果，农民大众从其中得到很多益处。

土壤测试和配方施肥系统解决了农民眼中准确的土壤测试配方施肥技术"走进村庄"和"推广"的难题。引入GIS、Web地图等先进技术手段，融入现代服务业等先进理念，提供测土配方施肥服务、施肥模型管理服务、数据管理服务等服务。系统平台的建设有助于推广测土配方施肥技术，使广大农民实现肥料减施、增产，降低成本、提高收入，可大大节约国家投入，提高施肥水平与效率，对推动耕地养分管理、节约资源、保护环境等都有较大的现实意义，并带来更显著的经济，生态和社会效益，这非常有利于农业生产的可持续发展。

目前，农民普遍存在大量化肥和重载土地，不利于土地耕作的可持续发展。农民群众在施肥方面有很多"病"，如注重氮肥，轻钾肥，注重化肥，忽视有机肥。重视大量元素化学肥料轻视中微量元素化学肥料，养分投入比例不合理，作物施肥量不均衡，施肥方法不科学等。为此，我们提出并开发测土配方施肥专家系统，对于改善上述现状具有良好的示范、实际作用，力争做到农业农村的全覆盖，并深入老百姓的生活。

当前存在各种测土配方施肥专家系统的单机版软件，随着互联网络接入的普及，老百姓文化水平的提高，为了适应我们国家农业信息化建设的需要，同时为农民提供方便施肥配方，我们下大力气开发了基于 Web 的测土配方施肥专家系统，该系统支持手机、PAD、PC、触摸屏等各种设备。

三、系统总体构架

服务对象大众化。系统以农户地块档案信息为基础，以村级地块信息为管理目标，以农户为重点服务对象，建立农户个人的专用账户，它还为个体农民提供个性化和有针对性的土壤测试和施肥技术的专业指导。系统的功能、交互等方面首先考虑农户的技术水平和应用能力，降低农户使用系统的知识水平门槛限制，实现系统推广的大众化、普及化。

服务内容个性化。每个农户均有自己的个人账户，可享受专门的有针对性的服务，土地管理部门检测的土壤养分情况可下发到每位农户，专家可针对农户提供一对一的技术指导，农户可针对自己的土地情况向各位专家进行更加有针对性的咨询。

服务媒介多样化。系统终端包括电脑、触摸屏、手机等常见设备，农民朋友可以通过电脑、触摸屏、智能手机等查询自己地块作物种植建议和施肥配方。当农户通过个人账户登录系统后，结合智能手机的定位功能，系统显示其用户地块的土壤养分含量数据及其评价，给出作物种植建议，与农户进行互动选择种植作物，给出施肥指导建议，极大的方便农民朋友的使用。

测土配方施肥专家系统的主要服务对象包括农业管理人员、农民大众、肥料经销点工作人员、农业技术专家等。系统里面我们集成了管理农业部门的土壤养分检测数据和农业技术专家测土配方的施肥模型，对农户的地块进行建档及管理、配方施肥模型管理及农户生产过程的服务和管理，针对不同的角色，系统可提供相应的服务。对于农户，系统为其建立私人账户，提供针对农户自家地块的测土配方施肥指导，为其建立施肥日志，便于日后查询与统计分析。对于农业管理部门，该平台提供数据管理和维护功能，包括空间数据库管理、土壤养分信息管理等，便于管理人员掌握土地基本信息，掌握该地区农村土地的情况，也可更新维护地块属性数据；对于农业专家，利用自己掌握的专业知识，结合土地的基本信息，构建各种测土配方施肥模型，给农民朋友提出合理的种植与施肥建议，同时对农户提出的问题进行了及时有效的解答，与农民朋友形成了良好的互动关系。

系统可与各级政府农业主管部门进行系统对接，为其提供农户地块档案信息，便于对农户地块土壤养分等信息进行管理。农业专家可以通过系统平台查看农户的施肥日志，根据农户的实际种植作物情况和农户的地块档案信息，给出有针对性的指导。基于现代网络技术和智能手机土壤测试配方施肥专家系统，对操作控制的简单性和易操作性进行了专门研究，做了专门的统计分析，针对广大农民朋友的操作习惯、行为方式等进行了专门的详细设计，同时提供电脑、触摸屏、智能手机等多种查询终端系统，从而大大的方便了农民朋友的使用，受到农户的热烈欢迎和赞誉。

四、系统建设内容

（一）系统功能

根据土壤测试和施肥工作的实际需要，我们精心策划和设计了土壤测试配方施肥专家系统的功能。

系统通过建立农户地块档案，配合政府农业管理部门采集测土数据，结合农业专家面向区域特征的施肥配方模型，研究和建立土壤测试和配方施肥专家系统，可以显著提高土壤测试和配方施肥技术在农业部门管理中的使用效率，大大提高了土壤测试和配方施肥工作的准确性，为农民提供了更有针对性的农业生产技术指导。系统包括以下管理功能。

作物肥料需求管理：对于不同的作物，以及不同农田，它们对肥料的需求是不同的，针对当地气候、土地等因素的影响，结合基于地理信息系统的作物肥料需求，精准的管理每块土地上的作物各个阶段对肥料的需求种类、需求量、需求时间，确保作物对肥料的需求能得到及时适量的供给。

作物施肥管理：结合地理信息系统，作物施肥管理对每块土地上作物的施肥时间、施肥操作、施肥量等进行管理，作物施肥数据是对作物施肥研究的重要数据来源，可对数据进行综合分析和数据挖掘，提取出有价值的信息给农户以指导和建议。

作物阶段施肥管理：作物各个不同阶段施肥后，作物的生长呈现出阶段性的特征，据此确立施肥效果指标，建立施肥阶段需求数学模型。对指标数据的采集、管理、分析，反馈于作物施肥的施肥管理，作物的施肥更加科学、精准。

地块信息管理：我们研究突破了以前的测土配方施肥管理系统，它们的管理单位不能达到地块级，还有不能到达农户级的限制，以农民土地为最小管理单位，实现了村级土地作为管理单位和土壤测试公式施肥指导，大大提高了土壤测试和配方施肥的准确性，对广大农民大众来说更加具有一定的针对性。同时提供可视化的网络地图形式，引导农户选择种植的地块进行直观化的查询。

施肥模型管理：可以根据区域、气候等因素的不同，这些因素导致作物种植种类可能存在着比较大的差别，从而农户种植作物的需肥规律肯定是不同的。通过我们设计的本功能模块，我们对施肥的模型进行各种各样的管理，方便在不同的情况下选择，我们最优的模型来计算施肥的配方，这样就可使得设计的系统有良好的针对性和适应性。

（二）数据库建设

建立土壤测试配方施肥管理系统数据库，可以对土壤测试配方施肥数据库数据进行更有效的管理，组织和实际应用，掌握耕地资源养分状况，优化配置土肥水资源，为科学配方施肥服务。我们可以根据农业部提供的土壤测试和配方技术规范的要求，探索建立适当的操作模式，来完成对现有采样数据的录入等管理，使得数据的分析、数据的处理、数据的管理、数据的汇总等相关操作更加便捷，为土壤测试配方肥料专家系统提供一些支持（图1）。

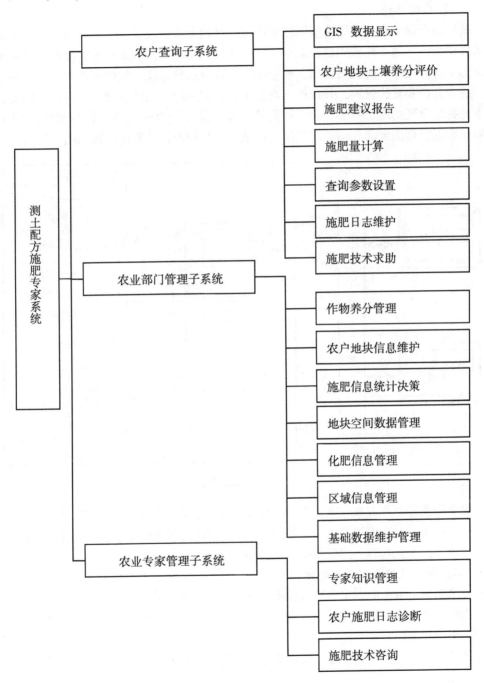

图1 土壤测试配方施肥专家系统的功能图

五、系统建设方案

（一）系统结构

我们结合测土配方施肥的实际应用情况，管理系统可根据各个区域内各个地块的土壤采样样品，土壤肥料施用试验数据，农民常用的土壤测试和施肥配方模型，建立每个地块的空间数据库、属性数据库、模型库等。通过这种方式，我们可以对各地区的地块进行综合分析和评价管理，并根据土壤测试公式应用施肥公式模型，它可以为农民朋友的土地提供合理的施肥指导。我们开发的管理系统基于 B/S 模型，该模型使用 Web 服务器、数据管理服务器和浏览器三层体系结构。我们设计的系统结构如图 2 所示。

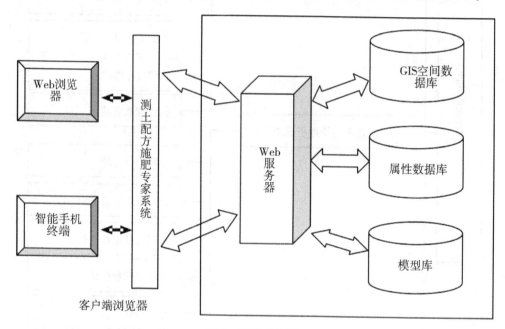

图 2 系统结构图

农业部及各个地方农业主管部门开发出很多测土配方施肥专家系统，均得到一定程度的推广应用，根据使用的效果，我们选取较好的系统进行对接，可缩短建设周期，减少建设成本，在此基础上建设数据库服务器。

（二）配方施肥专家系统的建设

土壤测试配方施肥专家系统数据库的建设和管理是土壤测试施肥工作和农民土地评价工作的重要环节，探索建立土壤测试标准化数据库和制定专家，我们能够实施有效的管理，组织和应用系统。通过该系统，我们可以查询科学施肥指导的相关数据，该系统可以评估耕地的基本土壤强度，为农业结构调整、新农村建设、化肥减施等提供重要的数据支撑。

（三）土壤测试数据库的建设

系统部署完成后，首先需要将辖区内地块的采样数据、地图数据等信息导入系统，完成地块数据的建立和更新。

数据库建设主要有两方面的任务，首先将土壤属性、地图采样数据等信息录入数据库，二是根据农户地块的种植情况，进行数据采样更新。系统为了方便数据的管理操作，系统需建设可以存储数字地图信息的空间数据库以及可以存储土壤地块信息的土壤属性数据库，同时，可以执行空间数据库和属性数据库之间的相互连接。

（四）实施保障

1. 加强组织领导，统筹合理推进

中国各级农业主管部门要切实加强组织领导，建立健全土壤测试和施肥技术指导小组和工作领导小组。我们需要加强组织的协调与领导，实现并进行强化行政推动，做到统筹农户、科研、企业等方面的各种力量，这样就将部门的行为变为国家各级政府的行为和社会的行为，合力进行统筹合理推进。

2. 提高业务水平，稳定队伍建设

我们应该进行切实加强测土配方施肥队伍的建设，可根据实际工作的需要配备相关专业的土肥技术人员。同时，要继续加强对基层农业技术人员的科技培训，积极调动全社会的力量。扩大与土壤测试和配方施肥技术相关的培训范围，不断加大培训管理力度，深入开发土壤测试配方施肥专家技术队伍，为土壤测试和配方施肥服务。

3. 加强信息宣传，营造氛围

我们可以利用有线的电视、计算机网络、传媒报纸以及各种各样的新媒体，采取电视机讲座、手机短信/微信、新闻报道、明白纸印发、有奖知识竞赛、现场会等相关方式，积极开展土壤测试配方施肥专家系统实际应用的广泛宣传和专业技术指导。积极开展土壤测试和配方施肥相关技术和生态低碳环境施肥，进行各种公益宣传。因此，可以形成上下联动的相互动态潜力，并且可以产生良好的气氛，并且土壤测试配方的施肥专家系统是积极的持续深入的在实际当中进行应用创作出来良好的社会环境。

农业物联网管理系统

一、项目背景

在传统的耕作方式下，耕地信息主要通过人工测量获得，采集过程需要大量的人力资源。农民大药、施肥、浇水全靠感觉、凭经验，导致化肥过量或不足，影响蔬菜和水果的质量，盲目使用农药，往往会导致农药残留过多。随着农业产业化、规模化的发展，以蔬菜瓜果温室大棚为代表的现代设施农业在农业现代化生产中发挥着重要的作用。温室内的温度、湿度和二氧化碳浓度等参数与果蔬的生长发育直接相关。温室栽培的农产品种类越来越多。实施环境自动监测，实现温室生产的自动化、科学化管理，通过统计分析监测和分析数据，并结合作物生长发育规律。建立控制模型，以调节环境条件，以实现高产、高质量和高效率的作物。

设施农业通过人为调控设施内的环境，来达到有利于农业生产和保护农业生产的目的，因此，环境的监测和控制已成为温室等温室环境中需要解决的关键问题。为了达到高效农业生产科学化、现代化的目的，提高农业科研的准确性，为了促进农业的快速发展，有必要对农业生产环境中的一些参数进行调节和控制，如空气温度和湿度、二氧化碳浓度、土壤含水量和光照强度。目前，大多数农民使用手工方法来检测温室中的温度和湿度，二氧化碳浓度和土壤含水量，控制更是凭感觉经验，这样必然导致劳动强度大、测控精度低、测控不及时等问题，容易造成人力资源的浪费，成本增加，甚至是不可弥补的损失，且很难达到预期的控制效果。

农业物联网管理系统使用各种传感器来监测和管理农业、林业、畜牧业和渔业的增长过程。以蔬菜种植为例，通过传感器检测感知土壤成分在生产过程中是否添加农药、化肥、各种生长添加剂（调节剂）等物质；结合二维码和RFID电子标签、存储和管理蔬菜苗木、种植地点、生长环境、日常管理、质量检验、运输等过程的数据。物联网系统以物联网技术平台为载体，提升农产品的质量和标准，让老百姓吃上放心的农产品，用现代手段实现菜篮子、米袋子等有利于广大人民群众的项目。

以蔬菜种植为例，利用物联网技术在设施农业中，温室内空气温度和湿度、土壤温度和湿度、CO_2浓度、光照强度、叶面温度、露点温度等环境参数的实时采集通过控制中心根据模型进行分析，自动采取措施对某（几）个参数进行调控，比如感知到温湿度过高，则需要通风或采用喷雾方式进行降温，这些操作是由物联网数据平台根据监测数据做出的控制策略。通过监测作物不同生长期所需的温度，湿度，光照和二氧化碳浓度，自动调节水肥投入，减少人力消耗，帮助农民实现集约化耕作。可根据种植户的需求，对作物进行实时监测，根据监测数据采取相应措施，为设施农业综合信息、环境自动化控制和管理决策的智能监控提供科学依据。

通过传感器采集各种需要的数据，经有线无线网络传输至数据中心，分析数据给出控制决策，对温室大棚进行远程控制。它是以人力为中心的农业转型，转到依靠机械生产，以现代信息技术为中心的模式。通过推动农业物联网系统和自动控制系统在现代农业生产中的应用，我们将积极提高现代农业生产设备的智能化和数字化水平，促进我国现代农业快速发展。

运用现代信息技术手段，实现温室大棚等设施农业生产管理的可视化、数字化、信息化、电子化，建立农业物联网服务平台，做到各节点信息的互联互通，形成完整的农业信息管理、科技服务链条，有利于提高安全责任意识，科技应用在农产品生产，加工和销售中的应用意识，增强科技的配套作用。加强监督和保护措施，形成问责机制，提高农产品质量要求；有利于增强农业主管部门的技术服务能力，提高科技示范作用，提高对问题农产品的发现处理能力，推进农产品质量追溯体系的应用，提高农产品质量安全监管水平和农业生产管理服务水平。

二、项目意义

农业物联网管理体系在保障农产品质量安全，引领现代农业发展中发挥着重要作用，可以改变广泛的农业管理和管理方式。提高作物质量，让人们吃上放心安全的农产品。

（一）改变农业管理管理模式，增加收入的必要手段

在作物种植的准备阶段，通过在温室中安排部署各种传感器，综合分析土壤和气候信息，以及选择适合种植的作物。

在种植和生长阶段，利用物联网技术采集温湿度等各种信息，进行管理与调控，使作物处于理想的生长环境中。如果需要对温室大棚进行浇水施肥，降温，卷帘等管理操作，应用物联网技术，只需简单操作即可完成控制管理，而且，控制区域基本上不受限制，取代了烦琐的手动操作，节省了大量人力，从而提高了工作效率。

在农产品收货阶段，利用物联网技术，综合整个作物的生长周期内的各种感知信息，进行综合分析，进行更精准的测算，同时可给出建议供下一轮种植时参考。

（二）发展现代农业的重要支撑

农业物联网技术的应用可以实现对农药、化肥、水、水产养殖和畜牧业的精确控制，减少资源浪费，节约资源，减少污染，保护环境；可加强疾病防疫与疫情控制，促进农业生产增产，效益提高，实现可持续发展；同时，利用物联网技术及时反馈控制，可以提高农业生产决策管理水平，加快实现农业现代化。

（三）适应当今信息技术发展的需要，改善工作管理

该系统采用多种传感技术、无线通信技术、云计算技术、射频识别技术、互联网技术、农业机械和农艺技术构建农业物联网管理系统。为智能手机、微信、手持设备等提供多终端支持为普通农户、温室大棚管理员、企业生活经营者、政府管理人员提供功能各异的服务功能，打造现代设施农业生产管理模式，实现以物联网技术为核心的集管理、控制、预警、监测、展示等多功能于一体的综合服务平台，提升工作的管理水平。

（四）加强农产品质量安全的有效措施

农产品的质量和安全是当今社会普遍关注的问题。为农产品质量和安全追溯系统提供支持，确保农产品的安全，如利用多种传感器对蔬菜水果生产生长过程进行全程监管，数据自动保存，真正做到追溯数据真实可靠，保证蔬菜生长过程的绿色环保。

三、建设原则

农业物联网管理系统的建设必须遵循以下原则。

（一）实用性原则

为了方便用户使用，系统必须在设计时考虑到实用性。在数据组织和系统性能方面，有必要以最终用户为目标并尽可能轻松地实施系统。在功能方面，满足服务、计策、控制、预警、报警、监测等功能要求和高效、可靠、稳定等性能要求。

（二）先进性原则

系统的建设尽量采用先进的，适用于农业物联网的信息技术、网络平台、硬件、软件、方法，确保系统的先进性，同时兼顾成熟度，使系统成熟可靠。在满足完整性和整体性要求的同时，系统可以适应未来的技术发展和需求变化，从而使系统可持续发展。

（三）可靠性原则

可靠的系统和稳定的运行是系统实用的先决条件。系统在建设时应通过增加冗余感知设备、冗余数据处理/转发节点、服务器、数据库等使系统具有高可靠性。当某些设备发生紧急情况或故障时，必须确保系统能够正常运行；系统还应具备一定的自我修复与诊断能力，当系统出现突发事件或故障时，能进行故障自我诊断并启动修复机制，最大程度的保障系统的正常运行。

（四）低成本原则

系统在达到设计目标的前提下，尽量降低系统的运营和建设成本，以最高的性价比投入使用。根据"逐步实施"，尽可能使用现有资源条件（硬件、软件、人员、数据），统筹规划"的原则在规定的时间内高效率、高质量的完成系统的建设目标。

（五）易用性原则

系统应尽可能的减少维护人员的工作量，给过短期培训后，一般工作人员均可掌握系统的基本操作。系统在交付使用后，应便于进行系统日常维护，能方便的进行软件的配置、软件系统升级、硬件备件的更换、系统预警的监测等。

（六）可扩展性原则

可扩展能力是农业物联网管理系统的重要原则，平台需具有灵活的适应能力、自动升级能力和可扩展能力。项目建设所需涉及面相对较广，项目建设周期一般来说相对较长，在这个过程中，硬件、软件以及集成设施等都有可能做出了调整，因此需在设计时留出必要的扩展余地。

（七）可管理性原则

项目建设在统一目标的指引下，进行整体设计，统一规划，具备完整的管控、预警、分析、统计等功能。该系统依靠统一的业务管理模型、统一的数据管理系统和统一的电子政务平台，实现统一的运营和部署。系统采用统一的网络互连技术标准、统一的计算机技术标准、统一的数据标准、统一的业务标准,开发出的系统具有可管理性。

四、系统总体设计

（一）应用架构

整个服务系统利用物联网技术,传感技术和通信技术整合太阳能温室和温室：通过各种传感器收集诸如空气温度和湿度、土壤湿度、CO_2 浓度和光强度的数据,并使用有线/无线网络将数据发送到数据中心。设施农业经理和种植者可以通过智能手机或手持设备和计算机查看不断增长的作物环境。及时采取控制措施,及早预防,提高作物产量和质量,提高工作效率。在系统环境参数超过规定的阈值或者根据后台系统模型检测出应采取动作时,系统可远程对微喷灌/滴管装置、卷帘机、放风机、供暖设备、降温设备等进行控制,实现农业生产的精准管理和智能化管理,促进现代农业发展,提升农业现代化水平。

（二）技术架构

农业物联网平台采用四层架构,即传感层、网络层、数据层和应用层,如图1所示。

1. 感知层

感知层的任务是通过各种感知方式收集现实世界农业生产的各种信息并将其转换为计算机可处理的数字信息或数据。感知层的各个感知节点感知采集的信息,通过有线或无线方式传送到上层网关,由网关将收集到的感知数据通过网络层发送至支撑层的数据中心。感应层用于在农业设施的温室环境中部署各种传感器,如气温、湿度、CO_2 浓度、光强度、土壤湿度传感器等。实时监测作物生长环境,如空气温度和湿度、二氧化碳浓度、光照强度和设施农业环境中的土壤含水量。

农业物联网收集的信息主要有以下几类。

作物环境信息：气温、空气湿度、CO_2 浓度、土壤湿度、光照强度、土壤温度等。

农作物基本信息：作物品种、名称、种植时间、特性等。

其他信息：温室结构（长度、高度、宽度）、管理设备等。

农作物生产信息：种植作物名称、品种、种植时间、生长期、管理措施等。

感知层的主要任务是通过感知手段收集现实世界的物理信息,并将其转换为数字信息,通过有线/无线方式传输。感知层涉及的技术有：各种传感器、RFID 标签和阅读器、QR 码标签和阅读器、相机、传感器网络等。

2. 网络层

网络层的主要任务是通过各种网络技术（有线/无线、ZigBee、adhoc 网络、GPRS、4G 等）收集传感层收集的各种传感数据。数据被传输到数据层的数据处理中心以进行存储集成,并被移交给应用层进行相应的处理。网络层包括各种有线网络、无线网络、通信与互联网、网络管理、信息中心、处理中心等。

3. 数据层

数据层起着应用系统的支撑作用,基于 Hadoop 的云计算技术可用于构建农业物联网云数据中心,支持在物联网中存储、分析和挖掘大量数据,为应用层提供数据支持。

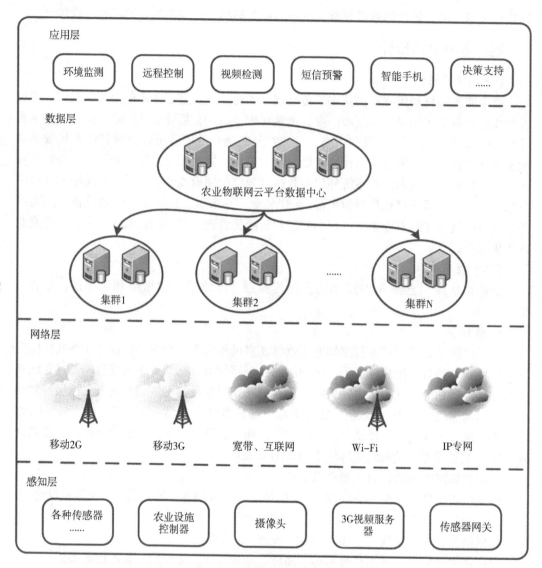

图 1 农业物联网管理系统架构

4. 应用层

应用层主要任务是对数据层的汇总存储信息进行查询、分析，进行基于感知数据的应用和决策，实现对现实世界的实际情况形成数字化认知，应用层是农业物联网管理系统的用户体验。

（三）部署架构

农业物联网管理系统的部署架构如图2所示。

（四）建设内容

农业物联网管理系统的建设包括以下两部分。

1. 农业物联网管理系统基础设施建设

包括面向种植户（温室管理员）等用户的农业物联网监控终端和物联网网关节点。

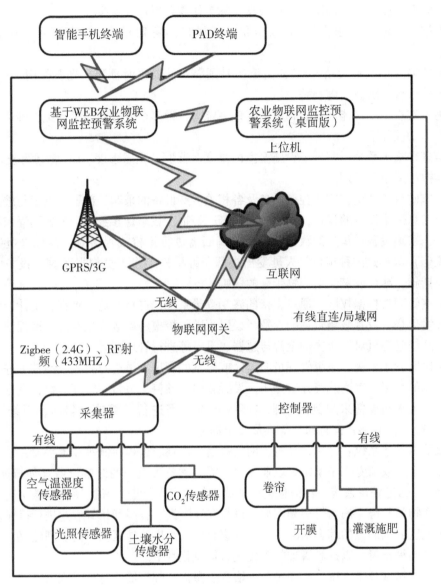

图2 农业物联网管理系统部署架构

员工可以通过智能手机、PAD和计算机等联网设备随时获取授权设施的环境监测信息。温室的自动化设备可以由智能电话、PAD和计算机等联网设备控制。农业物联网监测终端设备包括空气温度、空气湿度、光照强度、CO_2浓度、土壤含水量、土壤温度等传感装置；物联网网关可采用基于WiFi、GPRS、4G、NB、ZigBee技术的网关。以上物联网监控终端设备、网关设备可根据需要自行进行组合，构建针对农、林、牧、副、渔等领域的不同场景的监控系统。

2. 基于Web的农业物联网监控管理系统，实现了农业物联网应用中的N测五控五警功能

N测：设施环境温度、设施环境湿度、光照强度、CO_2浓度、土壤含水量、风速、

风向、叶面温度、露点温度等，可根据实际工作需要增加或减少。

五控（以温室大棚为例）：卷帘机、通风口、灌溉开关、施肥开关、地温控制开关。

五预警（以温室大棚为例）：高温实时预警、低温实时预警、光合作用条件预警、地温控制预警、灌溉实施预警。

为了实现以上功能，农业物联网管理系统需建设多个业务功能子系统，各子系统包含的基本功能如下。

信息采集子系统：实现了设施温室内温度、湿度、光照强度、CO_2浓度和土壤含水量等参数的采集和管理。

设施监管子系统：完成控制设备及各相关生产设备的采购、部署、设置与维护。实现环境信息采集设备的设置，设置各种预警、警报、栽培管参数等，配置远程控制监控功能。当采集设备采集到的数据，经决策判断需进行预警或报警时，系统以短信（声音、微信）等形式向种植户等人员发出提醒，种植户工作人员根据预警、报警内容采取相应措施，减少或避免给作物带来的影响。

温室管理员根据权限，通过支持的终端登录系统，进行浏览、查询、统计、分析温室的环境信息，设置栽培管理的参数，设置预警、警报的参数，配置进行远程管理的参数等；系统管理可对系统内的全部传感器、设备等进行管理配置。

园区监管子系统：系统管理员向园区或温室的管理人员进行授权，使其可进行温室环境信息的浏览、查询、统计、分析，设置栽培、预警、报警等管理参数，进行远程控制管理。当采集设备采集到的数据，经决策判断需进行预警或报警时，系统以短信（声音、微信）等形式向种植户等人员发出提醒。

农情专家子系统：运用计算机、网络通信、视频、远程控制等技术手段，结合专家知识库，开发农情专家子系统。农业专家仅需要配备可上网的终端设备（如智能手机、PAD等），通过登录系统，对信息进行浏览，与生产人员进行音视频的沟通交流，实时查看农作物的长势、病情等视频图像，进行在线咨询、答疑和指导农户进行操作。也可根据更大区域内温室的综合情况，进行专家会诊，给出该地区温室的管理建议，它以各种方式推向种植者，以提高管理的现代化和信息化水平。

信息推送服务子系统：实现短信、语音、微信、邮件等信息推送服务管理，为温室生产管理者提供预警报警信息的推送服务。温室生产者仅需一部入网的设备（手机、PAD）加入该管理平台，就能实时接收预警报警信息及生产指导等信息。

大数据分析决策子系统：使用大数据技术分析和挖掘温室环境信息（气温、空气湿度、土壤含水量、光照强度、CO_2浓度等）。结果供农业专家进行决策的依据，也是进行反馈控制的重要依据，为生产管理、预警、报警等提供科学参考数据。

专家系统：根据平台采集的基础数据，实现各种农作物生产参数、病虫害预防等知识的分类存储与管理，利用条件分析、数据挖掘等技术，开发专家系统，为种植户提供信息咨询服务。

图像采集子系统：实现了在设施温室中收集和传输网络摄像头监控节点的视频图像数据的功能。采集的视频图像发送给网关节点，网关节点汇总各个采集点的数据，发送给物联网数据中心；由数据中心进行视频图像数据的存储，各授权用户可登录系统，浏览、查看图片和视频信息。

云存储子系统：实现各类物联网数据的存储、管理与维护工作，并提供对外接口供其他分析决策系统使用。依托农业物联网管理系统，实现农业物联网数据库服务器、Web 服务器、FTP 服务器、应用服务器、存储服务器等的构建，实现系统资源的集成和统一管理，通过标准数据接口提供外部数据服务。

（五）建设目标

利用计算机技术、网络技术、传感技术、物联网技术、4S 技术、无线移动通信技术、云计算技术等现代信息技术，结合现代农业技术，构建基于云的农业物联网管理系统，满足各级农业生产经营组织（农业企业、农业合作社、农业科技园、种养植大户等）现代化、智能化、精准化管理的需要，建设服务支撑与监控平台，对于增加农产品产量、提升农产品品质、降低成本、节约资源、保护环境具有良好的作用，可明显提高经济效益与社会效益。

1. 多种应用服务模式

现场服务模式：农业合作社、种植企业、种植大户等管理员可通过智能终端在设施温室现场进行环境监测、设备控制。

远程服务模式：系统管理员通过权限分配开通相应账户，各级用户根据自己的权限进行远程控制操作。

平台管理服务模式：系统平台部署在农业大数据云平台上，为农业生产提供物联网监测、控制、预警、咨询及决策等支撑服务，实现传感设备的分布式部署和数据的集中统一管理。

2. 支持多种客户类型

大客户：农业园区、农业龙头企业、家庭农场、农业合作社等，建设和生产规模适中的客户。

集体客户：以村为单位形成的生产基地或农业专业合作社等，形成一定的建设和生产规模，进行农业信息化生产经营。

单一农户：分散独立的普通农户，需要政府部门进行组织和支持。

3. 支持分层和统一管理模式

管理平台支持地市、区县、乡镇、村四级管理，各级政府或农业技术部门的会员对平台进行统一管理、统一分配。

4. 支持平台式农技服务

支持地市、区县两级建立、更新、完善农业技术专家知识库，农业技术专家通过网络实时了解辖区内设施大棚内的物联网数据和其他综合信息，评估生产信息，在线接受咨询，利用云平台分析挖掘信息，提供生产决策建议，对需要预警的情况通过多渠道及时进行发布，以便于对生产进行指导。

5. 支持综合生产分析和管理

支持地市、区县政府或农业管理部门对辖区内的设施农业大棚物联网示范点进行分布分析、市场信息分析、生产情况分析、设施状况分析，辅助实现宏观管理与决策。

6. 建设温室物联网应用示范

建设设施温室物联网应用项目，项目要实现标准化的环境监测、控制、预警和报警，起到良好的示范作用。

7. 平台具有可靠性和可扩展性

可靠稳定的运行平台是系统实用性的前提，系统建设时通过增加冗余设备（装备）来提高系统的可靠性和稳定性，当部分设备出现突发事件或故障时，可保障系统仍然能继续工作。系统还应具备一定的自我修复能力与诊断能力，当系统出现故障或突发事件时，系统可启动故障诊断机制或修复机制，进行自我诊断、自我修复，并最大限度地提高系统的正常运行。

扩展性是农业物联网平台的重要原则之一，平台应具备灵活、充分的适应能力、自动升级能力、可扩展能力来支撑农业园区的发展。农业物联网项目建设涉及的范围面较广，建设周期一般来说比较长，这样硬件、软件技术等有可能发生较大的调整，因此系统设计时必须考虑可扩展能力。

五、系统建设方案

（一）技术路线

该系统使用部署在设施温室中的传感器节点来收集各种数据，例如空气的温度和湿度，土壤水分含量等。传感器节点以 adhoc 网络的形式形成传感器网络，并传输数据，并且数据最终聚合到网关节点。网关节点只通过外部网络处理从多个传感器收集的数据（GPRS、WiFi、4G、Internet 等）将数据传送至农业物联网云平台数据中心。数据监测中心负责对用户提出的各种数据查询请求进行答复，对采集的数据进行分析处理，为农业专家决策及制定农田管理措施提供依据。农业管理者、农户可通过手机、电脑、手持设备等对农作物生长环境进行现场采集，上传至数据中并采取相应的控制措施，预防病虫害，提高作物产量和质量，提高劳动生产率。同时，它可以远程控制通风，卷帘，施肥，灌溉等设备，实现现代，智能，精确的农业生产管理。

（二）关键技术

为实现本系统所需解决的关键技术如下。

1. 高性能农业物联网基础平台

以无线传感器网络为代表的大多数物联网感知设备都是电池供电的。发送和接收数据消耗大量能量，因此如何延长无线传感器网络的寿命成为一个关键的技术问题，它限制了它是否可以广泛用于农业生产。当节点失效时造成查询处理过程的中断，无法获得查询结果，提高传感器网络的可靠性也是本平台部署时需研究的关键技术；设施农业中存在多种控制设备，如何跟物联网结合，实现实时、可靠的控制平台，是农业物联网管理系统基础设施建设的一个关键技术。

2. 大数据的智能处理技术

如何分析和处理物联网感知层收集的海量数据的数据和信息，可通过数据中心提供的云计算与大数据处理技术，实施智能化控制。云计算是一种新兴的计算模型，它通过网络聚合和管理大量计算资源，形成计算机资源池，为用户提供按需服务。云计算主要使用数据挖掘、数据分析、搜索引擎、模式识别、人工智能等技术，为物联网提供高性能，大容量的决策和处理控制功能。

3. 安全的云计算服务

本系统部署在公共服务云上，系统中的数据可能包含大量的用户敏感信息，所以本系统需避免隐私泄露的安全风险，保证授权用户的合法权益。在特定实现中，可以针对不同的应用场景设计不同级别的安全和隐私保护技术。在分布式数据挖掘中实施隐私保护和隐私保护也是一项需要解决的关键技术。

（三）网络拓扑

农业物联网管理系统使用各种方法进行网络通信。近距离采集传输采用 RF 射频、ZigBee 技术、或与有线方式进行结合，保证网络系统的运行稳定；系统管理部分（系统运行管理、人员权限控制、数据增加修改删除查找、安全隐私保障）采用专线传输或 VPN 的方式，此类数据接入云平台；远程通信采用 Internet、GPRS、2G/3G/4G 相结合的方式。系统的网络拓扑结构如图 3 所示。

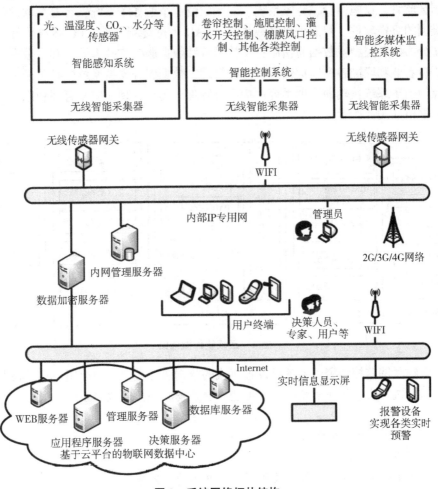

图 3　系统网络拓扑结构

（四）软件框架

通过系统应用软件平台，统一存储，处理和挖掘各种传感设备的基本数据，如设施温室环境传感设备，土壤环境信息传感设备和视频信息传感设备。通过云平台进行决策控制，生成有效的指令，控制执行机构直接进行调控，或以多种方式通知相关管理人员，设施环境中微气候的调节为作物生长提供了合适的生长环境。

从业务上来说，农业物联网服务系统包括农业感知数据采集、数据通信服务、数据支持服务、智能控制与管理、用户终端五部分，每部分包括业务范围内的一到多个分系统。为了更好的支持业务系统的运行，需建立一个系统低层基础服务平台，构建统一的运行管理和权限管理，根据不同的用户提供不同的数据通信服务和安全隐私保护。

在实施方面，农业物联网系统由三个主要部分组成：农业物联网监测与预警、终端监控软件、辅助功能。农业物联网监控预警系统部署在农业大数据中心，集监视、控制、管理、预警为一体的综合性服务系统。系统软件分为四个部分：前台应用软件系统，客户端系统，后台支持系统和辅助管理系统。其中前台应用软件系统包括信息采集、园区监管、设施监管、专家系统、专家农情5个业务子系统；客户端软件包括至少三种类型的终端软件：智能手机终端软件、PAD手持终端软件、PC终端软件；后端支持系统分为两种类型：数据通信服务支持系统和数据服务支持系统；其中数据通信服务系统支持农业物联网各类通信信道资源的管理，数据服务支持系统包括了云分析决策和云存储两个子系统；辅助管理系统包括系统管理、用户管理、安全管理和隐私管理。系统软件框架如图4所示。

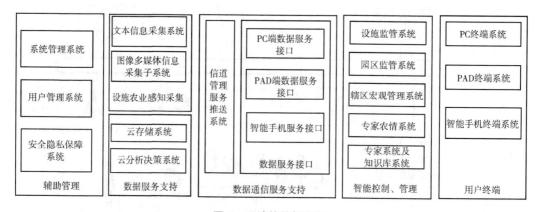

图4　系统软件框架图

农业物联网监控预警子系统可设置相应的数据服务接口，因此用户只需根据自己的终端类型，选择安装相应的软件，便可登录农业物联网监控预警系统，根据权限进行相应的操作。

（五）前台应用软件系统

1. 信息采集子系统

信息收集子系统包括图像采集子系统和文本感知采集子系统。

图像采集子系统：实现设施农业物联网环境下多媒体感知节点采集到的图像数据的

存储、查询和传输功能。

文本感知采集子系统：负责采集大气、土壤环境基本数据，视频监控数据及其他相关数据的采集与管理。

2. 设施监管子系统

控制中心可自由控制各种农业生产设备，管理员在授权的权限下对农业控制设备（放风机、卷帘机、灌溉施肥设备等）进行操控。设施温室管理人员通过本子系统浏览、查询、分析、挖掘温室环境信息、设置预警、报警、管理等各种参数，进行远程控制等管理任务。

3. 园区监管子系统

园区或基地的多个设施温室管理人员可通过本信息浏览、查询、挖掘、分析温室环境信息，设置预警、报警、管理等各种参数，进行远程控制等批量管理任务。

4. 辖区宏观管理子系统

各级政府主管部门，依据辖区内设施温室的各种信息，进行分析挖掘，结合专家意见，形成合理的建议，对辖区内农业生产进行宏观管理指导。

5. 专家农情子系统

通过农业物联网管理云平台，农业专家根据区域（地市、区县、乡镇、村）的各种数据，进行综合分析，足不出户，即可进行远程指导、在线答疑、咨询服务等，给温室管理提出合理意见与建议。

6. 专家系统及知识库子系统

存储和管理各种温室作物生长参数，病虫害预防技术等各种农业知识，实现网上专家系统，提供 24 小时的连续服务。

（六）客户端软件系统

1. 智能手机终端系统

工作人员可通过手机随时随地查看各个设施温室的环境监测信息，并进行授权权限的控制操作，它主要包括实时监控数据、变化曲线、设备控制、报警警告和历史数据。

2. PC 终端系统

对于最终用户，包括各级管理员和用户，可以实时监控设施的温室环境，查看历史数据、变化曲线等，各种物联网设备信息的查看，授权的操作控制等。

3. PAD 终端系统

包括各级管理员和用户在内，工作人员携带便携式 PAD 终端，用于环境监测和设备控制，主要功能包括实时监控数据、变化曲线、设备控制、报警警告和历史数据。

（七）后台支撑软件系统

1. 云存储子系统

存储、管理各类物联网数据资源。

2. 云分析决策子系统

利用云计算中间件对温室生产的相关数据进行数据挖掘等综合分析，为农业专家提供决策依据，为农业生产管理提供建议与参考，对紧急情况进行预警报警。

3. 信息推送服务子系统

将预警、报警等信息实时可靠的推送给各级管理员,让管理者及时采取措施,避免造成严重后果。

(八) 辅助管理软件系统

1. 系统管理子系统

负责各子系统的运行和管理,包括应用管理、运营管理、系统管理、网络管理等功能。应用管理负责管理工作人员使用系统;运营管理服务管理,监控系统运行,运行性能,故障情况等系统管理包括服务器、数据库、各种中间件等系统配置、运行性能的管理等,系统管理通常采用软硬件设备生产商提供的软件;网络管理是对网络安全性和性能进行全面控制和监控的过程。

2. 安全隐私保障子系统

安全隐私保障是物联网管理系统建设的前提,没有安全保障,数据毫无隐私可言,数据传输、管理等工作无法开展;没有安全隐私保护,就无法取得客户的信任。子系统需要确保管理、信息、系统、网络,物理等的整体安全性,并以有效性和应用为主。技术与管理并重,建立防范机制,保证物联网云平台可靠、高效、安全运行。对于不同的农业物联网应用场景,可以提供分级隐私保护机制以实现分级隐私保护。

以上各子系统软件采用组件化设计,统一集成在基于 Web 云平台下,便于集中管理和权限分配。

六、功能框架及各子系统

(一) 功能框架

农业物联网管理系统的具体功能包括:环境信息实时监测、农业生产基地管理、生产过程智能控制、预警报警管理、云数据存储与分析、专家知识库、数据服务接口、用户管理、流程控制等 (图5)。

1. 环境监测

收集温室中的温度和湿度、CO_2 浓度、光强度和土壤含水量等环境监测数据通过物联网网关经 Internet 或 GPRS/4G 网络传输至云平台进行存储,具体功能包括实时采集感知、可视化展现、数据查询统计、实时曲线等功能。

采集层支持单传感器节点独立采集和多传感器节点聚合采集。对于多传感器节点获取,可以使用 RF 无线电或 ZigBee 网络技术在设施温室中建立无线数据传输网络,并且可以通过外部增强天线实现无线信号的无限制传输。

通过农业生产环境中的温度和湿度、CO_2 浓度、光照强度、土壤温度和湿度等进行实时监测、图像视频监控、数据显示、数据处理、曲线显示等管理功能,基于监测信息的环境决策控制允许作物在适当的环境中生长。

2. 图像监测

利用在温室大棚内部署的视频监测设备,可进行视频采集和定时图像抓拍,各级管理人员可随时掌控设施温室内作物的生长情况。

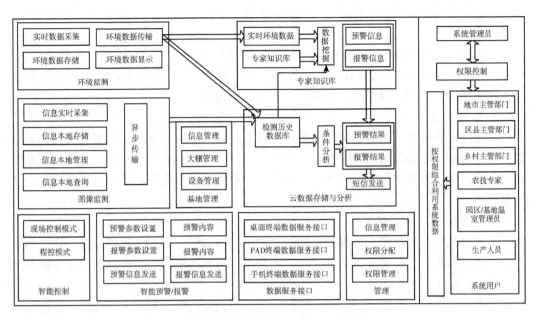

图 5 农业物联网管理系统的功能框架

3. 智能控制

智能控制主要由控制设备和相应的控制继电器组成，可自由控制各种农业生产设备，包括放风机、卷帘机、灌溉施肥、设施提温等。用户可通过智能手机、PAD 终端、电脑等对设备进行操控，也可通过现场手工的方式完成。

具体控制操作包括：卷帘控制、防风通风控制、灌溉施肥控制、设施加温降温控制、遮阳网控制等。

4. 预警报警

预警：对指定的监测指标，以及基于监测指标计算的二级指标，进行条件设置，根据预测模型，当预测到未来某个时刻事件发生的概论超过某规定值时，系统发出预警提示。

报警：对指定的监测指标，以及基于监测指标计算的二级指标，进行条件设置，根据计算模型，当某一个或多个条件达到时，系统发出警报。

5. 用户管理

为用户提供服务，实现用户信息的集中管理、权限分配、管理和控制。系统可为用户设置不同的管理、控制、监测权限，例如可以设置管理某个设施温室的权限。

6. 基地管理

包括基地新管理、设备管理等，提供直观的操作界面，方便用户对设备进行查找和管理，同时可按多种方式进行统计与查询。

7. 数据服务接口

通过数据服务接口，农业生产者和系统停靠，实现设施农业的智能监管、信息监测、控制管理、预警报警、远程查询、远程指导、生产指导、在线答疑等，我们将随时

农业物联网管理系统

解决农业生产者在种植过程中处于不同阶段的各种实际问题，全面提高设施农业的现代化和信息化水平。

为用户提供数据服务的包括智能手机终端数据服务接口、PAD 终端数据服务接口、PC 数据服务接口三类。通过智能手机终端、PAD 终端、PC 电脑终端可与系统进行连接，获得指定大棚的环境信息监测数据，进行远程控制。

8. 数据存储与分析

实现温室环境和历史数据构成的数据库数据的可靠存储，并基于大数据进行预警报警条件的分析，采取预警报警措施，以短信（微信、页面消息）的形式通知设施温室管理人员。为了预警更加准确可靠，还需采集并存储农业气象等相关信息进行综合分析挖掘。

9. 专家知识库及数据挖掘

基于对长期、大规模、多维数据的综合数据挖掘分析，形成了智能服务。通过整合专家系统知识库，与预警分析模型对接，实现基于专家知识库的生产服务的预警报警。

10. 系统用户及使用模式管理

为地市、区县、乡镇、村四级管理人员建立信息化档案，实现人员信息、企业机构信息的统一管理、统一查询。各级管理机构、人员的档案信息要相对独立，各级管理机构和用户采用与其授权模式相应的模式进行登录及系统操作。

（二）信息采集子系统

子系统是农业物联网管理系统的设施环境收集功能，主要收集设施的温室环境信息和温度环境信息。

收集和管理温室温室的各种环境数据，如环境温度和湿度、光照强度、土壤含水量、CO_2 浓度、温室外空气温湿度、风速、风向等文本型感知数据和图片、视频等多媒体数据。收集的数据通过各种互联网络传输到物联网数据库云平台。

1. 基于文本的环境监测数据实时采集子系统

该子系统为农业物联网管理系统的环境监测功能提供基于文本的监测数据。负责收集和管理温室的各种文本环境监测数据，如温度、湿度、光照强度、土壤含水量、CO_2 浓度、pH 值等传感器采集的数据。为实现物联网的预警功能，还需采集并存储设施温室当地的气象数据，如天气状况、大气状况等。将采集的环境信息和其他相关数据通过物联网传送至物联网数据中心云平台。

感知数据采集：支持单传感器节点独立采集和多传感器节点聚合采集。

单传感节点独立采集：传感器节点采集数据发送给物联网网关节点，网关节点接收到数据后，通过 Internet 或 GPRS 等方式将数据传送至农业物联网数据库云平台。

多传感器节点汇聚采集：多个传感器节点借助于 ZigBee、adhoc 网络形成传感器网络，并且多传感器节点首先将收集的数据聚合到多个传感器中的一个。然后，聚合数据由传感器发送到网关节点。网关节点将接收的数据通过 Internet 或 GPRS 等方式将数据传送至农业物联网数据库云平台，可通过外置加强天线实现无线信号的稳定传输。

传感器节点的功能设计：传感器节点直接与物联网的目标测量相关联，将农业信息转换为数字信息以进行传输。传感器节点有四种工作状态：节点休眠、节点唤醒、采集信息、发送数据。传感器节点通常处于睡眠状态，并且当唤醒所接收的命令时发送加入网络的请求。在等待来自聚合节点的响应之后，加入网络，然后开始感知收集信息并将感测收集信息发送到聚合节点。在接收到汇聚节点的确认成功信息之后，传感器节点再次进入睡眠状态，并执行循环。

汇聚节点的功能设计：汇聚节点从传感器节点接收数据，并将聚合的传感器感知数据发送到网关节点。聚合节点始终处于工作状态，或者由普通传感器根据一定的概率选择；等待监测命令、发送唤醒信息、等待接收数据、汇总接收的数据、发送数据至网关节点。

网关节点功能：网关节点主要负责建立并管理网络，将收集的传感器节点数据发送至互联网，物联网数据库云平台接收数据并存储。网关节点始终处于工作状态，其工作过程分为：等待监测命令、建立网络、发送监测命令、等待数据、发送数据。

环境信息数据存储管理：系统管理员对温室大棚的监测信息进行管理，具体包括温室编号、温室名称、监测数据类型、监测值、监测时间等信息。系统支持数据库标准化格式数据导入导出及持久化存储，支持报表打印、历史曲线打印等。有多种查询方式可供用户选择：按数据进行查询、按时间查询等。数据查询可直接显示在网页上，可直接打印或导出 Excel 格式，方便保存查看。

系统提供通过输入查询时间，可查询被测点对应时间内的数据记录和相应曲线；可统计某个时间段内监测数据的平均值。当监控数据达到报警状态时，及时发出报警信息；为用户设置不同的管理权限，具有数据报表、实时曲线、实时数显等多种数据显示方式。

实时曲线可切换 [环境温度] [环境湿度] [土壤温度] [土壤湿度] [光照强度] [CO_2 浓度] [风速] [风向] 等参数。当鼠标在曲线范围内移动时，您可以在鼠标位置看到相应时间段的值。

2. 视频图像监控信息采集子系统

基于视频信号传输技术和网络技术，对温室作物生长进行视频监控。部署在设施温室内的网络型视频监控节点进行数据采集并传输至数据库云平台。系统的授权用户可以登录系统查看图像信息。

图像信息的采集和存储：图像采集点通过有线的方式与网络型视频监控节点相连，视频监控节点采集的视频图像信息，数据通过互联网或 GPRS（4G）传输到物联网云平台，用于数据检测和存储。

图像浏览、查询功能：授权用户登录系统，可根据时间、视频图像的内容标注、类型等浏览图片和视频信息。还可按照指定的条件进行查询：根据视频图像的内容查询、按类别查询、按时间查询等，查询结果可以直接在互联网上显示，也可以由用户下载保存。

(三) 设施监管子系统

温室生产生产管理人员通过设施监管子系统进行浏览、查询、分析温室环境信息、进行远程控制、设置预警报警及栽培管理参数等。

浏览查询功能：设施温室生产管理人员通过智能手机、PAD 终端、电脑等与数据库云平台进行连接，获取指定大棚的环境监测数据。通过设施监管子系统，可实时浏览当前大棚的各项环境数据及历史记录。

该系统提供数据查询、数据聚合、曲线分析和环境数据综合管理等功能。

系统提供多种查询方式供用户选择，按温室名称查询、按时间查询、按数据类型查询等。数据查询可直接显示在网页上，也可保存成 Excel 文件或直接打印。

例如可输入查询时间，根据不同权限查询输入时间内授权的温室的监测环境数据记录和曲线图等；可统计某个时间段内监测指标的平均值；当监测值达到报警条件时，改变相应的数据颜色给出提升（当时已实时报警过，现在只是查询功能，给出提示即可）；根据用户设置不同的权限，如管理员、温室管理员等；子系统具有各种数据显示模式，如实时数据显示，实时曲线和数据报告，数据显示更直观，生动。

实时曲线可切换［环境温度］［环境湿度］［土壤温度］［土壤湿度］［光照强度］［CO_2浓度］［风速］［风向］等参数。当鼠标在曲线范围内移动时，您可以在鼠标位置看到相应时间段的值。

温室环境信息分析功能：用户通过各类终端，如智能手机、PAD 终端、计算机等与数据中心云平台系统连接，获取指定设施温室的环境监测数据。用户可以在线分析指定时间段内某个监控指标的平均值、最大值、最小值和累计值。实现数据的在线分析功能，为生产技术人员做出决策提供数据支持。借助数据中心云计算平台，用户可进行复杂的分析与数据挖掘功能，如对数据进行分组、过滤、聚合、聚合、关联等操作；根据数据挖掘结果设置合理的阈值；对异常信息提供分析判断及自动处理功能，如当温度高于某个数值时，首先综合各种因素判断是否可靠，如果可靠，将根据型号提醒，提醒相关人员采取相应措施。

参数设置功能：该子系统与专家系统、云计算决策子系统和专家知识库进行对接，用户为管理区域内温室作物建立相应的指标，设置相应的报警和控制参数，根据参数实现对设施温室的通风、采光、灌溉、温度控制等设备实现远程控制。

预警报警参数设置：收集各种传感器的环境信息：温湿度、光照强度、风速、风向、CO_2浓度等，通过设置相应的预警报警阈值来启动预警报警管理。如果可以在温室中设定温室温度，则温度采集的下限为 20℃，上限为 35℃，当测量到的监测数据低于 20℃，或高于 35℃时给出预警、报警信息。当然我们不能只依靠单个数据值的监测给出简单的预警报警信息，我们可结合其他指标，如大棚内空气的湿度等在后台进行大数据分析，假如当前温度是 33℃，但湿度超过 80%，CO_2浓度偏高，仅按照单个设置的阈值来说达不到预警报警条件，但运用大数据模型进行判断则必须给出预警报警提示，并给出操作建议。

预警/报警处理的主要过程如下。

温室作物生长环境特征数据的实时采集：通过物联网数据中心或数据采集子系统获取当前设施温室的环境数据。

为作物生长设定合适的生长环境条件：通过参数设置功能来实现。

确定预警报警参数：对指定的单个或多个监测指标及基于监测指标二等二级指标，进行条件设置，当预测（监测）到某个指标（多个指标）达到阈值时，生成预警报警信息。

云计算大数据中心根据当前监测的数据，虽然各个指标均未超过设置的阈值，但根据预警报警模型判断必须进行预警报警，也生成预警报警信息。

发布预警报警信息：预警报警信息主要通过手机，以短信的形式发送给责任人，也可通过微信、Email、广播、在线通知提示的形式给出，便于用户采集及时措施进行处理。

栽培管理参数设置：不同类型的作物对生长环境有不同的需求，如各种蔬菜，水果，花卉等需求；在不同生长期，不同作物的生长环境要求不同，如苗期，生育期和结果期。用户需根据作物的生长生理周期，制定相应的生产管理周期，用于指导农业生产过程中各阶段生长生理期的管理。本系统可根据作物种类、当前天气状况、所处生长期等，对设施温室内的放风机、卷帘机、微喷灌设备等进行操控，温室中的作物在合适的环境中生长。

控制功能：用户通过移动终端或计算机控制相应的设备，完成日常生产管理工作。控制设备主要由控制设备和相应的继电器控制电路组成，可根据需要自由控制各种农业生产设备，如卷帘机、放风机、灌溉施肥设备等。

系统根据作物种类、植物营养状况、生长期等及当期天气因素等信息给出管理控制建议，控制功能具体包括：卷帘控制、放风控制、灌溉施肥控制等。

卷帘控制：对卷帘的程度进行控制，可全卷、全放、按百分比开放。

放风控制：对放风（通风）设备进行控制，可全通、全闭、按百分比通闭。

灌溉施肥控制：通过系统控制为微喷灌开关和流量；结合灌溉系统，对营养液、酸液进行自动调配，控制 EC 值、pH 值，实现作物水肥一体化控制。

遮阳网控制：对遮阳网进行控制达到某种效果，如展开、收拢、按百分展开。

控制功能支持多种控制模式，如现场控制模式，可手动控制；非现场模式（程序控制模式），可采用智能手机、PAD 终端、电脑等进行控制，实现远程自动控制。

（四）园区监管子系统

设施温室管理人员可对授权内的温室大棚进行信息浏览、查询、管理等工作，预警、报警、栽培等管理参数的设置工作；远程操控及相关设备的管理工作。设施温室管理人员通过智能手机、PAD 或计算机登录系统，进行园区内设施温室基本信息管理。

其中数据浏览、查询、管理以及预警、报警、栽培管理工作，远程操控工作在前面已进行详细介绍，在此不在赘述。

设备管理功能：对网关、监测设备、控制设备、设备参数等进行综合管理，如添加、修改、查看、删除、维护日志等。

添加设备：物联网设备的添加，包括设备名称、数量、型号、购买日期、生产日期、位置信息等，被添加到系统中。对于自动控制设备，根据需要添加各项控制参数，便于远程自动控制。

设备信息管理：管理、查看、修改和删除现有设备信息。

设备日志维护：记录设备的维护，例如发生故障时的特定维护情况。记录类型、时间、维护人员、位置结果或故障状态等信息。

（五）辖区宏观管理子系统

农业政府各级主管部门和领导，对辖区内设施温室情况进行宏观管理，对农业生产进行宏观分析和管理决策指导。各级农业主管部门根据权限设置通过智能手机、PAD、计算机等终端设备，实现辖区设施温室信息的在线浏览、查看，了解温室的基本信息，环境监测信息，以及对农业生产的直观和实际的了解。

其中数据浏览、查询、管理以及预警、报警、栽培管理工作，远程操控工作在前面已进行详细介绍，在此不在赘述。

宏观分析决策功能：该功能与专家农情系统、云分析决策子系统、专家知识库系统进行对接，实现信息的过滤、聚合、关联、分组等数据挖掘操作，对数据进行宏观分析，制定农业生产管理决策，并以短信、Email、通知公告、文件的形式通过相关人员，使得指导工作更加有效。

（六）专家农情子系统

农业专家根据系统提供的地市、区县、乡镇、村温室的综合情况，分析并提供温室管理建议。农业专家通过电脑、PAD 终端或智能手机终端登录系统，实现在线浏览、查看、咨询、答疑和管理、用药指导等操作，还可进行远程视频实时指导、视频会议等形式，支持多领域、多行业、多部门的专家，多用户同时参与视频会议集成服务功能。

其中数据浏览、查询、管理以及预警、报警、栽培管理工作，远程操控工作在前面已进行详细介绍，在此不在赘述。

远程实时指导功能：农业专家和生产管理人员可进行双向的音视频交流沟通，进行远程实时农业生产指导。例如根据农户提供的农作物病情实物，专家根据农作物生长、病虫害防治依据，给出农药喷施种类、时间、剂量的建议，用于指导农业生产，准确使用农药，减少农药残留，减少农药滥用，种植绿色健康食品。让老百姓吃上放心的农产品。

系统支持通过短信、Email、在线提问等方式，将农业生产实际中遇到的问题及时向农业专家咨询，获得及时解答。系统支持专家咨询视频会议，多位专家、多位农业生产管理人员可同时在线沟通、交流，实现多用户综合服务功能。

（七）专家知识库子系统

管理和存储各温室作物的生产参数、病虫害防治技术等形成知识库，代替专家实现 365 天×24 时连续信息服务。对温室作物生产参数、病虫害防治技术等信息进行处理和存储，利用大数据处理技术、数据挖掘技术为农业生产管理者提供分析决策依据，进行农业生产指导。

在线咨询功能：对于用户没能通过浏览或查找解决的问题，提供在线咨询功能。用户在线填写并提交咨询问题，专家将及时回答问题，以便及时有效地解决问题。

（八）智能监控终端子系统

开发基于 B/S 结构的综合服务平台，用户可浏览查看数据，掌握各种农作物不同时期的生长情况。用户可将历史数据导出或打印，进行分析与挖掘，找出其中的规律，作出有效的决策，对农业生产管理者提出有指导性的建议。针对不同的用户终端，提供设施温室的功能展示和服务，具体包括：数据监控、数据查询、数据分析、预警和报警、远程控制等功能。

监控终端智能手机版：基于移动互联网，方便工作人员随时随地查看授权日光温室的监测信息，包括实时监测、设备控制、变化曲线、报警控制、历史数据等功能。智能手机版监控终端主要面向园区内管理人员、种植人员、农场主等（图6）。

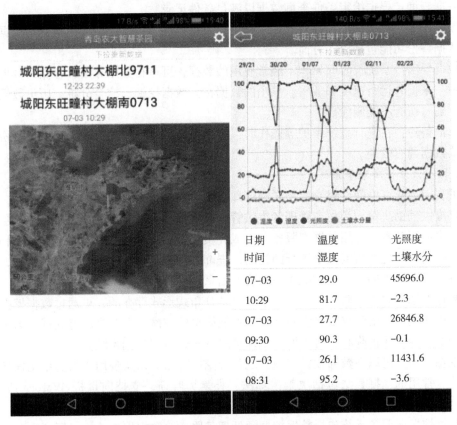

图6　监控终端智能手机版

监控终端PAD版：基于移动互联网，实现工作人员便携式近距离可视化进行现场环境监测与设备控制，包括实时监测、设备控制、变化曲线、报警控制、历史数据等功能。智能手机版监控终端主要面向园区内管理人员、种植人员、农场主等；PAD版相较于智能手机版由于功能相对单一，应用少，专用性强。

智能终端PC版：基于互联网，面向种植大户，实现辖区内设施温室的管理，包括基地管理、温室管理、设备管理、控制管理、预警报警管理等，设施温室的实时监控，设备控制，曲线，报警控制，历史数据和其他功能。同时，它具有数据上传功能，与数据库云平台互联，实现数据共享，为云平台提供基础数据。

（九）后端支持软件系统/云存储子系统

后台支撑软件系统（云存储子系统）管理、存储各类物联网的数据资源，实现对系统内温室环境数据、系统各级用户的管理，为用户查询、数据挖掘、在线分析等提供支持，用户可通过智能手机、PAD、计算机等随时接入进行管理查询；系统管理员通过

Intranet 或 VPN 管理权限更改。存储和管理功能包括如下。

数据添加：添加系统管理、各级用户、设施温室基础数据、设施温室环境监测数据、控制数据、农业相关数据等。

数据管理：实施修改和删除添加数据等操作。

数据查询：根据给定条件查询数据以满足条件记录。

数据排序：根据某个条件对查询结构进行重新排序，也可创建视图来管理数据。

数据备份功能包括以下方面。

数据导出：系统长时间累积不经常使用的数据，可进行数据导出，存储到其他存储介质中，来提升运行速度和效率。导出的数据在大数据挖掘、综合分析、历史数据查询时还会到，所有不能删除。

数据备份：在进行系统维护等操作时，为避免发生意外情况，对数据进行备份，确保系统数据的有效性。

（十）云分析决策子系统

可利用云计算中间件对农业物联网管理系统数据进行大数据挖掘、综合分析，为农业专家提供辅助决策，为生产管理者提供管理科学参考。云分析决策子系统由上层应用程序、决策服务器、专用数据库服务器组成，负责对云平台上的大数据进行挖掘分析，同时接收用户通过手机或电脑进行的访问、查询，对采集的信息进行存储和处理，为用户提供分析和决策的依据，并把用户需要的信息返回给用户。

数据挖掘：农业物联网数据在云平台上是分布式形式存在的，可通过数据挖掘实现在线分析功能。但由于权限和隐私问题，不能简单的在网络上共享所有数据，为了保证数据的安全性，保护隐私，可采取隐私保护的分布式数据挖掘系统。

挖掘算法：可以在数据预处理中实现各种挖掘算法，以实现用于数据挖掘任务的调度算法。常用的挖掘算法包括决策树算法、朴素贝叶斯、支持向量机 SVM、期望最大化、FP-Tree 算法等。

数据挖掘云服务系统包括数据挖掘预处理云服务，数据挖掘算法云服务和工作流子系统。数据挖掘算法云服务包括关联规则云服务、集群云服务、分类云服务、异常发现云服务等；工作量子系统可以在数据挖掘任务上执行多任务组合。

在线分析和决策支持：决策支持是一种多策略数据挖掘工具，包括数据挖掘工具，如描述、聚类、分类、关联、演化和偏差分析以及时间序列分析。提取的知识存储在具有不同抽象级别的知识库中，适用于不同决策级别的数据分析和决策。从农业物联网中的大量信息中挖掘出可能有价值的信息，以实现在线分析、预测和决策支持。对物联网中的设备等进行有效的反馈控制，达到决策的有效性。

云分析决策日志：所有操作均进行记录，并对记录日志进行管理，支持日志的备份、导出、打印等操作。

（十一）信息推送服务子系统

信息推送服务子系统实现短信、邮件、通知等信息的可靠传输，为设施温室管理者提供预警报警等信息的推送服务。通信介质采用有线（Internet、WAN、LAN 等）或无线（WiFi、2G/3G/4G）。信息推送服务子系统针对不同信道及相关通信协议资源需求情况，调度和分配合适的信道资源，保证系统及用户通信的正常进行。也能实现向特定用户或用户群推送特定的预警信息，实现某些控制效果。

有线网络信道管理及推送服务：以有线网络为通信介质，根据通信类型确定采用的通信协议，如 FTP、TCP、UDP、HTTP 等，根据采用的协议预留信道带宽和容量。在系统部署阶段需考虑系统的可扩展性，预留冗余信道带宽和容量。

2G/3G/4G 网络信道管理及推送服务：用户以 2G/3G/4G 作为接入物联网的通信介质，通信过程的信道管理由 2G/3G/4G 移动通信供应商服务。需解决在系统部署时，根据通信类型和数据量大小，对可靠性的要求等选择合适的 2G/3G/4G 移动通信网络，并租用足够的带宽。

WiFi 网络信道管理及推送服务：用户以 WiFi 接入物联网时，系统根据接入点 AP 的带宽、容量、通信类型、数据量大小，并选择适当的通信协议和通信标准，例如 802.11a/b/g/n，以满足可靠性要求。根据通信的类型确定信息发送的优先级。同时考虑到 WiFi 接入用户一般具有移动性，信道管理模块还需具有快速网络切换和断定续传功能的支持。

（十二）系统管理及安全保障管理

管理每个子系统的操作，包括应用程序管理，操作管理，系统管理，网络管理等功能。建立信息化档案，实现人员和机构信息的统一管理和统一查询。

应用程序管理负责管理应用程序系统各级的用户信息，包括以下方面。

各级农业管理部门的用户管理。

本部门人员的管理：人员的增加、修改、删除、查询等。

个人用户的管理：系统管理员、农业专家、设施温室负责人、生产管理人员等。

个人用户信息的管理：姓名、性别、所学专业、职称等信息的管理。

数据应用统计功能：全面的数据访问统计分析和访问次数，形成数据报告，便于查询分析。

系统日志功能：系统操作日志，可以全面管理和统计日志，并可以打印，备份和导出功能。

系统运行管理：管理和监控系统操作、操作性能和故障情况。

系统管理是服务器、主机、中间件、数据库等的系统管理、配置管理、操作性能管理等。它是每个软件和硬件系统的相对独立的管理系统，通常由硬件和软件供应商提供。

网络管理：对网络的安全性、性能、品质等进行全面的控制和监测，生产并维护网络监控、管理日志。

农产品价格预警管理系统

一、项目背景

农业部要求进一步加强农产品市场信息的收集和发布，完善市场体系，形成更加透明、开放的农产品价格机制，将使市场信息的传播更加便捷。在全国农业信息化发展规划中，提出加快农产品批发市场信息化进程，开展农产品批发市场信息化示范。重点支持大型棉花、粮食、禽类、油脂、蛋类、肉类、蔬菜、水产品、茶叶、花卉等重点农产品批发市场的信息化，加强农产品物流、配送、交易、管理、市场等信息化建设，减少交易中间环节，降低交易成本，提高交易的效率。

智能农业是建设现代农业的重要组成部分，各级政府提出了建立农产品供求和价格行情信息的采集、发布、分析系统，及时对农产品价格行情信息、农产品供求信息进行分析，为领导决策提供依据与参考。

控制宏观调控，改善农产品和初级产品供需监测预警机制，制定市场供应和价格应急预案，防止总体价格水平，是政府的一项重要任务。目前农业部已经建成农产品价格采集、市场动态信息发布、农产品价格趋势分析一体化专业性网站，在政府宏观调控决策、微观生产指导、搞活农产品流通、稳定农产品市场等方面发挥着重要作用。各省市还建立了农产品价格监测预警系统，分析和发布国家和地方农产品批发市场的农产品价格。促进农产品价格信息查询服务，为政府部门和农产品交易提供宏观数据，促进有效决策。

二、项目意义

农产品价格不稳定，准确把握农产品市场状况并不容易。加强市场信息的分析利用，开展农产品市场产销，围绕市场做文章，是农产品生产获得良好经济效益和农民增收的有效途径。因此加强对市场农产品价格信息的分析处理，利用信息化技术手段进行处理、挖掘、决策，是开展农业信息化服务的有效途径。

为及时收集农产品市场价格信息，为农产品生产者和消费者提供有效的价格信息，为决策者提供准确、及时的决策依据，需要建立一套集农产品价格信息采集、分析、农产品价格信息管理系统，集处理、发布、查询、管理、预警等功能于一体。

农产品价格预警系统的建设不仅可以为种植者提供及时准确的市场信息，而且可以合理地指导农业生产经营。为农业部门管理人员提供宏观调控决策支持，指导农业和农村产业结构升级调整，降低农产品市场价格波动，有利于社会稳定和繁荣，具有重要的实际和社会意义。

三、系统总体设计

(一) 总体构架

1. 业务流程

农产品价格收集和预警系统业务流程：建立农产品批发市场，合作社，大型种植户，主要生产基地等价格信息采集点，价格信息采集员应当当天报告农产品价格信息；收集方法可以基于智能手机终端或计算机，价格信息上传到服务器进行存储。农业部门管理人员应当对报告的价格信息进行审核，批准和批准的价格信息应当正式存放；针对不同的需求做不同的统计分析，可按照年、季、月、周等周期进行数据的查询与统计分析，提供给农产品批发市场、种植大户、合作社等下载感兴趣的价格信息，也可以通过各种媒介进行发布。同时能够分析某一农产品的价格走势或预测涨跌，生成价格简报，由管理人员结合各种内外部因素，进行价格预警，方便进行决策，并参考那些关心农产品价格的人了解的各种数据。

价格信息采集员的业务流程：采集上报每日价格信息，接受反馈的审核信息，包括价格报送成功的反馈和价格审核情况的反馈；查看农产品价格统计分析数据；获取农产品价格分析报告数据。

农业管理部门人员业务流程：收到价格收集者报告的价格信息，审核并处理，并存入农产品市场价格数据库进行数据汇总；农产品价格数据分析：价格市场分析，价格数据提交分析，农产品价格指数分析等农产品价格信息通过门户等渠道发布。

2. 该系统面向的用户群

(1) 农业管理部门人员，职责为：接收、审查和处理农产品批发市场报告的价格信息；对价格数据进行统计分析；对异常价格进行重点审核并预警；发布价格信息到农业类综合门户网站；在本系统内部发布通知、公告等信息；对本系统的管理工作。

(2) 价格信息采集员，主要由批发市场、合作社、种植大户等负责人组成，职责为：收集和报告收集点的农产品价格信息；询问农产品贸易情况和价格走势；查看农业管理部门发布的通知公告。

(3) 公众可以查询批发市场中农产品的价格信息；查看农业管理部门发布的通知公告等。

3. 系统建设的原则

(1) 先进性原则。该系统依托互联网，物联网，云计算等先进技术和手段，通过升级保持系统的先进性。系统在体系结构设计上应充分考虑系统的扩展能力，使系统在功能扩张、容量扩展、性能扩展方面都有足够的空间；该系统实现农业资源数据共享，建立农业信息资源数据库，消除农业信息孤岛的出现，使农民能够存档信息，农业企业数据、市场供应数据、农业种植数据、价格行情数据、科技信息数据等完全互融互通。

(2) 标准化原则。系统的建设按照国家发改委颁布的相关标准，统一编码、统一分类，应充分考虑系统的开放性、兼容性、安全性、可靠性，形成价格预警信息系统建设的标准和标准体系。

(3) 经济性原则。充分利用现有技术，依托现有的资源（硬件、软件、人员、数据）进行建设规划，使系统实用、方便，它不仅可以减轻业务人员的工作量，还可以使系统适应需求的变化，避免重复建设。

（4）可扩展性原则。系统需要具有强大的可扩展性，以便在应用程序环境发生变化时可以轻松扩展系统功能。在设计中，应充分考虑系统功能，系统数据和业务可扩展性，坚持开放式系统结构设计；采用最先进的架构设计，为将来的构建保留各种接口，以确保系统的可扩展性。在系统的设计中，将采用业界流行的当前组件化技术和XML技术来确保系统的可扩展性。

（5）易用性原则。根据软件工程质量标准，良好的系统界面应整洁简洁，一目了然，层次清晰，易于使用。

（6）统一性原则。随着企业和政府信息化建设的深入，信息资源的整合越来越激烈。它不是自下而上的分散式构建，而是应用程序增强，无缝集成，物理分散和逻辑集中，非结构化数据和应用程序软件系统。通过门户整合、内容整合、数据整合、应用整合、流程整合支持到统一的门户管理系统，通过个性化、安全可控的门户页面提供给各类用户。

（二）建设内容

1. 价格信息采集功能

将各信息采集点当天的价格信息、交易量、市场基本信息等内容通过手机采集终端或电脑进行采集录入，为用户提供在线报告、在线修改、离线报告等功能。价格信息可通过智能手机价格采集终端发送到云平台数据库服务器，也可以在用户登录本系统后通过在线填报页面进行价格报送。

在农产品批发市场、合作社中选择多处建立价格信息监测点，并配置相应的软硬件设备。为了便于收集农产品市场价格，可以使用智能手机终端通过GSM收取价格，GPRS、4G、WiFi和其他网络传输技术确保收集信息的高效和实时传输。

2. 基础数据处理功能

系统的基础数据包括系统的用户信息、采集终端设备、采集点的数据管理、价格数据信息等，以及农产品数据管理、日志管理等。价格数据处理主要对接收到的数据进行审核、分类、存储，可由系统进行自动审核，或者管理员进行人工审核。为了方便对价格数据进行分析，系统还需要接收到的数据，包括手机端接收的数据和在线填写的数据，分别进行预处理，处理后存储到数据库。同时，管理和维护系统基本数据，以确保系统的正常运行，为系统管理和业务管理人员提供系统维护、用户管理、价格维护等功能。

3. 价格信息发布功能

通过各种门户网站，价格信息将实时公布，并发布农产品价格变动信息，相关国家政策措施和综合价格统计信息，让人们了解价格变化。能够为用户提供各种条件的综合查询功能，这样用户可以很方便的查询与市场相关的各种信息，如市场基本状况、价格行情、成交量等。查询信息以曲线、专题图、表格等形式直观地显示农产品的价格信息。价格信息以各种方式下载和查询，例如农产品类别、变化趋势、交易时间和地点。

4. 价格信息分析功能

对才加的信息和历史数据进行对比分析，利用专用公式进行换算，计算加权系数。将价格的变化转化成可视化的比较模型，展现出市场价格的变化趋势，作为决策依据。可以按照季度、月、周、日查看价格走势图，分析平均价格（平均价格需考虑交易价格和交易量等因素），不同地区的价格比较，以及相关政策和专家文章、新闻报道等，方便用户及时了解农产品的价格走势。进一步掌握市场行情，可以基于数据分析的结果生成相应的报告或价格简报，供各级用户使用。

5. 价格预警功能

根据采集入库的价格信息，依据分析数据得到的各种结果，对农产品价格的变化进行实时监控；当价格的变化超过某个规定的阈值或者价格变化与某模型严重偏离时，系统发出相应警告，并触发应急预案，提出相应的干预手段和建议策略；警告可利用弹出窗口等方式，通过日常监控终端进行提示。

农产品的价格受市场供应、市场需求、天气变化、农产品的产品、农产品品质、节假日等因素的影响，均会发生一定程度的波动。价格预警主要在系统自动监测预警的基础上，结合专业预警分析师的分析，决定是否发布某种级别的预警，供各级用户参考，供主管政府部门进行适当的宏观调控的依据。

四、系统建设方案

（一）网络拓扑架构

农产品价格预警管理系统主要供政府农业部门管理人员、批发市场管理人员、种植户、合作社、经纪人使用。其中批发市场管理人员、经纪人、种植户通过智能手机价格采集终端和电脑进行市场价格信息的采集，系统审核存储处理后，价格信息可由农业管理人员查询，给出分析和预测，通过微信、手机短信等平台发布给种植户、合作社、经纪人等人员使用，它也可以通过报纸和电视台等媒体向公众发布（图1）。

（二）技术架构

系统总体技术架构采用基于 B/S 的三层体系结构，将所有的服务组件化设计，通过 Web Service 和 C/S 方式向各类应用和管理模块提供服务。三层体系结构分为表示层、应用层和数据层。表现层主要指的是用户界面，要求用户界面尽可能简单，最终用户不需要进行任何培训就能方便的使用访问；应用程序层是应用程序服务器，通常称为中间件，数据层存储大量数据信息和数据逻辑。系统的三层架构如图2所示。

表现层：即农产品价格预警管理系统门户。通过门户为用户提供视觉信息服务，如数据发布和数据查询。提供各种数据分析和图表生成功能，为用户提供农产品价格信息查询功能的各种复合条件，并支持多种类型的图形化、图表方式的直观展现给用户。

应用层：实现业务功能，如信息审查、信息发布和数据分析。应用层的功能主要包括价格收集、价格分析、价格预警、价格发布和基础数据管理。

数据层：收集农产品市场价格信息后，通过数据转换、数据汇总、数据清理、数据过滤、数据提取等数据处理进入价格信息数据库，其中数据转换和数据清理可由应用支持平台采用专门的工具来实现。

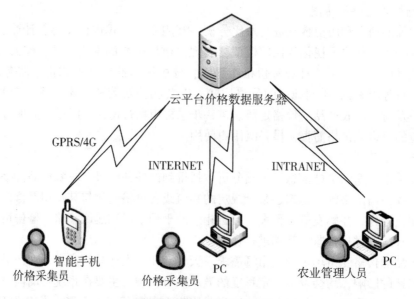

图1 系统网络拓扑架构图

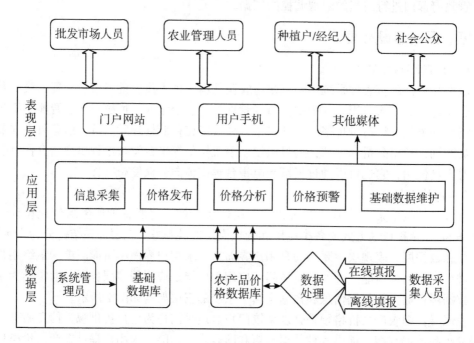

图2 系统三层架构图

(三) 系统架构

该系统分为前端管理子系统、价格收集子系统和后台管理子系统。前台管理子系统用于查询和显示各种农产品市场的价格信息；价格采集子系统运行在智能手机端或电脑上，由价格采集员进行农产品市场价格的采集和上报；后端管理子系统有农业政府管理人员进行系统管理，数据审查和分析。系统的功能架构图如图3所示。

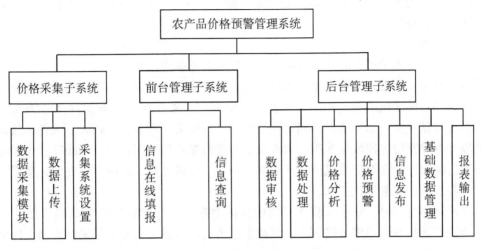

图3　系统功能架构图

（四）价格采集子系统

价格收集子系统针对价格收集点的收集者，开发在智能手机（PAD）侧运行的价格收集程序，实现价格信息的收集和传输。采用 GPRS、4G、WiFi 等方式进行信息传输，完成价格信息的相关数据、表格信息的定时或实时传送。

在价格收集过程中，根据不同的收集点设定不同的价格收集标准。为了使采集信息具有统一性和全面性，减少误差，方便分析和预警，在价格信息采集子系统中，对农产品种类、等级、交易市场等的编码应符合农业部颁布的"农业全息市场信息分类和计算机编码"（NY/T 2137—2012）和"农产品全息市场信息收集标准"（NY/T 2138—2012）。子系统包括用于数据采集，数据上传和采集系统设置的模块。

农产品价格采集系统的数据自动采集功能，基于智能手机终端（或 PAD）进行市场价格信息的采集上报，用户采集的数据经过现场处理后，通过 GPRS、4G、WiFi 等网络传输技术，数据被传输到后端数据库以实现数据收集过程。

价格数据采集处理过程如下。

1. 数据初始化

根据用户特性和业务逻辑，初始化本地数据库，下载相应数据至移动终端。

2. 数据采集

采用智能手机（PAD）到业务现场进行数据采集，数据输入可采用自动录入、手写输入、键盘录入等多种输入方式，还可以使用其中数据电缆连接到专用设备以读取数据的形式。价格信息采集员在上报信息时，根据设计好的表单填报采集数据，并根据需要采集该农产品的图像资料，数据填写完成后系统将采集时间、采集人等信息附加上，完成一次价格信息的采集过程。

3. 数据上传

用户采集完数据且经过现场处理后，采用 GPRS、4G、WiFi 等网络传输方式，以各种数据接口将采集的数据上传到后台数据库云平台中，采集的价格数据可以进行在线上传或离线上传的形式，若采用离线上传则采集终端中需要临时数据库的支持，便于对采集的价格数据备份。

4. 数据处理

在根据业务需求处理收集的数据之后，完成数据获取过程。

通过收集系统设置功能，用户可以指定当前收集人员信息和当前收集数据的存储模式。考虑到智能手机终端（PAD）所在的网络环境可能不稳定，数据存储分为在线存储和离线存储模式：当采集点的网络信号状态良好时采用在线存储模式，采集的数据实时传输并存储在服务器数据库中；当没有信号或收集点信号不好时，离线存储模式将收集的数据临时存储在智能手机的本地数据库中；信号恢复后，临时存储在本地数据库中的数据将上传到服务器数据库。

当网络不可用时，信息采集人员将收集的数据临时存储在智能手机的本地数据库中，然后在连接网络后将价格信息采集数据上传到服务器。当需要将采集数据上传至服务器时，首先读取手机中暂存的未上传成功的数据记录，将数据读取并上传至服务器数据库，系统需要清除已上传的数据或对已上传的数据进行标记，以免导致数据的重复上传。

（五）前台管理子系统

前台管理子系统为价格采集员提供价格在线填报功能，并将价格信息对外发布，面向价格采集员和社会公众。价格收集者可以通过网络登录系统，填写价格收集信息的在线信息。管理人员通过门户网站，向社会公众，市场管理人员等展示农产品的价格信息、走势分析等，并提供农产品价格的对比分析，生产按年、季、月、周的价格走势图，发布价格行情，提供各种类型的查询功能。

前端管理子系统包括用于信息收集器的在线填充和信息查询模块。

1. 信息在线填报

当智能手机（PAD）价格收集终端的价格失败时，收集的价格数据无法及时报告。手工完成价格采集数据的上报工作（此时采集人员通过 Web 表单页面完成价格采集信息的填报），防止因为设备故障延误数据的上报，导致数据不连续。

2. 信息查询

信息查询模块主要提供农产品价格信息的显示。

（1）每日行情信息。显示每日农产品交易信息，您可以根据产品名称，时间，市场等不同条件，选择查看各批发市场各种农产品的当前价格和历史价格信息。

（2）价格走势查询。查询某一周期（年、月、周、日）某种农产品的价格走势情况。

（3）市场行情价格简报查询。查看农业部门发布的农产品价格分析和农业相关政策信息。

（4）价格预测查询。查看某种农产品未来价格趋势的预测分析，上升和下降警告情况等信息。

批发市场、合作社、种植户等人员还可以下载价格统计分析报表、图表，便于了解市场行情。

（六）后台管理子系统

后台管理子系统主要面向农业政府部门管理人员，实现价格信息的审核、分析、预测、发布，对系统的数据进行查询分析，同时根据需要发布价格预测预警信息，以及涉农政策等相关信息。

系统数据管理由系统管理员管理和维护，主要包括上传的价格信息的查看和摘要、维护和备份系统用户、市场信息和农产品等数据。数据信息存储在云平台中，并为其他系统调用提供开放接口。

该子系统包括数据审计、数据处理、价格分析、信息发布、预警发布和系统数据维护等模块。

1. 数据审核

数据审核功能对手机终端（PAD）、在线填报的价格数据进行审核、存储，提供自动审核和人工审核两种方式。

自动审核：系统指定价格的有效范围，无效信息自动反馈或提示收集者确认，收集者根据情况重新获取或确认报告。在查看确认的价格信息后，系统通过后台软件将数据写入数据库。

人工审核：建立在自动审核基础上，系统自动审核后，价格数据写入数据库。登录系统后，农业政府管理部门可以查询负责区域的价格数据信息，进一步确保数据的准确性，纠正并防止错误被自动审查。

数据的审核结果情况（是否通过审核），也需要及时反馈给价格信息采集员，以供价格信息采集员参考。

2. 数据处理

收集的价格信息量通常很大，需要在存储到数据库之前进行预处理。同时为了便于数据挖掘模型的价格分析预测，借助专门工具软件对数据进行清洗、抽取、装载、转换等操作，数据的数据挖掘有助于进一步的统计分析

3. 价格分析

对每日上报的价格进行统计分析，主要分析农产品价格、供求关系的变动等情况，为价格预警提供数据支撑。为实现基于数据挖掘的自动智能分析功能，需通过选择某个时间段和（或）某个品种的农产品，自动生成一定时期内农产品的价格变动曲线。

可显示短期均价曲线，如 5 日、7 日、10 日曲线，也可显示中长期价格曲线如 15 日、20 日、30 日等价格曲线；提供均价曲线与预警阈值线的对比；提供某种农产品在某一段时间内的最低价格、最高价格等信息；通过鼠标等操作可直接定位到某明细情况，通过键盘上下左右键移动、旋转曲线，实现从不同维度进行观察其动态变化；将分析结果按日、周、月、季生成大类农产品价格简报，重点监控农产品要提供专题简报分析。

在计算农产品平均价格时，可以通过销售加权算术平均法计算，并对某些农产品在不同时间和地点的销售进行加权，计算农产品在特定时间或某个时间点的平均价格。

4. 价格预测和预警

对某一时间段内（年、季、月、周）的农产品价格数据进行统计分析，结合权数、

指数、均价的数据信息，建立科学的数学预测模型，利用数据挖掘技术预测农产品未来的价格走势。也可重点监测农产品的品种及各大类农产品未来的涨跌情况，建立农产品价格中长期变化趋势预测分析决策支持系统，提供对未来农产品趋势的分析和预测（如1~3个月）。

价格预警可通过设置预警参数对农产品特别是重点监控农产品进行价格预警，预警的级别、类型、预警的内容等预警信息也需设置。当某种或某类农产品价格超过或接近预警设置参数时，系统需自动发出预警警报提醒用户。

根据系统自动预警分析结果，有预警管理人员结合各种内外部因素，如农产品产量、市场需求、市场供应、节假日效应、天气气候情况等，发布预警信息，发布预警信息的功能需要支持自动和手动通知方法，根据设置的预警参数自动通报到相关负责人。

5. 信息发布

信息发布功能将经过汇总处理后的价格信息以各种形式发布到门户网站系统、电视等媒体，为公众提供信息服务。发布的信息主要有以下五类：农产品价格信息、农产品价格分析、市场分析、政府相关政策措施、价格简报等。

6. 系统基础数据管理

基本数据管理功能完成系统基础数据的维护，备份和日志管理功能，由系统管理员使用。基本数据主要包括系统用户信息、智能手机（PAD）采集终端信息、农产品数据信息和农产品价格信息。

系统管理员登录系统进行系统管理。系统用户分为4类，即系统管理员，负责系统设置，数据维护等；农业政府管理人员，负责采集信息的审核统计，决策的制定；预警师：分析价格走势，确定是否预警并发布；信息采集员：采集发布价格信息。另普通消费者无需注册即可访问公共页面，查看发布的公共信息。

价格信息管理实现对各种农产品价格信息的管理，包括价格采集点、农产品名称、等级、交易时间、交易量、采集时间等。

智能手机终端（PAD）信息包括采集终端信息的维护，包括本地数据库的管理、市场信息、价格信息、产品信息、等级、数量等。

农产品数据管理完成了产品、蔬菜、水产品、粮油、药材、畜产品、茶叶等各种农产品主要数据信息的管理。

7. 报表输出

农业政府部门管理人员将采集汇总处理后的价格信息，以各种形式发布或上报主管部门。系统需能够按照多条件组合查询结果导出报表，针对不同的格式要求，输出不同的报表格式，也可以导出常用的文件格式便于数据报表的处理操作。

农产品电子商务服务平台

一、背景和意义

农业电子商务是指基于农业的农业生产管理、农产品网络营销、电子支付、物流管理和客户管理等一系列电子交易活动,基于信息技术和网络,完成了一系列农产品生产,供应和营销管理活动。

各级政府高度重视电子商务的发展,在我们国家的中长期科学和技术发展规划和信息产业部的信息产业科技发展规划和2020年中长期规划纲要中,都将"电子商务应用平台技术""农业信息化技术"列为发展重点。

电子商务得到了很大发展,并已广泛应用于各个领域。随着信息化推进,提高信息化发展水平,让现代科学技术服务于农产品销售,利用现代化技术手段,增加农产品的销售渠道,提高农业相关部门的管理服务水平,满足市民日益增长的对食品安全、健康和可追溯的要求,依托信息技术和现代技术的优势,我们通过现代手段为农业生产和产品销售提供服务。现代农业中,信息的实时交流已成为一个关键因素,如何利用现代信息技术,实时了解农资信息,利用现代科技信息服务农业生产,提供准确及时的农资信息,提供高效的农资交易平台已经成为发展的必然。

农业发展新阶段农业电子商务的实施可以作为精准农业和绿色农业的手段,也可以作为农业科技发展的基础工程。电子商务在农业部门的应用具有重要意义:首先是速度优势;其次是客户资源的优势,电子商务系统可以通过各种方式收集市场和客户信息,为企业提供最直接,最有价值的资源;第三是成本优势和个性化产品的优势。

二、发展现状

农产品电子商务模型有三种主要类型:首先是主要模式,包括虚拟社区、信息经纪人和在线黄页,主要用于提供服务,不进行实物交易;第二种是先进的模式,包括电子商店、电子采购等,其特点是在线交易;第三种是第三方交易模式,其特点是使用合约在互联网上进行交易。

目前,电子商务在农业领域的应用和研究已取得一定成效,形成了一套农产品电子商务销售平台的应用程序,具体形式包括:企业自有农产品销售电子商务平台+自有物流配送系统;农业企业发展自己的电子商务平台+第三方物流平台;第三方农产品销售电子商务平台+自营物流配送;第三方农产品销售电子商务平台+第三方物流平台等发展模式。

2003年5月,中国第一个辣椒专业网站——中国辣椒网络启动。目的是更好地为种植辣椒的农民和相关企业服务，从而带动整个产业的发展。此外，农业网络、中国农业科技信息网、农业装备企业电子商务系统、中国种子集团公司、中国农业信息网、金龙网、北方种子产业信息中心等，都是比较典型的电子商务平台。

新的农业商业指数信用评估体系启动了由商务部中国国际电子商务中心建立的诚信新农村商务平台。用于评估农产品贸易实体在电子商务活动中的信用状况，以提高农产品交易实体在交易中的标准化和信用透明度，便于建立良好的农产品交易环境。

三、总体目标

建立农产品电子商务平台，将农民、消费者和配送物流有机地结合起来。对农民来说，他们可以增加产品附加值，扩大销售渠道，增加收入；对于消费者来说，可直接面向农户，可了解农产品产地、安全标准、食品安全回溯等。

电子商务平台依托智能终端和通信网络，最大限度地支持移动互联网终端的接入。您可以将 PC 用作接入终端或将智能手机用作终端。农户通过手机发布农产品信息，消费者通过手机进行查看和购买，都是该系统提供的功能的一部分。

通过电子商务平台，可以改善物流配送系统，优化电子商务交易环境，发布企业信息。处理业务信息查询、在线订购、在线支付、物流和其他服务。搭建企业与企业、企业与政府、企业与消费者之间的信息交换渠道，实现产业链上下游协同作用，促进涉农企业产业的大发展。

（一）规划原则

该系统的设计目标是满足用户的业务需求，构建功能齐全，使用灵活，操作高效，易扩展，维护方便，可靠性高，投资少的系统，为达到这些目标，需遵守以下设计原则。

1. 经济性原则

在系统设计和设备选择中，应充分考虑资源配置的合理性，使系统具有较高的性价比；在系统设计中，软硬件都以开放式为原则，用户可以方便的在原有基础上增加设备，灵活的扩展系统。电子商务平台所需的软件和硬件环境可以在线租用或单独购买，并可根据具体情况决定。

2. 可扩展性原则

系统设计时应预留多种接口，为将来需求和功能的扩展做好准备，良好的扩展性设计可保证后续系统功能不断完备。

3. 高可靠性原则

系统的高可用性、高可靠性是企业成功与信誉的关键，通常建议用户使用高可靠、高可用的软硬件产品和技术，如 RAID 技术、内存分页、进程管理等，本系统采用 RAID 技术和双机技术，来保证数据和服务的高可靠性与稳定性。

4. 高安全性原则

数据的完整性、安全性也非常重要，所以客户的系统必须有安全可靠的数据备份、恢复系统与措施，同时，在本地建立完整的网络防御系统，以提高安全性。

5. 易升级性原则

信息领域发展迅速，应用环境、软硬件系统都可能不断更新，计算机多层体系结构使我们可方便的扩充升级新系统，达到性能最优的"数据集中，分部处理"的模式。

6. 可管理性和可维护性原则

在分布式计算环境中，系统的可管理性已成为系统成功的关键。设计的系统采用图形化的操作管理界面，抽屉式、模块化的结构，维护方面，同时，通过集中管理平台，用户可以完成系统，数据库和存储备份的管理和维护。

（二）主要功能

网站的栏目分为以下五类：信息类、业务类、服务类、功能类、移动终端功能，如图 1 所示。

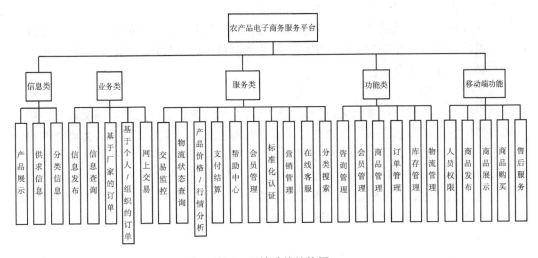

图 1　系统功能结构图

其中，

信息类：提供信息显示和沟通，以及相关的组织和业务信息。

业务类：提供电子商务平台的交易功能。

服务类：提供后端支持服务和系统使用帮助。

功能类：提供给个人、商家、其他组织等系统管理功能。

移动端功能：为了方便用户，将其提供给商家和消费者以进行移动访问。

1. 信息类功能

信息功能主要包括分类信息、供求信息、产品展示等。

产品展示：根据产品类别，商店和其他类型显示各种产品。

供求信息：图像和文本的组合用于显示商家的业务和页面上的最新促销供求信息。

分类信息：分类显示业务分类信息和促销以及其他相关信息。

2. 业务类功能

商务功能提供一系列功能，如商品购买、商品交易和在线支付。电子商务平台的主题是业务流程，可进行用户认证产品、选择产品、确认交易、确认在线支付。

信息发布业务流

经分析概况总结，信息发布业务流程如下：会员登录信息发布后台审核审核通过后，前台显示信息匹配，匹配成功，信息推送后台记录，前台显示同步。

信息查询业务流

信息查询业务流程如下：会员登录信息订阅按订阅内容推送信息根据需要查询信息；信息搜索发布查询内容背景点播查询回复推送，信息查询（非会员）网站搜索，类别搜索信息显示或部分显示（指导成为会员）。

基于厂家的订单服务

企业登录（成员）根据需求生成订单选择订单交付对象接受订单并对其进行处理。非会员单位在规定时间后可看到订单，若想参与必须先注册成会员。

基于个人或行会组织的订单农业

个人或行业组织登记（会员）提交订单购买要求后台接受订单购买需求，经过审核后推迟，会员公司立即收到订单需求；同样非会员单位在规定时间后可看到订单，若想参与必须先注册成会员。

网上交易功能

用户登录后，可进行网上交易，买卖双方可进行交流，达成交易后提交订单，购买方可直接付款或选择基于第三方的支付通道进行网上支付，卖方发货，买方可查询快递的状态，买方收货后确认付款或超过规定时间自动付款（若快超过规定时间未收到货物，及时联系卖家）。该平台支持 B2B，B2C，C2C 和其他交易类型。B2C 建立了农产品企业与农民之间的沟通桥梁，以满足双方的需求；C2C 满足了消费者和农民的需求，并在双方之间建立了便捷的沟通渠道（图2）。

3. 服务类功能

服务类功能提供系统支持服务和使用帮助等，包括交易监控、物流查询、价格行情分析、语音接口、支付结算、使用帮助、呼叫中心的等功能。

交易监控：提供交易时间、金额、数量等内容，具备查询、删除、作废等功能。

物流查询：用户可以根据时间、订单号、名称和状态等各种条件查询物流信息。

价格行情分析：结合市场的实时情况，提供农产品供需的实时分析报告。

支付结算：支持银行支付结算界面，支持在线支付和货到付款。

会员管理：它包括会员认证、会员权限、会员级别、会员升级制度、会员折扣、会员奖惩规则等，为会员提供全面的服务。

产品认证管理：为农产品提供标准化的认证管理体系，提供有机食品认证管理体系，建立绿色有机食品区；用户可以根据提供的有机食品认证信息选择有机食品和绿色食品进行系统交易。

营销管理：提供宣传信息、促销信息管理功能，为各类用户提供最新动态信息，其中登录用户可在专区查看最新促销宣传信息。

信息交流管理：提供供需双方沟通的平台，交换交易中各种问题和相关信息的信息。可根据不同的用户类别（企业、个人等），设置不同的交流便捷功能，方便用户的使用。

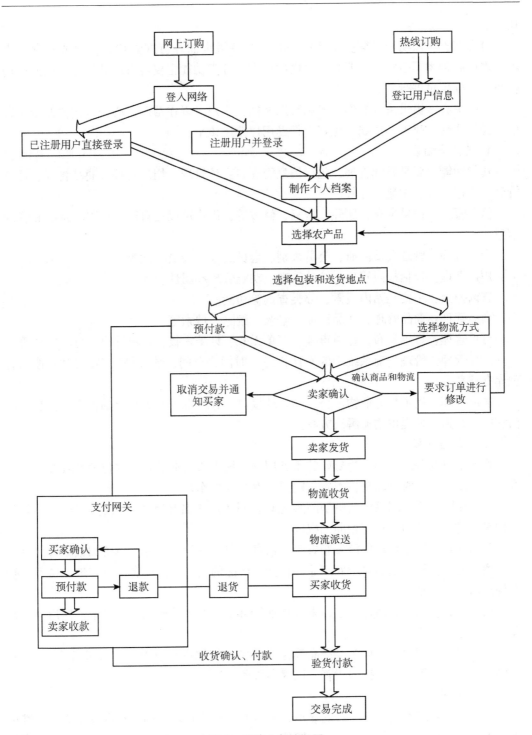

图2 系统业务流程图

在线客服：包括在线客服、客服留言、FAQ手册等。在线客服提供实时的交流解答疑问，客服留言实时性稍差，FAQ手册主要是针对常见的问题进行归纳总结，方便用户。

分类搜索：该功能贯穿整个系统平台，在各级页面均有搜索功能，比如说批发市场、农产品种类等搜索，主要是平台内的搜索。分类功能主要对农产品信息，批发市场信息和基础信息进行分类。

帮助中心：提供系统操作指南和帮助文档，用户可以在帮助中心查看相关的交易说明，操作帮助，政策和法规，更好地帮助用户使用该平台。

4. 功能类功能

功能功能主要是系统管理，包括信息管理、会员管理、用户管理、商品管理、订单管理、库存管理、物流管理等。

信息管理：信息发布、删除、搜索、修改等，用户可以在自己的权限内维护和管理信息。

会员管理：管理会员注册、会员级别、会员属性、会员功能等。

用户管理：包括权限管理、统计管理、功能管理和属性管理。

查询搜索：主要是站内搜索、分类查询等功能。

商品管理：商品管理、产品添加、修改、删除、搜索等

订单管理：订单发布、订单审核、订单无效、订单审核、订单导出、订单审核等。

库存管理：管理当前系统中的库存状态（时间、空间、数量等），并实时监控和管理库存信息。

物流管理：提供与专业物流公司的系统接口，方便电子商务平台的查询、交易、运输和配送，为用户提供物流解决方案。

5. 移动端功能

智能手机的普及应用，为更好的服务用户，提供移动端功能，这样消费者和用户均可通过智能手机对农产品进行查看、购买、发布等操作。

人员权限：手机端系统与电脑版系统有相同的用户名和密码，仅需注册一次即可，两种登录终端方式用户具有相同的权限。

商品发布：通过移动电话发布农产品信息，包括图片、名称、单价和数量等信息。

商品展示：移动电话显示农产品，包括农产品的名称、生产者、原产地、数量、折扣信息和促销活动。

商品购买：消费者通过查看搜索等将他们喜欢的农产品放入购物车中，然后进行结算、支付等。

售后服务：用户可对订单、产品等存在的各种问题进行反馈，卖方进行售后服务；一些平台问题也可以及时得到改善，以提高服务质量。

6. 后台管理功能

任何信息系统不仅需要前台的显示，还需要后台的管理。如何建设方便简单，维护人员可及时维护、更新信息，投诉反馈能及时转交，处理过程和结果可追踪，这些都是后台管理的功能。

除了页面设置、目录维护、用户管理、日志管理、信息发布、个性化信息管理等基本功能外，还需提供诚信管理、在线支付等功能模块。

7. 物流管理

通过平台购买农产品后，消费者自动生成物流管理任务。从成本的角度考虑，可根据农产品的位置、消费者的位置进行通盘考虑，将相近位置的农产品、消费者进行汇总，生成任务单，自动填写入系统中。

8. 会员管理

农产品电子商务系统，各级用户想进行交易必须注册，本系统提供了会员注册、管理、升级、奖惩等方面的制度与措施，包括会员注册，会员资格认证，会员升级，会员发展，会员活动等，以管理各级会员，提供优质贴心的服务。

9. 诚信服务

对于农产品电子商务服务系统，信息的真实性尤为重要。需要保证平台数据中个人信息，企业信息和产品信息的真实性。本系统通过多种方式确保信息的真实性，对每一个用户进行交易的用户进行严格的审核，确保每位用户的资金安全、交易安全。平台对于普通消费者，至少需验证银行卡、姓名等基本信息，通过转入或给消费者银行卡打入额度的费用进行验证；对于商户，至少需进行营业执照、银行公户的认证，这些步骤虽然麻烦，但能更好的确保用户交易的安全性。

10. 在线支付

电子商务和在线支付是不可分割的，在线支付必须涉及资金的安全。本系统引入银联、支付宝平台，为确保交易安全，构成三方支付的监管模式。银联和支付宝拥有广泛的用户基础，具有良好的品牌知名度，借助银联和支付宝两个支付平台，对目前国内各个层次的消费者能做到基本覆盖；下一步是引入更多支付平台和方法，为用户提供更方便快捷的服务（图3）。

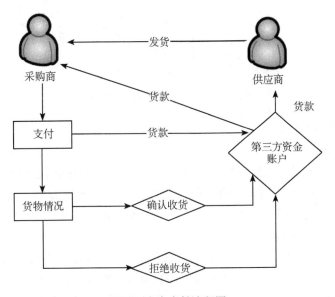

图3　安全支付流程图

四、技术方案

（一）基于 SOA 的总体架构

SOA 是一种面向服务的体系结构，可根据需要在网络上分布式部署，组合和使用松散耦合的粗粒度应用程序组件。服务层是 SOA 的基础，可以由应用程序直接调用，从而有效地控制与系统中软件代理交互的人为依赖性。

系统采用基于 SOA 的架构，可与第三方物流公司、支付平台或其他电子商务网站进行交互，兼容各种通信协议，方便部署和应用（图4）。

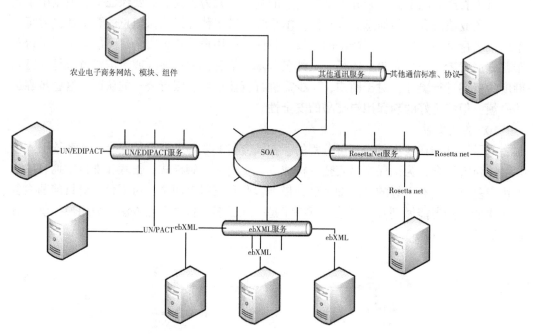

图 4 基于 SOA 的平台总体架构

（二）硬件/拓扑结构

农产品电子商务平台的硬件拓扑结构简单，易于维护，具有良好的可扩展性。能应对一定程度的突发流量，当然性价比越高越好，系统的拓扑结构如图 5 所示。

农产品电子商务平台硬件方面的设计，需充分考虑平台建成后系统的可扩展性，随着云计算技术的普及应用，价格的降低，本系统可部署在云平台上，为整个电子商务系统平台提供稳定的运行环境。

应用服务器：对于电子商务平台，最广泛使用的应用程序是 J2EE 体系结构，可用于典型的应用程序服务器软件，如 Apache、Tomcat、WebLogic、JBoss 等。

数据库服务器：整个电子商务平台的数据存储和数据吞吐量的基础，数据库服务器可以根据具体情况选择，我们选择 MS SQL Server 数据库服务器，数据库服务器可以满足大中型数据吞吐量的要求。

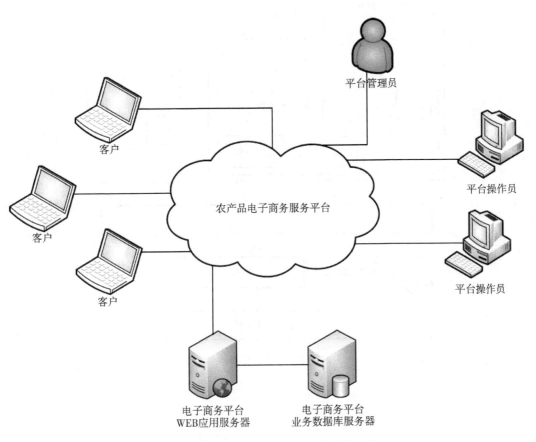

图5　农产品电子商务服务平台拓扑结构图

（三）软件架构

使用基于 J2EE 的 B/S 多层体系结构，J2EE 是企业多级应用程序开发的标准之一，提供了一种基于组件的标准开发方法。提供了相关的服务功能，如负载均衡、消息序列、事务处理等机制，且跨平台操作，与具体的硬件和操作系统无关，这样可较容易做到对多种平台的支持，兼容性好。

J2EE 是当前主流的企业应用程序架构，具有可扩展性、高可靠性和高安全性。以及易移植性等特点，被广泛应用在各个领域尤其是电子商务领域。农产品电子商务服务平台采用 Spring+Hibernate 作为主要框架结构，并结合 SOA 面向服务的体系结构。农业部门电子商务平台的总体结构是通过结合必要的组成部分和外部扩展来实现的（图6）。

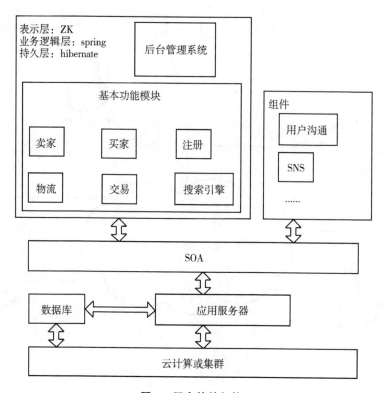

图6 平台软件架构

农产品质量安全可追溯管理系统

一、项目背景

随着我们国家经济社会的持续快速发展，人民的生活水平得到显著提高，农林牧副渔产品种类与数量得到了很大的丰富，极大地满足了各消费者的需求。但同时，我国农产品的质量与安全事故频繁发生，农产品生产、加工、流通、消费等各个环节的安全隐患不断被暴露出来。各种假冒伪劣农产品而引发出来的恶性事件不断涌现，对消费者造成了极大的伤害。从"有害食品"到"有毒食品"再到"杀食"，农产品质量安全危机激发了消费者的心，并不是一件令人震惊的事。随着这些食品安全事件，网络已经开始"全民化学知识扫盲活动""鲜肉火腿"让我们再次了解敌敌畏。这些食品安全事件/问题的曝光对消费者造成不良影响，已经引起了各位消费者的不安与恐慌。

现在各个国家均在集中国家的各方资源，努力控制与提高农产品的安全，主要包括生产、加工、流通和消费等各个环节与领域。所采取的措施一方面是通过不断制定和更新食品安全法律法规和标准，建立和实施可追溯系统，农产品和食品受到各级控制，一些国家的可追溯管理制度和管理得到改善；另一方面，要实现新的贸易保护主义，设置各种障碍，实现对外国农产品"高要求"的高标准。对外来食品的市场准入实现严格管控，要求必须达到或高于其质量安全要求与标准。这样导致的结果是，国际农产品的贸易环境变得越来越复杂和多样化。

我们国家农业监管有关部门和各级地方有关政府相继出台的一系列关于农产品质量安全的管理办法、法规和法律，以此来确保消费者餐桌上的农产品安全，利用这些办法和法律条文来应对贸易环境变化，督促各位加强农产品质量安全体系建设，做到确保我国人民大众的食品质量安全。我们国家自21世纪以来，农业部相继实施了"无公害食品行动计划"，这样带来的结果是我国农产品认证检测取得了比较好的效果。目前，已建立国家、行业、地方和企业农产品标准体系，初步形成覆盖国家、省、市、县的农产品检验检测体系。已经逐步的建立了以无公害认证为基础的认证体系计划，有机食品和绿色食品的认证作为辅助体系计划。

中国的农业产品质量管理随着信息化、现代化的深入不断完善和制度化，我们国家拨出专项资金用于农产品质量管理试点建设。2008年的中央一号文件提出了"加快我国农产品质量安全标准体系的完善，建立农产品质量安全追溯体系"。全国相继开展了农产品质量追溯体系建设的试点工作，选择北京、上海、海南、山东、山西、天津、陕西、广东、广西、江苏、辽宁、河南等省市为试点。在中共中央的积极号召下，全国各地进行了探索和实践农产品可追溯系统的开发建设和示范点的建设工作。

现在，有些部分地区已经初步的制定了一些基于当地实际情况对各个地区的农产品质量进行监督监测检验的措施，如利用 IC 卡、RFID 或条形码来记录和传递农产品信息，我们国家农产品的可追溯性监管体系的原型已经逐步形成。我们在可追溯系统建设要求和形势的推动助力下，将采取积极办法进行严格监督生产、加工、流通和销售的全部过程。要求农产品从现场到生产、加工、包装、运输、流通、消费等方面，做到有迹可循、有据可依，进行有效管控。

我们国家农产品质量安全追溯体系的建设，不论是各级政府管理部门还是学术界，都将重点关注"农产品质量安全可追溯制度"。在我们的建设和实施过程中，积极进行探索农产品质量可追溯系统的内容和框架建设，以及如何进行有效实施，来确保我国农产品质量安全的全程可追溯性管理，保证全国人民的食品安全。

二、项目意义

通过农产品质量安全追溯系统，我们可以进行有效传递农产品的全方位信息，以此来提高农产品流通环节的透明度，这是限制或减少农产品贸易当中的欺诈行为，监测农产品质量和安全的有效手段。农产品质量安全可追溯系统的建设是一个系统工程，包括"从田间地头到餐桌"的多个环节。通过生产信息、产品信息、操作规范、产品认证、检验检疫和法律法规限制，如果发现问题，我们可随时进行召回，实现对老百姓餐桌上的农产品质量安全的有效控制和监督。

通过建立农产品质量安全追溯体系，可较容易的实现从原产地到消费者的质量监督全过程，以及农产品生产区的环境监管，进行有效监督和管理生产和加工、产品测试、包装和标记等工作。如果发现质量问题，尤其是那些危害消费者健康甚至生命的问题，可由负责人和监督管理部门根据相关的可追溯系统追究责任。以此来提高广大生产者的安全意识和责任感，确保老百姓餐桌上农产品的质量安全。

建立农产品质量安全追溯体系，实现农产品可追溯管理，是进行农产品质量安全监管的有效手段，这不仅仅是对消费者负责，同时它也是实现"绿色，安全，环保"农产品的必要保障，响应农产品国际贸易环境的有效途径和措施。农产品质量安全的可追溯是未来农业贸易和消费需求的必然趋势，将得到极大的推广和认可。

农产品质量安全追溯系统为农产品"生产—加工—流通—消费"建立起了可追溯的监管数据环节。可方便容易的实现信息查询、监控和追溯，同时可以链接各级（省、市、县）检测监管网络，通过农产品批发市场监测网络来实现省、市、县（区），基地和市场监测数据的共享和互通，及时有效的控制农产品质量安全动态信息，加强预警信息发布，充分发挥信息技术在农产品质量安全监督管理中的作用，切实提高农产品质量安全监管水平，确保普通老百姓餐桌上的食品安全。

我们一定要坚持提高（保障）农产品质量安全、提高农业效益、增加农民收入的目标。坚持"政府推动、龙头带动、统筹发展、整体推进"原则，建立和完善农产品质量安全标准，从获取农业物资，监管执法进行质量可追溯性管理，全面提高农产品生产标准和质量安全标准。建立质量安全放心的农产品品牌，使之在国内享有盛誉，深受消费者信赖，并得到市场的认可。

坚持"统一规范、重点突出、分步推进"的原则，以信息技术为手段，以法律法规为依据，以现代信息技术为基础，建立以生产技术档案为管理平台，产品可追溯性代码用作信息传递工具，产品可追溯性标签（RFID、IC、条形码等）用作表达形式，以查询系统为市场服务手段的农产品质量安全追溯系统。不断丰富完善可追溯体系，创新运行机制，建立统一共享的农产品质量安全管理体系，积极探索农产品质量安全监管全过程的新途径和新方法。

三、需求分析与可行性

（一）总体需求

在应用方面，农产品质量安全追溯管理系统满足了对主要农产品、农业投入以及农产品的前期生产、中期生产和后期生产过程实施质量和安全监督的需要，进行从"农场桌"的全过程实现农产品的质量控制。

在应用能力方面，农产品质量安全追溯管理系统满足国家标准、行业标准、地方标准和企业标准的需求，用于检测农产品的质量、安全性和性能，做到数据准确、及时的传输和共享。

在管理层面，国家、省、市、县（区）和乡镇可以实现农产品质量检验的分级管理。检测数据实行逐级上报，上级对下级进行指导、监督、管理，能够实现政府的实时监控，辅助决策。

在服务水平上，能够满足广大农民群众的农产品质量安全知识的普及与提高，为广大消费者提供方便快捷的信息查询平台，提高农产品质量安全意识，对违法者进行投诉举报。

（二）性能要求

1. 系统安全性

（1）系统应防止非法用户进入系统并采取用户登录安全认证措施。

（2）系统应严格控制用户的操作权限，引入合理、有效、灵活的用户及其权限管理机制。

（3）系统应拥有日志管理功能，对整个系统的运行情况进行跟踪、检查和分析。

2. 系统扩展性

系统需具备可扩展性，可方便动态的扩展系统功能，如可方便进行功能升级、硬件升级等。

同时提高系统的稳定性，系统应能够不间断地运行稳定，并保证系统运行的效率。

（三）安全需求

系统安全风险包括黑客入侵、病毒攻击、拒绝服务攻击、内部信息泄露以及入侵网络的不良信息。系统安全需要采取措施，以确保网络的稳定运行和数据信息的安全性。

1. 网络可用性

防止对网络设施的入侵和攻击，防止通过利用网络爬虫和带宽消耗的方式消耗网络，确保网络的可用性。

2. 访问的可控性

必须有效地控制对网络、数据和系统的访问,跟踪和记录任何访问信息,确认访问者的身份,并仔细授权以确保访问的可控性。

3. 数据机密性

确保信息在存储和传输过程中保密,确保信息不会受到损害。

4. 可管理性

系统应具备审记和日志功能,对重要操作提供方便可靠的管理维护功能。

(四) 可行性

目前,在种植业中,以青岛为例,制定了158项农业地方标准,推广了218项无公害绿色食品生产技术法规,涵盖了蔬菜、果茶、粮油等有利的农业领域。建立了市级农产品质量检测中心为龙头,区县农产品质量检测站为骨干,基于城镇(企业、基地等)农产品快速检测点的农产品质量检验系统;建成农业标准化生产基地258个,农产品标准化生产示范区面积达到200万亩。通过实施农业生产档案统一登记制度,重点农业生产基地档案的建立和完善率达到95%以上,为建立农产品质量安全追溯体系奠定了良好的基础。

我们可以充分借鉴青岛奥帆赛和残奥会帆船比赛中食用农产品质量安全的经验,选择名优特色农产品,特别是在标准化的农业生产基地,在这些农产品生产基地建设农产品质量和安全追溯系统的试点项目。建立胶州大白菜、马家沟芹菜、大泽山葡萄等生产基地的农产品鉴定体系,并在其他标准化生产基地进行应用推广。区县已结合当地实际情况,进行了以条形码、二维码、RFID标签等为主要内容的农产品质量安全追溯体系的构建,通过这些实践和探索,我们为农产品质量安全追溯体系的建设积累了丰富的实践经验。

目前,北京、上海、深圳等市已建成并开辟了农产品质量安全追溯体系;潍坊寿光市实施"蔬菜安全可追溯信息系统"建立了城市无公害水果和蔬菜的质量和安全追溯系统;杭州市根据农产品规范,建立了农产品质量安全追溯体系。无锡市开展了农产品质量IC卡管理系统建设,并通过以上实践,特别是农业部农用土地复垦系统试点方案,为农产品质量安全可追溯体系的构建提供参考。

四、功能设计

(一) 系统总体设计

关注农产品质量安全监督管理业务和信息化建设管理的需求,构建农产品质量安全可追溯管理平台,以基础认证管理、质量检验管理、编码应用管理、农产品信息可追溯为核心,提高农产品生产基地的效率和农产品质量安全监督管理,从源头上监督农产品的质量和安全,建立健全市、区、生产基地、农业资金管理点参与的农产品质量安全监管体系,构建农产品质量安全可追溯管理系统,包括信息收集、系统应用和综合信息服务管理。

整个系统采用基于J2EE标准的多层体系结构,分为数据访问层,业务层和表示层。同时采用B/S架构,普通用户通过浏览器进入系统安全框架,从而访问每个子系统。

系统采用 J2EE 标准，可以在 Unix、Linux、Windows 等系统上移植，提供良好的系统兼容性。数据访问层使用我们自己的数据访问层中间件，它支持 Oracle、DB2、Sybase、Mysql、MS SQL Server 和其他数据库。系统的总体架构如图 1 所示。

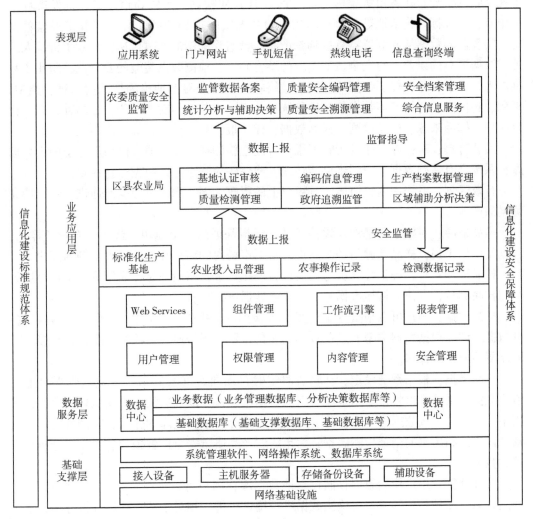

图 1　系统总体架构图

（二）农产品质量安全监管平台的功能

1. 建设区县级农产品质量安全监督分平台

区县级的农产品质量安全监督分中心是区、县监管数据的管理中心。各区县监管用户通过分中心平台管理当地农产品质量安全监管信息。对区县农产品质量安全监督中心的数据进行支持，以提高农产品质量安全监管平台的绩效和管理效率。

2. 建设农产品追溯微信平台

目前追溯平台已开通查询热线、门户网站、短信平台等查询手段，在此基础上建设农产品质量追溯微信平台，改善多渠道查询渠道，为消费者提供更多途径方便消费者查询。

3. 建设农产品质量检验检测体系

农产品质量检验检测系统是检测终端数据的收集，通过与检测仪器、计算机连接，依托互联网将分布在各市级检测中心、区（县）质检站、乡镇街道政府、批发市场集散中心、各种标准化基地等试验样品的试验数据实时报告到区（县）级和市级监督系统。实现监测数据从检测仪器直接到农产品质量监管部门的实时传递，消除了传统人工方式数据汇总的误差，与安全认证系统的结合确保了检测数据传输的安全性，防止了数据的人为修改，并确保了数据的真实性。终端检测用户对本地检测样品的品种、类型、来源、超标率、合格率等数据信息进行统计分析，它包括的功能模块，如品种管理、行政区划、样品来源、样品检测、样品数据、样品报告和统计分析等。

品种管理：对检测样品进行管理维护，建立检测样品库，通过网络汇总至中央数据库，系统具有基础数据下载同步更新功能，确保各检测点基础数据保持一致。

行政区划：系统中内置市属区县级以上的行政区划，用户可根据产地进一步维护完善样品产地。

样品来源：对标准化基地来说，样品来源指基地内各个地块，或各个农户的样品；对批发市场来说，样品来源指各个商户，包括市场内的商户和外地进入市场的商户。

样品检测：首先确认计算机与仪器设备连接正常，然后开始样品检测，用户可进行空白样品测量，方便对待检数据和本批检测数据进行查看。

样品数据：用户可以根据数据类型（未检测、未报告、报告）、检测时间等执行各种数据查询。根据数据类型通过设置不同颜色显示，还可通过对检测样品选择直接进行样品复检或终检操作。

样品上报：根据系统设定，在联机时定时自动向区（县）级、市级监管平台上报检测数据，系统可以将市区示范试验点的数据汇总并存储到区级农产品质量安全监督数据中心，并与上级数据库互为备份。上级数据库起着中央仓库的作用，存储着主管部门辖域内当前和历史的全部检测数据，为市级农产品质量安全监督预警、决策和分析提供基础数据。

统计分析：用户检测数据存储在本地数据库，用户可以根据检测时间、检测类别、样品来源、样品品种等条件对检测数据进行统计分析。相对于传统统计分析方法，运用计算机进行数据统计分析，科学性和可靠性更高，能保证所统计数据的准确、正确、可靠。

（三）农资执法监督管理平台建设

农业生产资料直接影响农产品的质量和产量，与农民收入和农村社会稳定有关。近年来，由于农资生产不均，动植物病害频繁发生，农业生产资料价格波动剧烈，导致农业生产资料市场秩序不稳定，影响农业生产，损害农民利益。我们使用移动终端的便携性，物联网设备的智能，并从农产品开始，建立农民监督制度，引导农业生产资料企业以"诚信，规范，合法"的方式运作，进一步规范农业生产资料市场的经济秩序，建立可靠的农业生产资料流通体系，确保农民对消费的信心。

农业物资综合执法管理平台为相关行政执法部门提供全面的信息管理，确保辖区内农业生产资料的安全。管理以种子、农药、兽药和化肥为主的农业生产资料管理商店，包括他们的基本信息等。

系统权限控制和角色，各部门可以实时总结管辖区内监督管理执法的详细检查记录。对有问题的农业生产资料的生产经营单位、监管部门应当按照相应的措施予以处理。

通过任务、信息、预警等信息管理功能模块，可以对监督管理部门的监督主体进行监督检查，或者对案件进行查处。

通过信息统计功能，实现监督管理数据按年份、乡镇、农业生产资料种类进行统计、分析、比较，同时为可视化提供各种统计曲线和图表，为辖区内农业行政主管部门和服务中心的监督管理决策提供科学，全面的依据，做到有法可依。

1. 农业生产资料销售管理系统

农产品质量安全管理与农业化学品投入监管密切相关，农业化学品投入准入门槛实现严格控制。

农业投入监督管理系统包括两个部分：业务系统和公共服务系统。业务系统主要面向政府职能机构相关工作人员，作为系统授权用户，根据不同的分配权限，可具有申请管理、登记备案管理、标签申领管理、标签溯源管理、高（剧）毒农药实名制管理、系统管理等功能模块。公共服务系统主要面向农业生产资料投入品的生产企业、各级代理商、各级经销商、消费者，可进入系统进行查询农业投入品的准入信息，提交备案申请，查询受理情况，进行信息反馈，修改备案资料等操作。

准入申请管理：经销商、代理商在线提交、打印农业生产资料投入品市场准入申请表，对已提交的申请信息进行管理，如果系统提示未通过审核的申请信息，则给出原因并可以返回修改。可在线实时查询申请审核情况，政府职能机构人员可以各种组合查询管理准入申请信息，进行审核并回复处理结果。

登记备案管理：已通过准入申请的产品，经销商、代理商可到相关部门办理登记备案手续，产品经过审核后对备案信息进行公示，申请人可根据要求打印生成的备案通知。

标签申领管理：取得备案通知书的经销商或代理商方可对备案产品进行采购，据实填写采购数量、品名、规格等进货管理记录，系统根据采购的批次自动生成追溯电子标签，代理商（经销商）可到相关部门领取或打印农业生产资料电子标签。

标签溯源管理：系统根据国际先进的 QR-Code 矩阵式二维码，为每一个物品生成唯一的二维码，用户可以通过扫描二维码来查询产品的归档信息、购买记录、销售记录等，及时了解产品的流向并进行有效监管。

高（剧）毒农药实名销售管理系统：通过对购买人身份信息进行识别，将高（剧）毒农药销售信息录入系统，记录消费者购买高（剧）毒农药的用途。经销商和代理商不需要手动录入农药信息，只需要扫描条码，系统就自动识别读取并显示产品信息，方便经销商（代理商）进行进销台账记录，确保高（剧）毒农药可追溯。

系统管理：政府管理部门授权工作人员，通过系统管理模块添加系统用户，添加完善经销商、代理商信息，办理准入备案手续，对违反准入备案规定的经销商、代理商资格进行吊销操作。

2. 农资流通监管移动执法系统

开发智能手机 APP 用于监管移动执法，结合二维码（条码）扫描、无线通信、RFID、NFC 等多种技术，实现农产品监督员对农产品经营企业的现场检查，取证，执法，处罚，改变过去人工手工记录的执法方式，使监管更便捷、高效、准确。

通过手机等手持终端设备与监控服务中心之间的数据交互，农产品安全监管部门可以实时下载执法任务，上传执法检查记录，实现农产品安全监督管理人员现场工作的考勤记录，确保监督管理行为的真实性。

3. 农资执法管理信息系统

为提高执法管理的信息化水平，便于检打联动管理。该系统包括功能模块，如投诉、报告、执法检查、执法活动、执法人员、文件管理、案件管理和统计报告。

执法走势：根据辖区内投诉、举报、执法检查、案件数据等情况，自动生成区域执法情况走势图，具体包括投诉举报走势、执法检查走势、案件走势等，便于各级用户实时掌握辖区内执法情况。

投诉举报走势显示辖区内投诉举报情况的走势情况，用户可根据归属地、执法单位、投诉来源、涉及的产品、涉嫌单位及类型与名称、受理时间等条件进行组合查询。

执法检查的趋势是辖区内执法检查的趋势图，用户可以根据执法情况对该区域进行检查。执法人员、检查机关、检查人员、任务发布单位、任务来源、产品类别、产品名称、查询对象类型、检查单元的类型、检查对象的名称以及检查时间。

执法案件走势是辖区内执法案件情况的走势图，用户根据所属辖区、处罚部门、执法人员、案件性质、案件来源、案件地位、当事人、当事人的类型、涉案财产，对行政诉讼现状、行政复议和受理时间等条件进行综合查询。

投诉举报：为用户提供投诉举报记录，执法检查任务分配功能。政府相关部门接到群众举报，根据投诉来源、投诉内容、所属辖区、涉及产品类别、涉嫌单位、投诉人（或匿名）、联系方式等信息后，根据具体情况部署执法检查任务，农业生产资料经营单位的投诉举报可作为单位信用评价的重要指标之一。

执法检查：执法检查人员根据投诉举报分配的任务进行执法检查，详细记录执法检查情况，填写执法检查任务名称、检查产品类别、检查依据、执法检查重点等信息，执法人员也可自行建立执法检查任务，检查记录可以与移动监视系统共享和结合，以实现移动执法检查。

执法活动：为用户提供对执法宣传、农资下乡等需要农业生产资料生产经营单位参与执法活动的管理功能，记录执法活动的详细信息，各单位的参与情况等。

执法人员：显示用户管辖范围内执法人员的详细信息，包括姓名、性别、年龄、身份证号码、工作单位、所属部门、职务职称、执法区域、执法年限、执业资格、取得时间、截止时间等信息。执法人员的信息涉及执法监督和案件管理相关联，可只选择或输入比如证件号码，系统可自动读取该人员的详细信息。

案件管理：主要实现执法部门在行政执法过程中执法案件的登记，审批和处理。制度必须严格遵守农业部颁布的"农业行政执法文件生产条例"和"农业行政处罚程序管理办法"等有关规定。完成案件处理，如案件登记、案件报告、案件审批、案件调查、案件移交、案件备案、复议诉讼、案件备案和举报信息。

文书管理：根据案件处理的不同环节，系统根据用户填写的信息生成执法文书。文书按照农业部相关规范提供的标准文书模板，根据需要选择适当的模板，查询、打印按照文书模板原样显示。文书模板主要包括："现场决定""行政处罚审批表""查询记录""现场调查（调查）""取样取证证书""产品确认通知书""证据登记保存清单""注册存放物品通知书""封印（扣留）通知""解锁（扣留）通知""案件处理意见""行政处罚通知书""行政处罚实现通知书（适用一般案例）""行政处罚实现通知书}适用听证案件）""行政处罚听证通知书""听证录""行政处罚听证报告""行政处罚决定书"等28个执法文书。

统计报表：包括统计分析和报表管理两部分，用户通过统计分析功能对辖区内投诉举报、执法检查、涉案产品类别、问题类型、投诉来源、投诉单位、执法时间（年、月、日）等进行分类统计；报表管理功能根据系统数据生成农业生产资料打假统计报表，包括农业生产资料打假情况统计表、农业生产资料打假大要案（6万元及以上）统计表，该统计表需支持常用的格式如Excel等。

4. 实时在线监控系统

实现农业生产资料销售监督管理电子化，同时实现了与重点监管对象及各个操作环节终端视频信息的互联互通。

通过电子监控技术与GIS技术的整合，生成电子监控视频位置分布图，实现各种农业生产资料销售点的视频监控和管理。

视频监控系统基于用户网络，并使用用户网络布线来部署监控点。系统集成度高，前端设备可直接将视频信号和控制信号转换为IP数据包，并在IP网络上传输数字信号。系统可靠性强，操作简单，图像质量高，不受保存时间影响。

视频显示界面可以多种方式显示，如单张图片、双张图片、四张图片、六张图片、九张图片等，可以远程控制高端高清相机。

用户可以在电子地图上标记监控点的位置信息，可以通过直接点击地图进行标记。该系统可实现城市农产品检测点的实时监控视频（默认），为领导决策和现场控制提供实时视频支持。

（四）示范基地建设

1. 新建80个农产品质量追溯示范基地

根据项目建设需求，扩大追溯体系的覆盖面，新增80处农产品质量追溯示范基地，覆盖区域内主要农产品生产基地，配备电脑、条码打印机、二维码识别器、农田数据采集设备、相关制度、实用手册等，提高生产基地的信息化管理水平，保证基地生产档案等基础数据的及时上报，加强对生产基地生产过程的有效监督管理。另外，根据项目建设需要，配备信息查询终端，方便用户查询信息。

2. 新建18个农产品质量追溯示范基地

依托农业现代化园区，建设18个农产品可追溯质量示范基地。建立智能生产管理控制系统，建立土壤养分分析检测系统，实现示范基地温湿度、土壤水分等相关信息的同步实时更新；建立农产品质量速测室，配备功能齐全的速测仪，通过平台与上级平台进行对接，实现检测数据的同步传输；建立农业生产全过程视频监控系统，收集种植、生产、成长、收获、仓储、加工、包装、销售等产品的视频信息；农业投入管理系统实现采购和投入使用等信息的收集；建立健全园区农产品质量追溯制度，并贯彻落实执行，为其他基地提供示范借鉴。

3. 试点建立46处示范农资销售管理信息系统

示范农资销售单位（以下简称"示范点"）是农业投入品准入备案及产品追踪管理数据的上报采集端，示范点用户通过PC进行产品准入备案、进销存记录、电子标签管理，通过农资监管设备或PC记录产品信息，向当地区（县）级农产品质量安全监督管理分中心报告，实现农产品进口可查询，销售可追踪，为用户带来真正的便利。

（五）运行管理

操作管理包括操作员和权限设置，以控制操作员的操作范围和操作模块。统一管理平台为整个系统的安全统一认证提供基础管理，包括角色管理、用户管理、单元管理、安全管理、系统监控和日志管理。

管理运行还涉及人员培训、耗材、设备维护更新、宣传材料印制等。

五、技术路线

（一）技术手段

该系统基于B/S多层体系结构构建，所有服务都是组件化的，并且通过Client/Server和Web Service为各种应用程序和模块提供服务。

系统的数据访问层独立于业务系统和数据库，并使用统一的数据库连接和访问对象从根本上解决数据操作对数据库平台的依赖，系统对Oracle、MS SQL SERVER、MySQL和其他数据库的访问是完全透明的。

上述技术符合农产品质量安全监督管理体系的要求，具有良好的可扩展性和适应性。

（二）多层构架技术

多层体系结构是传统客户端直接连接到后端数据库的模式，改变为客户端通过业务逻辑层连接到后台数据的访问模式，这种核心业务流程中的业务逻辑可以是一个独立的部分。可以根据业务逻辑的实际情况修改分离的业务逻辑层，而不会导致软件其他部分的修改，大大提高了软件的可扩展性和可维护性。

在传统的三层体系结构中，用户层、应用层和数据服务器是三层。用户层主要是指用户界面，尽可能简单，最终用户无需培训即可方便快捷地访问。应用层是应用服务器，即中间件、所有应用系统、应用逻辑、控制等都处于这个层次，系统的复杂性主要在这个层面；数据服务器层存储大量数据信息和数据逻辑，并且数据相关完整性控制，数据一致性，并发操作，安全控制等被放置在数据服务器层。

在系统中，我们采用三层结构，但传统的三层结构被修改，并提出了四层结构的概念。

1. 用户界面集成层（展现层）

每个应用系统都开发自己的用户界面，每个用户界面可以使用有限数量的终端设备（或者界面效果大大降低），导致跨系统业务流程的操作必须使用不同的用户界面，受终端设备可用性的限制，并显著的降低工作效率，对于客户端的 Web 界面也是如此，导致在线服务系统的满意度和利用率的显著降低。将用户界面集成在顶部，并通过集成多个业务系统的用户界面来构建跨设备和应用程序，统一，集成的交互式用户界面，为用户提供实用，灵活，适应性强，实时，舒适的体验，客户可以从任何设备随时随地获取所需信息。

2. 业务流程集成层（业务集成层）

业务流程是一组业务功能，它们根据各种业务规则进行连接，以便在特定时间段内实现特定目标。业务功能是抽象定义的：业务功能的具体实现受到业务功能操作所需的可用资源的限制，包括业务人员、客户、应用系统等。业务功能的组成由目标决定，并且建立任何活动，任务和操作以实现目标。在业务流程中，业务规则要么是受限制的，要么是有条件的，或者表示为序列化和并发等过程中的行为节点。

在还没有实现业务流程集成的系统中，业务流程的实现分布在应用程序代码中，需要跨部门的人工协作来执行业务操作，从而降低了快速流程更改的灵活性。使用业务流程驱动的体系结构，可以从应用程序中释放业务流程的逻辑，专注于业务流程管理器，形成一个新层，我们可以将其称为"业务流程集成层"。

3. 逻辑功能层

逻辑功能层抽象业务功能，并以各种组合提供接口和表达方法，它允许我们将精力集中在使用逻辑接口作为通信对象的业务流程模型上。

4. 应用数据访问层

应用数据访问层主要是指集成系统内部的数据模型，该层与特定系统负责的功能区域密切相关。但是，从整体数据集成需求的角度来看，有必要在各种系统数据模型之外定义和规划模式形式，其定义和规划主要基于统一数据视图的需要。

(三) 开放性设计

1. 可扩展性（开放性）原则

社会越先进，业务越发达，需求变化越快，可扩展性也越来越重要。需求的变化将导致扩展（修改）的软件功能，并且如果软件的可扩展性较低，则扩展（修改）的成本将非常高。

如果软件的可扩展性较低，则开发新版本软件的成本将特别高。根据系统的稳定要求设计架构的稳定性。从表面上看，稳定性和可扩展性似乎是矛盾的，但两者之间存在辩证关系：如果系统不可扩展，则没有未来的发展，因此您不仅要注意系统的稳定性且不能忽略系统的可扩展性；另一方面，软件可扩展性的前提是保持结构稳定，否则软件很难按计划开发，更不用说可扩展性、稳定性是系统继续发展的基础。从上面可以看出，稳定性和可扩展性是架构设计的要素。

2. 组件式开发技术

组件开发的重用已经从原始源代码，目标代码和库类演变为当前的组件开发技术。

组件是具有特定功能的软件模块，以硬件为中心，以芯片为中心的过程思想完美地融入软件的设计、分析和开发中，使得组件形式的软件开发与堆叠木材一样简单。组件技术是目前发展最快、最好的软件重用技术，它完全解决了软件开发中的适应性，可重用性和循环性问题。我们在系统中使用组件技术来集成OA、MIS、Web等。实现整个系统的功能，结构和接口的集成，真正集成图形和文本。

业务处理将随着时间的推移而变化，作为业务处理平台的信息系统的灵活性将发生变化。尽管由于业务需求的变化，"参数驱动"方法可以减少对应用软件的各种调整，但是参数驱动存在许多限制。在系统设计和开发阶段考虑未来的业务需求，处理规则和变更可能性既不现实也不可能。实际上，当业务需求大大调整时，应用软件必须进行相应的变更和调整。另一方面，应用程序软件的更改不会对正常的业务处理产生影响。如何使应用软件适应业务需求的变化，将信息系统从"技术导向"转变为"面向业务"。组件技术和软件总线技术用于解决上述问题。

组件技术使用软件架构作为组装蓝图，使用可重用的软件组件组装预构建的块，并支持组装软件重用。软件体系结构是对系统总体设计的描述，为构建组件提供了基础和上下文。软件组件是可以在应用程序系统中清楚识别的组件，可重用组件是指具有相对独立功能并具有可重用功能的组件。

组件技术有利于探索不同系统的高层次通用性，确保系统设计的正确、灵活、分析、规定、验证和管理整个系统结构和全局属性。使用架构作为系统演化和构建的基础，实现系统化、大规模的软件复用，大大缩短应用软件的开发周期，提高软件产品的质量，应用软件的灵活性，以及软件对业务变化的适应性，这也是大型信息系统软件的发展方向。

3. 面向构件技术

组件技术是一种类似于"组件组装"的集成组装软件生产方法，它将零件、生产线和组装操作的概念应用于软件行业，打破手工作坊式的软件开发方法是一项革命性的技术。

组件具有一定的结构和功能，符合一定的标准，是软件的组成元素，可以完成一个或多个特定的服务。该组件通过接口提供外部服务，隐藏特定的实现。通常，组件是软件系统中相对独立的软件实体，接口由规范指定，对上下文具有明显的依赖性，可以组装，并且可以独立部署，并且可以重用。从广义上讲，组件可以是数据，也可以是文档、软件体系结构、对象类、测试用例等。

软件组件库是软件资产管理工具和支持软件重用的基础结构，提供用于描述、存储、分类和检索软件组件的功能，有效支持基于组件的软件开发，提高软件产品的质量和软件开发的效率。

组件的技术标准逐渐成熟，目前使用的软件组件的技术标准是：Microsoft COM/COM+、SUN 的 JavaBean/EJB、OMG 的 Corba，这些标准为应用程序开发提供了可移植的异构环境和平台。它结束了面向对象的开发混乱，解决了互操作性和通信等异构环境中软件重用的瓶颈问题。

（四）统一门户技术

随着企业和政府信息化的深入，对信息资源整合的需求也越来越强烈，它不是自下而上的分散式构建，而是物理分散，逻辑集中，无缝集成和应用程序增强。既要无缝地访问各种现有的或新构建的各种结构化数据信息，非结构化数据信息和应用软件系统。通过门户整合、内容整合、应用整合、数据整合、流程整合到统一的门户管理系统，通过安全可控、个性化的门户页面提供给各类用户。

统一门户技术是聚合和管理异构和分布式信息资源，通过单点统一门户页面为用户提供个性化内容服务，支持分类分级式的可控访问。统一门户就是各个分散的政府应用集成在一起，发布在门户上，用户可以从单一入口进入浏览，使用所有政府应用系统提供的服务。对用户来说使用方便，至少无须记忆或注册大量的用户名和密码，目前在各级政府、大型企业中应用较为普遍。

（五）XML 数据交换技术

XML 是一种很有前景的全球数据标准，将在许多领域和领域发挥重要作用，例如，实际上不可能搜索多个不兼容的数据库，但 XML 技术允许轻松合并来自不同来源的数据结构。软件代理可以集成来自后端数据库的数据和中间服务器上的其他应用程序数据，可以将合并数据传递到客户端或其他服务器以进行进一步处理，聚合和分发。

XML 的灵活性和可扩展性使得描述网页和应用程序中的材料等许多不同类型成为可能。由于 XML 数据可以自我表达，因此可以在不需要内置数据的情况下交换和处理数据。

XML 存在众多优势，如独立于任何语言的数据格式；独立于任何体系机构的数据格式；开发更灵活的 Web 应用程序；XML 便于不同系统之间信息的传输；不同方式的资料检视；已经被广泛使用。

XSLT 是一种描述将一种 XML 结构变换为另一种结构的标准方法，是 W3C 提出的标准，通过 XSLT 对 XML 数据进行格式转换操作，可进行如下应用：将数据转换为浏览器的显示格式；在不同的内容模式之间转换商业数据。

（六）安全身份认证

互联网具有开放性，方便快捷与外界沟通，加强内部管理、提高工作效率，同时互联网存在很多安全问题，因此需要系统采用一定的安全措施进行防护。

基于本系统的实际情况，我们拟采用静态密码加动态认证结合的方式，系统对用户身份进行审核，如果静态密码过于简单，则不会通过密码审查，从而提高系统身份认证的安全性；对于动态认证，拟结合手机短信或动态令牌的方式，由登录用户输入令牌中的动态密码或手机动态短信密码，检查服务器端动态密码以验证用户身份的有效性。若采用 U 盾登录方式，登录用户必须统一发放 U 盾，然后才能在电脑上进行登录；若采用手机密码认证的方式，则用户注册时必须提供可用的手机号，便于接收短信验证码。

智能农业机械物联网系统

一、背景和意义

农业机械化作为我国现代化的重要内容和组成部分。近年来随着我国现代化、城市化、工业化目标的提出，国民经济的迅速增长，农业机械化的性能和科技含量得到了明显的提高。利用现代信息技术充分发挥农业机械设备的作用，可以大大降低劳动强度，降低成本，提高生产效率和效率。随着现代农业机械化作业范围的不断扩大，由于缺少有效的农业机械调度手段而导致的农业机械分布不均匀，农业机械化信息滞后，信息反馈时效性差，可用性低，农业机械化作业组织和参与者对信息的详细、准确、快捷的要求难以满足等问题逐渐凸显。这些问题不仅降低了农业机械作业的质量和效率，而且造成了农业机械的不合理分配，这导致了资源的严重浪费，给农业机械作业的进一步发展带来了很大的困难。

国家强度利用信息化手段加强农业机械调度，及时收集、分析和发布农业机械供需、作业价格等市场信息，为农业机械企业，农业经营者和农民提供有效的信息服务，促进经营机械的合理有序流动；另随着购机补贴的进一步推广，相关的监管、管理工作得到重视。

农业部强调利用信息化手段加强春耕等重要农时农机调度工作，及时收集、分析和发布机具供需、作业价格等市场信息，为农民、农机手和农机企业免费提供有效的信息服务，促进作业机械的有序流动；并且随着购机补贴的进一步推广，相关管理、监管等工作也受到重视。要求各级政府加强技术创新，提高机械作业的自动化和智能化水平，促进信息化与机械化的融合；要求创新形式的农业机械化服务和发展农业机械合作社，发挥农业机械合作社的作用，组织开展跨区域经营、订单经营、承包经营、收集和替代服务。完善农机服务组织，要加大对农业机械研究、农机维修、安全监管、信息服务、教育服务等公益性设施建设的支持力度，提高农业机械化的公共服务能力。

农业机械是农业生产的重要生产资料，它的管理水平从某个侧面反映了农业现代化的水平，农业机械的发展趋势我们可以总结为以下几个方面。

（一）智能化

农业机械的高科技一体化和农业机械的一体化已成为提高农业装备制造业竞争力，并最大限度地发挥作物和土壤潜力的需要。满足作物生长需要，减少农业生产资料投入，从而减少消耗，保护生态环境，节约资源，增加利润，实现农业的可持续发展。

（二）精准化

以环保，经济农业为发展理念，开展精播农艺生产，如播种、精准定位、施肥、精准喷洒等。对精制农业机械和设备的需求不断增长，农业机械产品精准农业系统的提供已成为世界农业机械发展的趋势。

（三）节约化

环境退化和资源短缺是人类面临的重要问题，也是农业现代化建设和促进农村经济可持续发展过程中不可忽视的问题。积极推进节约型农业机械的发展，实现节能、种植、医药、化肥、燃油经济、节约用水的目标。同时，它可以减少环境污染，获得良好的经济、生态和社会效益。

（四）大型化

考虑到中国的大规模、土地集中、双工作业、节约成本等问题，农业机械化向大规模发展的趋势非常重要。原因是大型农业机械具有效率高，质量好，平均运行成本低，联合作业有利的优点，大型农业机械是农业机械未来发展的趋势之一。

中国正处于从传统农业向现代农业转型的关键时期。逐步向经济、智能、精密、规模化方向发展，注重产品质量与配套技术的整合。其中，农业机械化进程离不开农作物信息的认知、农业生产的智能决策、农业数据的传递、农业系统的优化。农业机械物联网技术可以在这些方面发挥重要而独特的作用，因为物联网技术是利用互联网技术全面整合信息感知、数据传输、智能处理和优化决策等关键技术。利用现代信息技术建立农业机械管理现代化系统，将有利于提高野外作业水平，提高农业机械利用率，快速调度农业机械作业。建设的必要性包括以下三点。

（1）系统建设，以满足改善工作管理和信息技术发展的需要。近年来，国家各级都高度重视信息管理和信息技术应用，要求各级农机管理部门加强信息系统管理，完善管理和监管工具。一些农业先进地区先后开发建立了农业管理调度指挥系统，并首先在全国范围内实施。基于北斗定位导航系统，地理信息系统，移动通信等技术的农机信息管理。建立和构建农机操作指挥调度监控系统，实现信息的共享和上传，全面提高农机的管理水平和监管手段。

（2）制度建设是完善和加强农业服务体系建设的需要。随着农业机械的发展和普及，农业机械管理系统建设中信息系统的运行越来越明显。但是，由于多年来对工程机械信息系统的投入不足，信息系统分散，规模小，信息终端数据采集不完整等问题和现象尤为突出。农业机械技术推广体系，技术创新体系，流通服务体系和教育培训体系四大体系信息系统分离或缺乏功能。它限制了农业机械安全监管，农机购置补贴监管，农业专业合作社发展的有效实施。保护现代农业的发展，建立一个高效，高水平的农业机械作业调度指挥控制系统迫在眉睫。

（3）系统建设是实现农业机械安全生产监控，应急响应和指挥调度的需要。为了实现对农业机械动态的准确，快速和直观的理解，必须使用电子地图技术和卫星定位和导航技术，使指挥官和决策者能够直接了解农业机械的当前位置、速度和路线。做好准确科学的指挥，提高农机运行调度、防洪、抗旱、苗木养护等应急响应能力。建立农机安全运行监控，农机专业合作管理，跨区域运营管理，购置补贴，以及年检率、上市率和许可率"三率"管理的协同制度，充分发挥农机指挥监控中心在有效，及时处理农机调度和安全监管中的作用。

二、建设原则和目标

建设现代智能农业机械物联网系统，需要遵循农业现代化的一般规律，且需要考虑同其他农业信息化系统间的相互关联与通信，智能农业机械物联网中的数据是农业信息整合的重要组成部分。同时信息化建设必须照顾行政管理、农户使用等各个方面，达到便于推广，农业方便使用和数据录入，管理部门可顺畅获取数据，实现后期统计分析。具体建设原则如下。

（一）数据标准化

系统建设实行整体设计、统一规划，在统一目标的指引下，采取分步、分期建设的方式。该系统可以依靠统一的农业云平台，统一的业务管理模型和统一的数据管理系统来实现统一的运营和部署。系统采用统一的互联网络技术标准、统一的计算机技术标准、统一的数据标准、统一的业务标准。

（二）系统互联互通

农业信息化建设是一项全面、长期的工程，智能农机作为其中的一部分不是孤立存在的。系统要求其设计和运行阶段需要和其他各类系统进行数据交换，同时应专注于应用系统及现有技术，考虑到成本最低的网络技术的发展，当系统规模迅速扩大时，不需要重新设计和规划系统规模，使现有的系统能够与需求同步增长，避免因相同数据的重复处理导致的重复性建设。

（三）适用性和先进性相结合原则

项目建设是一项长期建设、长远规划的工程。在工程设计上应具有前瞻性，充分考虑到未来发展的需要，在技术上要保证在想当长的一段时间内不落后。与此同时还应当结合项目建设的实际情况，在保证实现系统建设目标的前提下，尽量选择成熟可靠、物美价廉的技术和产品，使系统的性价比达到最高。

（四）农户可直接使用

鉴于智能农机物联网系统用户的范围和用户的参与，需系统落地，农户可直接使用。在系统的设计中，应该考虑到维护应该是方便和用户友好的。系统设计中，系统管理方便简洁，系统功能强大，系统界面简单实用，系统架构易于理解，系统维护自动化，简便易行。

（五）数据安全

对于农业信息服务系统和农业管理服务，安全是不可替代的。因此安全问题是系统设计的重要原则，有时甚至超过系统的功能。整个系统架设在农业云平台之上，采用防攻击手段和多种容错措施，采用镜像备份和双机容错的工作方式，确保系统的正常运行；在内网和外网均采用防火墙技术，隔离非法的访问，保证系统不受网络攻击；使用多级权限管理系统，给系统管理员、应用管理员、领导、一般办公人员进行准确的权限分配，保证重要资料的安全性。

（六）管理便捷

为方便用户，需要考虑项目的方便性和实用性。在数据组织和系统表现方面，需面向应用人员，通过简单易用的手段实现系统应用。系统功能强大，系统界面简单实用，系统架构易于理解，系统维护自动简便，系统管理方便简洁。

智能农业设备物联网系统的建设分为软件和硬件两部分，网络系统，服务器系统，视频会议系统，各级调度指挥中心和子中心硬件结构。智能农业机械调度指挥控制系统中包含的子系统和功能包括以下两个。

1. 农机智能调度子系统

系统在地图上实时显示安装北斗终端的农机的位置和相关的状态信息。指挥调度监控中心根据作业需求的地点、时间、性质进行资源查找，系统运用时空智能调度模型，实时提供可用农业机械的位置和数量，协助指挥调度监测中心配置和调度农业机械。

2. 农机管理子系统

包括合作社管理、农田区域登记、作业供需管理等方面，应充分调研掌握本地作业需求，为统筹规划农机资源提供基础数据支撑。

三、建设内容

智能农业机械物联网项目针对管理者、农机手等提供信息化服务，主要建设内容包括以下方面。

（一）网络系统

网络系统建设采用分级结构，根据需要一般采用三级网络结构，采用电信（移动、联通）专线建设，保证视频会议及管理调度数据的实时高效传输，整个网络由专用线路组成，以确保数据安全性，可靠性和良好的传输。

（二）服务器及存储系统

服务器采用云计算平台，该平台借助当地政府管理部门已建设的项目，这样可大大降低成本，系统运行的专业化水平也得到明显提升。

（三）视频会议系统

视频会议系统采用成熟稳定可靠的高清视频会议终端，实现与上下级部门的互联互通，支持至少720P的高清视频会议，保留上行链路的视频会议系统接口以促进与国内或国外视频系统的互连。

（四）指挥控制监测中心和分中心建设的运行

普通农机运行监控调度系统分为三个层次：市级调度指挥中心、区级调度指挥中心、区域调度指挥中心。

城市调度指挥中心的建设目标：充分掌握城市农业机械的状况，以视听形式发送和指导全市农业机械，实现工作机械的有序合理流动，避免整个区域的盲目操作，支持急需的区域。建设市级调度指挥中心，包括中央控制系统、音响控制系统、大屏幕显示系统和装饰工程。

区级调度指挥分中心的建设目标：监测和管理各区县的农业机械信息，使用专用网络连接上级调度指挥中心，通过视频会议终端参加市级调度指挥中心的视频会议，与市级指挥调度中心建立高效，便捷，安全的信息交换连接。

区域调度指挥分中心的建设目标：监测和管理区域中心的农业机械信息，利用专用网络与区级调度指挥分中心建立联系，通过视频会议终端参与视频会议和远程培训，与区、县级调度指挥中心分支机构建立高效、便捷、安全的互动连接。

（五）智能农机指挥调度监控系统

该系统可实现农机运行监控、实时调度、农机定位、调度信息上报、信息展示以及当前农机状态查询，农机需求上报，合作社负责人、管理员、农机手、车主等可根据权限查看车辆的位置信息等，控制加入系统的车辆的当前位置和状态。其中农机定位装置安装我国自主研发的北斗定位系统，由北斗定位终端获取当前位置信息，并将农机运行信息、位置信息反馈给农机监控调度系统。

监控调度系统分为后台管理系统和前端系统，由农机智能调度系统、农机管理系统和系统接口模块组成。实现农机信息查询、需求报告、农机调度、农机管理、系统管理等功能，其中地市级调度指挥中心可实现对地市级区域内农机信息的监控与查看，区县级、区域中心只能监控或查看所辖区域内的农机，即上级可以管理（指挥）下级，下级不能管理（指挥）上级。

四、总体构架

农机运行调度监控指挥平台部署在云平台上，通过标准接口实现同其他各类农业信息化系统的连接，地市、区县、区域中心等级管理机构通过权限管理使用该系统，实现农机监管、信息展示、数据采集、数据分析、数据上报等全方位农机管理。该平台采用B/S架构，客户端无需安装专用软件即可通过通用浏览器登录和访问系统。

我们从图1中可以看出，系统分为地市级、区县级、合作社、农机户四个层面的管理平台，每个平台通过专用网络交换信息，并使用Internet实现信息显示。该系统架构的优点体现在管理和信息发布上，确保信息可靠与安全，分为地市、区县、合作社及农机户四个层面，使得系统可进行全局统一管理，又可直接为农机户服务。

应用服务器：实现业务逻辑处理，用户界面显示，初期可只部署一台；随着应用子系统、用户量的增加，可采用动态平衡技术，增加服务器的部署，满足客户端服务请求响应的需要。

北斗导航定位终端：北斗导航定位终端安装在农机上，提供导航服务，同时将农机定位信息、运行信息返回给导航服务器。北斗导航定位终端数据的传输需要依靠数据网络（如电信、联通、移动网络）的支持。

导航服务器：负责与北斗导航定位终端的数据通信，为北斗导航定位终端提供导航定位支持。

数据库服务器：该系统使用Oracle数据库，该数据库负责系统数据和信息的存储、检索和优化。

短信平台服务：借助和信息短信群发平台，实现调度指挥监控中心农机手的短信发送功能，包括各类通知、公告、注意事项等信息。

五、技术路线

智能农机物联网系统建立在云平台之上，可与各农业信息化系统进行信息交互，实现系统间的数据共享与互通。各级用户可采利用远程接入技术，移动计算技术，实现系

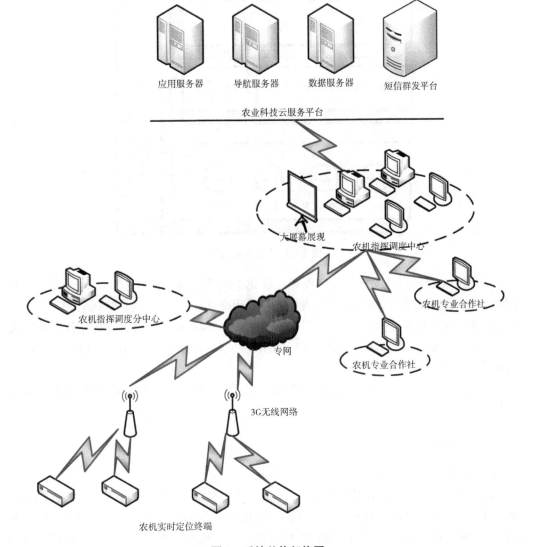

图 1　系统总体架构图

统的无缝接入，供各级用户随时使用。对于一般用户，系统开放统一的信息平台，对当前所有农业机械调度情况进行信息发布、需求征集等，利用互联网络即可实现；对于地市、区县、农机手等上报农机信息、进行调度等数据操作，则需通过后台管理平台来实现，此方面使用专用网络实现，以确保数据安全性和可靠性。

在上述网络架构上，系统软件功能分为地市级调度监控中心、区县级调度监控分中心、农机手、普通用户四个层面。这四个层面相互关联、相互配合、共同完成数据采集、分析、共享和监视，软件架构如图 2 所示。

主要采用的技术路线包括：

开发平台：系统采用 B/S 架构，采用基于 Microsoft.NET Framework 5.0 开发平台，采用 MVC（Model-View-Controller，模型—视图—控制器）系统架构。

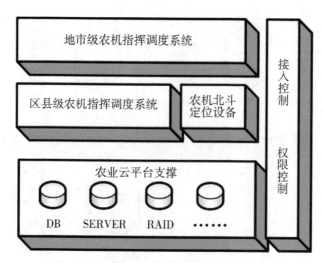

图 2　系统软件架构图

遵循 SOA 面向服务的体系结构思想，优化应用程序模型的粒度，并设计高可用性组件；采用通用高效的 XML 数据总线技术；采用交互式网页开发技术 Ajax。

支撑平台：服务器操作系统采用 MicroSoft Windows Server 2010，兼容各种浏览器（Chrome、IE 10）；采用 Oracle 数据库；使用统一身份验证技术，用户可以登录一次以访问所有授权服务。

位置服务平台：电子地图采用高德电子地图，导航定位系统采用北斗卫星导航系统。

六、关键技术

（一）北斗导航定位

北斗卫星导航系统是由中国自行开发和制造的全球卫星定位和通信系统（BDS）。GPS 和欧盟伽利略卫星导航系统（Galileo satellite navigation system）等之后的卫星导航系统。该系统由地面端、空间端和用户端组成。它还具有短消息通信功能，包括授权服务和开放服务。授权服务为具有高可靠性和高精度导航要求的用户提供定位、速度测量，定时，通信服务和系统完整性信息；开放式服务为世界免费提供定位，速度测量和定时服务，北斗在导航精度上并不逊色于欧美。

与 GPS 和 Galileo 相比，北斗导航终端具有结合短信服务和导航功能，增加通信功能的优点；快速定位全天候，减少通信盲点和 GPS 的准确性，为世界提供免费导航服务，提供时间服务和被动位置导航的用户数量没有限制，并且与 GPS 兼容；该系统适用于群组用户的广泛管理和监控，不依赖于区域数据收集用户数据传输应用等。另一个独特的中心节点定位过程和用户机器设计可以解决"你在哪里?"和"我在哪儿?"；系统独立设计，具有高强度加密设计，为关键部门提供稳定，可靠，安全的服务。

（二）电子地图

数字地图是使用计算机技术以数字方式存储和查看的地图。电子地图存储数据的方式通常存储在矢量图像中，并且地图可以按比例放大、缩小、旋转等，而不会影响显示效果，使用 GIS 存储和传输地图数据。国家测绘局在全国范围内拥有各种电子地图，各省市测绘和城市规划部门制作了大量的大型电子地图，可用于城市规划、航海、旅游、交通等多个部门。这些数字地图将各部门的日常工作频繁地转化为地图，转化为信息技术手段，直观、准确、科学，工作效率大大提高。

为保证地域数据的唯一性，农业系统应部署使用区域内专用农业地图，包括行政区划、地块划分等。在前期地图信息不能满足的前提下，考虑农机运行情况监控、作业范围及路径，可先利用高德地图来实现。高德 Maps 的 API 免费向公众开放，非营利性公共服务组织可以直接使用。

（三）SOA

Service-Oriented Architecture，针对智能农业机械物联网系统的特点和业务目标，采用面向服务技术架构分析和设计方法、项目设计、分析和开发遵循抽象、统一、合规和迭代、业务驱动的设计原则。基于 SOA 的分析构架如图 3 所示。

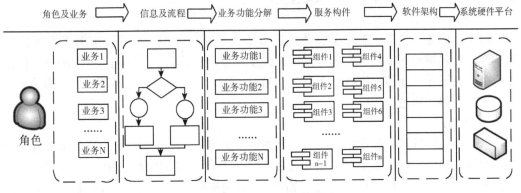

图 3 基于 SOA 分析构架图

分析步骤与方法如下。

一是确定系统建设目标和业务目标。

二是了解业务及角色。

三是根据现有业务流程了解关键业务流程并提出业务流程，并根据信息技术要求通过重构思路进行优化。

四是确定系统建设目标和业务目标的业务需求，并获得业务功能。

五是分解功能实现为服务构件，整合各业务构件，形成专业服务构件和公共服务构建。

六是确定系统硬件平台和环境。

七是进行系统架构设计。

（四）云计算

系统使用统一的农业云平台来实现后台系统的运行与响应，相对于传统的独立服务器架构，采用云平台可更好的实现数据共享和资源的重复利用。其主要特征如下。

1. 资源配置动态化

可以根据需要动态划分和释放物理资源。如果用户不再需要此部分资源，则可以动态释放资源。云计算为客户提供 IT 资源利用率的可扩展性，实现动态"无限制"的资源使用。

2. 以网络为中心

云计算的整体架构和组件通过网络连接，以通过网络向用户提供服务。客户可以使用不同的终端设备通过标准应用程序访问网络，从而使云计算服务无处不在。

3. 需求服务自助化

云计算为客户提供资源服务自治，用户可以自动获取自助服务计算资源，而无需与提供商交互。同时，云系统为客户提供特定的应用服务目录，客户以自助方式选择满足自身需求的服务内容和项目。

4. 服务可计量化

通过针对不同服务类型的客户进行计量，优化资源分配和自动控制，也就是说，可以使用即用即付服务模型来控制和监控资源的使用。

5. 资源透明化

对于云服务提供商，阻止了各种低级资源（网络、存储、计算、资源逻辑等）的异构性。边界被破坏，所有资源可以统一调度和管理，成为"资源池"，为用户提供按需服务；对于用户而言，这些资源是透明的、足够大，无需了解内部结构，只需关心您的需求是否得到满足。

（五）VPN

系统担负管理任务和用户访问，其管理数据应当和公网作出区分，来保障数据访问的安全性。VPN（虚拟专用网）是在公共网络上建立的专用网络技术，VPN 网络的任意两个节点之间的连接不具有传统专用网络所需的端到端物理链路，并且是由公共网络服务提供商提供的网络平台上构建的逻辑网络。用户的数据在逻辑网络链路中传输，因此称为虚拟网络，其特点如下。

（1）节省大量的通信费用。通过公共网络建立虚拟专用网络不需要大量的手动人力资源来维护和安装 WAN 设备和远程访问设备。

（2）数据传输安全可靠。虚拟专用网络产品使用诸如认证和加密之类的安全技术来确保用户的可靠性以及传输数据的机密性和安全性。

（3）连接灵活方便。如果没有虚拟专用网络，如果两个合作伙伴想要建立安全的专用连接线，则需要建立租用线路。

（4）网络的控制权。VPN 使用户可利用 ISP（Internet Service Provider）的服务和设施，又完全掌控着网络控制权；可以在企业内部建立虚拟专用网，满足企业信息传输的需要。

（六）MVC 体系架构

MVC（模型—视图—控制器）表示具有不同层的软件架构模式，其将软件系统划分为三个基本部分。模型（Model）、视图（View）、控制器（Controller）。模型—视图—控制器模式的目的是实现一种动态的程序设计，将应用软件与数据访问层、业务逻辑层、展现层、用户界面分离开，各层之间耦合松散；通过高扩展，对程序的后续修改和扩展进行了简化，并且其他层中的更改不会由一层中的逻辑更改引起。能够快速响应变化，支持分布式部署和跨平台；通过简化复杂度，程序结构更加直观，节省了应用软件架构设计的麻烦，提高了开发应用效率。

低耦合性。业务层和视图层的分离允许更改视图层代码，而无须重新编译控制器和模型代码。类似地，应用程序的业务流程或业务规则的更改仅需要更改 MVC 层模型层，因为控制器和模型与视图分离。更改应用程序数据层和业务规则很容易。

可适用性和高重用性。随着技术的进步和技术的进步，用户需要越来越多的应用程序访问，而 MVC 模式允许使用各种视图访问服务器端代码。包括任何网络浏览器（Chrome、IE、Opera、Firefox 等），并且因为模型返回的数据未格式化，同样的构件可被不同的界面使用。

生命周期成本低。MVC 可以降低开发成本并维护用户界面，使用 MVC 模式可以大大缩短开发时间，使开发人员能够专注于业务逻辑，而 UI（用户界面）设计人员则专注于演示。

可维护性。业务逻辑层和视图层的分离使得更容易修改和维护 Web 应用程序，因为每个层都执行自己的职责，不同级别的不同应用程序具有一些相同的功能，这有助于工具和工程来管理程序代码。

（七）构建快速响应的分布式应用程序并集成 Ajax 的 B/S 架构

B/S（浏览器/服务器）结构是浏览器服务器结构，在该结构中，用户工作界面由 WWW 浏览器实现，并且大多数事务逻辑在服务器端实现，形成所谓的三层结构，这极大地简化了客户端的负载，减少了系统维护和升级的工作量和成本，并降低了用户的总体成本。从现有技术的角度来看，建立 B/S 结构，通过因特网访问数据库以及建立基于 Web 的应用程序的应用相对容易掌握且成本低。该方法使不同的人能够从不同的位置和不同的访问方法（例如 WAN、LAN、Internet/Intranet 等）操作和访问数据库，它可以有效地保护数据平台并管理访问权限，服务器数据也是安全的（图4）。

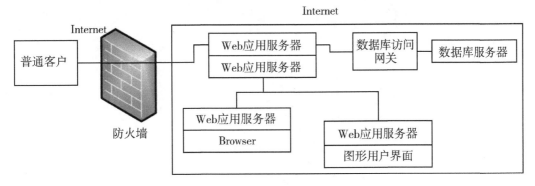

图4　软件结构图

Ajax 是一种可以更新网页部分而无须重新加载整个网页的技术。有些人认为 Ajax = 异步 JavaScript+XML，Ajax 是一种用于创建快速动态网页的技术。

用户在后台与服务器交换少量数据。更新了页面的某些部分，如果您需要更新内容，传统网页（没有 Ajax 技术）必须更新整个页面。有许多 Ajax 应用案例，如谷歌地图，新浪微博等（图5）。

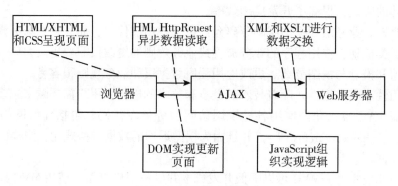

图 5　AJAX 技术的构成

农民教育培训系统

一、建设目标

一是建立新的农村社区居民技能教育机制和培训体系，将各个培训教师、培训机构、服务机构、业务管理部门、志愿者、社区学校、受训者纳入平台进行统一管理，建立新型农村社区居民技能教育培训管理信息系统。

二是实现农业科技知识资源共享，利用现代信息技术传播农业科技，用户可以通过手机和计算机获取农业科技知识和成果，建设具有在线考试、在线培训、点播、直播功能的农民培训网络学习系统和资源共享系统。

三是通过多媒体传播普及农业科技知识和成果，加快农业科技媒体资源数字化建设，建立电视制作系统。

四是利用报纸期刊进行农业科技知识的传播，可通过报纸期刊登共享利用农业科技知识和成果，建设报纸期刊管理信息系统。

五是引入先进农业科技培训资源，接入全国远程培训管理系统，促进农民科技培训事业的迅速发展。

二、建设原则

（一）先进性原则

采用主流信息格式、成熟的技术方法、先进的设计思想、采用严格的用户验证，紧跟信息技术的发展，顺应当前的发展趋势和未来的发展趋势，具有较强的使用价值和生命力。

（二）开放性原则

通过面向服务的体系结构，通过统一的数据交换、统一的身份认证和统一的信息门户，每个系统应用程序相互连接，具有良好的兼容性和开放性。

（三）标准化原则

采用业界主流的技术规范，系统建设严格遵守国家标准，系统规划设计及设备选型遵循标准化原则。

（四）统一性原则

与其他农业信息化系统采用统一的技术架构、统一身份认证、统一网络部署、统一信息标准，便于系统信息数据的互联互通。

（五）可用性原则

系统使用各种安全机制来确保系统可用性。

三、建设内容

（一）新农村社区居民技能培训体系

新农村社区居民技能培训系统主要包括：新农村社区居民技能培训体系和工作机制；将培训机构、服务机构、农村社区学校、业务管理部门、培训教师、志愿者、受训者纳入平台统一管理；实现新农村社区学校管理的统计分析和查询功能以及新农村社区居民的技能教育培训数据；建立高效规范的新农村社区学校管理流程，建立新型农村社区居民技能培训工作流程，包括：项目管理、任务管理、班级管理、教学计划、考试评估、培训实施、培训档案、总结验收。

该平台采用流媒体技术、多终端编程和发布技术，实时通信技术，并结合行业远程教育的数字教育标准。用户沉浸在体验和技术中，并为机构、企业和政府推出用户交互式网络培训系统；该平台侧重于个性化和谨慎的培训模式，这是传统培训模式的一次革命。该平台为培训师提供广泛时间分布、互动和创新的学习方法，帮助机构、企业、政府和其他网络培训任务，如技术培训、政策培训、技能培训、在线学习、评估培训等。提高人才培养效率，增强人才竞争优势具有重要意义。

该平台支持多级分布式部署，多权限资源管理和统一调度，采用开放接口，标准化管理中间件，统一多终端发布引擎。支持多屏互动，多服务扩展。该平台通过在线课堂直播，培训内容定制，按需培训材料，交互式学习交流，内容多终端推送，实现各种形式的网络培训和教育应用。该平台支持学习和管理功能，如用户管理、培训管理、课程管理、统计分析和系统管理。

多码流支持，是课件录制系统的采集模板中设置好相应的多个视频比特率等参数，录制完成的课件可以是多码流课件，供学生点播学习、教师授课时使用，确保了在复杂多变的网络环境下均能流畅的播放学习课件。

系统融合当前音视频主流编解码技术，采用高效率的视频渲染和码率压缩转换机制，支持高清、标清以及各种自定义的码流和格式输出，支持集群和单机式后台操作，提供批量、统一、集群式的专业化音视频媒体格式自动转换解决方案，它节省了大量的人力资源成本，效率极高。

音视频多种格式内容转码，系统支持将多种媒体格式进行批量、统一转码。

转码策略自定义定制，系统支持转码参数自由设置，可动态添加、修改、删除转码任务，可以设置代码转换文件名，目标文件的格式，分辨率，帧速率，音频编码格式，视频流等。满足不同用户对转码格式的不同需求。

高效率、集群式转码部署方案。系统支持多线程、多服务器集群式转码，设置好转码参数后，可选定多个任务、多个服务器并行进行集群式转码。用户可根据服务器的性能自由设置转码的并发数量，支持对转码服务器的自由扩展和临时调用；支持对流媒体应用用户提供网站视频格式整体批量转换方案，可迅速将用户网站已有的媒体文件进行统一无缝批量转换。

任务状态的实时监控。用户可以查看等待转码、转码和完成转码的所有音频和视频文件的详细信息。了解实时解码速率、转码次数、转码进度等。实时管理和控制转码，提高转码效率。

(二) 农民培训课件制作

名师教学培训过程录制成课件并进行后期制作管理，是培训工作的重要环节，是教育资源建设的重要措施，可供用户点播学习，完成知识学习与传承，建立数字共享资源库。课件录制和播放系统集成了多媒体技术、图像分析和处理技术、网络流媒体技术、自动控制技术、人工智能技术、智能跟踪和录制讲座、师生互动、计算机信号、课堂书籍等。自动生成课件资源（当然可对课件资源进行后期处理）。将教学方法、教学理念、教学设备等完全结合起来，实现教学活动的全面，完整再现，形成智能化、自动化、规范化的培训资源建设和应用模式。

1. 视频/图像采集

电脑屏幕、声音及图像的同步采集。课件实时录制系统同步实时采集授课教室的计算机电脑屏幕（幻灯片、操作演示过程）、声音、授课现场图像，实时保存压缩，授课结束或在规定的时间，课件能进行自动合成。

屏幕动态捕捉。课件实时录制系统完整记录计算机显示内容，包括幻灯片演示、Web文件、Word文档、PDF文档、鼠标移动轨迹、Flash动画、操作程序等。录制屏幕内容时，该软件具有以下三个特征：第一是CPU资源占用少，不影响其他程序的正常运行；第二个是系统使用算法的高压缩比，并且记录的过程占用磁盘上较少的空间；第三是可捕捉屏幕动态，图像流畅清晰，常见捕获帧率是每秒60帧。

2. 课件编辑工具

包括课件的编辑、剪辑和个性化定制功能，用课件编辑工具可以完成对录制课件的剪切、删除等操作，还可完成对视频的同步剪切等操作；可对录制的视频进行任意合并，而且可保证合并后的视频课件能够同步操作。课件编辑工具还提供如下功能。

图片水印/视频文字。课件录制系统可在视频上叠加图片水印或文字，是录制的视频更具有个性化。

照片加载。根据教师的授课需求，课件录制系统提供照片加载功能，在视频上加载个性化的照片，来达到某种效果，如活跃课堂气氛等。

静音检测功能。课件记录系统可以在记录过程中实时检测音频信号。

片头片尾。根据实际授课工作的需要，课件录制系统可自定义设置片头片尾。一般来说都需设置片头和片尾，如选择默认的片头片尾，则系统自动截取某段画面信息来作为片头片尾；当然用户也可以自定义片头片尾，这样更有针对性，一般来说效果会更好。

文字缩印自动生成。课件记录系统根据教师的幻灯片演示或教学文档自动生成文本索引，还可对生产的文字索引进行编辑，包括修改、添加、删除索引结构、索引条目的移动等操作；还可以创建缩略图索引等。

课件样式模板。课件录制系统提供丰富的课件风格模板，可根据个人喜好进行选择，当然用户可自行创建并添加自己的模板。

(三) 教学资源共享

教学资源共享系统集教学资源分布式存储、课程资源生产、资源管理、资源评估和多模式用户应用于一体。实现多形式，多渠道教学资源的快速上传，归档和检索功能，实现教学资源的多级分类存储管理。

系统支持动画、表格、图片、文字、流媒体音视频节目、网络课件、教学音视频内容、用户内容等不同形式的多媒体资源统一采编、入库、汇聚、加工、索引、转码、审核到发布的流程化操作，达到对流媒体教学资源的统一管理和发布，满足网络教育教学培训的实际应用需求，支持多终端、多码率、多制式、多格式、高清标清等，提高资源的综合利用和管理。

1. 上传管理

教学资源的上传管理可以多种方式来实现，支持用户权限管理下的大容量文件的上传与下载，支持断定续传、支持多格式、支持统计日志等。

在上传教学资源的过程中可以进行视频筛选、动画截图、缩略图等，实现自动分类操作。该平台支持用户将本地资源上传到数据库服务器，供管理人员审核和验收；也可以由管理人员将资源内容批量导入数据库，完成资源库的内容的动态添加。

该平台支持上传多种格式文件，如各种办公文档、PDF 文档、音频和视频文件、图像文件、HTML 文件以及 Authorware、Shockwave、Flash 等其他文件，也可支持 LOM、SCORM 等标准课件的上传与运行等。

2. 审核编目

系统将上传的网络教学资源按照一定的方式进行审核编辑编目，可进行统一检索和管理。

该平台支持审查和接受上传的教学资源，审查和存储教学资源，并进行必要的编辑和管理，如设置资源的应用范围、资源星评级、资源发布、下载点、日志管理、评估管理、资源清理和删除，资源目录导入和导出等。

3. 存储管理

支持音频、视频、文字、图片、幻灯片、Web 课件、Flash 动画等不同类型不同格式资源的分类存储管理。支持教学资源的分类管理，建立数据目录中心数据库，便于资源的检索和搜索。系统支持根据每个资源库建立相应的数据表以存储不同种类的数据元素，并且可以使用主题分类方法。专业分类、主题分类和部门分类用于分类和整理资源。

4. 资源共享

通过该平台用户可进行资源下载、检索、浏览、评价、收藏、个人订阅等多种资源共享应用，提高资源的利用率。

(四) 在线培训

按照农民教育与培训、农业技术人员进行继续教育的主要形式、主要项目、主要管理模式，构建网络教育与培训系统的基本架构。

1. 培训项目管理

培训项目编辑、培训计划增加、培训计划删除等。

2. 培训项目属性

培训班次、培训目标、学习要求、课程名称、考试要求、培训资金、使用要求等。

3. 培训课程管理

在线课程的特点是开放性、共享性、互动性、自主性和协作性。在线课程结合了学科的教学内容和通过网络实施的教学活动。根据一定的教学策略组织教学内容，达到一定的教学目标，其中网络教学支持环境指的是它是网络平台上实施的网络教学、软件工具和教学活动的教学资源。课程管理包括课程的创建、修改、删除和编辑；课程属性管理包括课程所属的类别、课程所属的科目、课程应用的科目、课程开放和课程学分；课程素材管理包括课程所需要的图片、文本、音视频、流媒体课件、试卷、作业练习等。

4. 培训证书管理

管理培训涉及的各种证书，如学历证书、职业资格证书、继续教育证书或其他一些证明。

5. 培训班次管理

参与者必须在一定时间内完成某些课程内容学习。培训班次包括新建、修改、删除、编辑等操作，培训班的属性包括课程、学时、适应对象、考试成绩、证书等内容。

6. 在线学习

学员按照分配的权限参加在线学习，选择课程进行培训、学习课程内容、做作业、查询学习过程、参加考试、查询成绩等。

四、技术架构

随着基于 Web 的信息管理系统业务拆分越来越细，占用的存储空间越来越大，应用系统的复杂度呈现出指数增长的趋势，部署维护系统越来越困难。所有应用和数据库的连接，在成千上万的服务器规模的网站中，连接的数目是服务器规模的平方，经常导致数据库连接资源不足，拒绝提供服务。

每一个应用系统常需要执行许多相同的业务操作，如系统登录、用户查看信息等，可将这些共用的业务提取出来，独立进行部署。由这些可以服用的业务连接数据库，提供共用业务服务，而应用系统仅需管理用户界面，通过分布式服务调用共用业务服务来完成具体的业务操作。

农民教育培训系统采用 J2EE 技术架构：服务器操作系统采用 Linux，应用服务器选择 WebLogic，数据库选择 Oracle，应用服务器和数据库服务器部署在统一的数据中心。不单独投资建设，建设规矩方案不含视频内容的制作，视频服务器系统需支持Flv+H264 高清网络视频应用，该平台的整体架构需要支持 PC 和智能手机多终端应用（图1、图2）。

农民教育培训系统

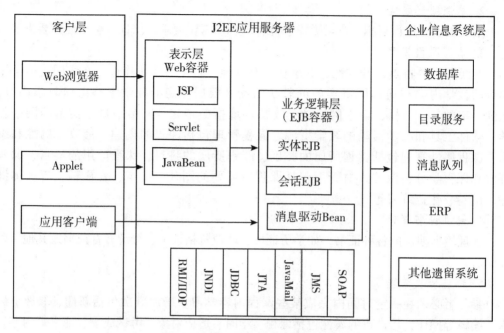

图 1 农民教育培训系统技术架构

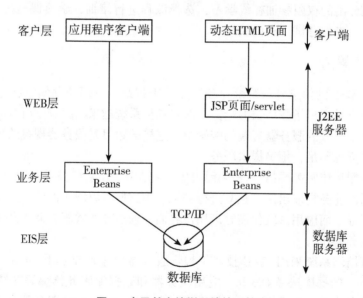

图 2 农民教育培训系统核心技术